Marine Cargo Operations

Marine Cargo Operations

SECOND EDITION

CAPTAIN CHARLES L. SAUERBIER, USNR

Master Mariner

CAPTAIN ROBERT J. MEURN, USNR

Master Mariner

Chief, Division of Nautical Science
Department of Marine Transportation
U.S. Merchant Marine Academy

A RONALD PRESS PUBLICATION

JOHN WILEY & SONS

New York · Chichester · Brisbane · Toronto · Singapore

Library of Congress Cataloging in Publication Data:

Sauerbier, Charles L.
  Marine cargo operations.

  "A Ronald Press publication."
  Includes index.
  1. Stowage.   2. Cargo handling.   I. Meurn, Robert J.
II. Title.

VK235.S26   1985   623.88′81   84-20920
ISBN 0-471-88616-5

Printed in the United States of America

10 9 8 7 6 5 4 3 2 1

To
Christine and Andrew

PREFACE

The safe and economical carriage of goods continues as an ongoing mission of the shipping industry. This second edition has a twofold purpose. It aims to help the shipping industry fulfill its mission by presenting formally and systematically the basic principles and techniques of cargo operations, and it seeks to acquaint merchant officers present and future with the principles of stowage and their implications as they apply to the responsibilities of a carrier.

The method of presentation grew out of our experience within the classroom, which we believe to be an excellent testing ground for the methodology used in the presentation of any idea or group of ideas. We slanted the discussions toward the individuals who make up the shipping "team" and whose work is one of the most important factors in accomplishing the shipping industry's mission. The "team" includes the operations manager, the terminal superintendent, the marine superintendent, the stevedore, the ship's master, and their staffs.

We have attempted to set forth and discuss as many of the tasks, concepts, and facts affecting cargo operations as we could encompass within a book of reasonable size. The discussion is directed toward the ship's master and his deck officers, but the substance of the discussion is no less valuable to other members of the team. These members of the team are often underrated, even though they are in an ideal position to observe and evaluate operating methods. They can do much to improve any given system. A lack of information among these men may prove very costly. Compliance with the three "C's" (Communication, Cooperation, and Coordination) is vital for members of the team. One need only study the litigation resulting from cargo claims for only a short time to be convinced that in many cases damage to cargo and the resulting legal action were needless.

Communication and cooperation should insure coordination in the movement of cargo. Much of the success of a shipping venture depends on intelligent coordination of movement among all individuals and groups participating in the business of overseas commerce. This coordination should be present from the domestic shipper's terminal in the hinterland, through the marine terminal, to the foreign consignee, and then, with other commodities, back to the shores of the mother country. We hope that this book will contribute to this coordination of movement by giving a better understanding of the full shipping cycle to many who are involved in it only partially.

Much of the text is based on scientific fact, but there are a number of recommendations regarding methods, procedures, and even philosophies that are based purely on our own opinion. Such opinion is based on much study and nearly a half a century's experience in the marine shipping industry. There is undoubtedly room for argument on some of these points, and there may be some omissions that will come to the mind of the careful reader. The authors invite written comment from all who care to discuss such points.

Much of the research alluded to in the first edition has indeed led to astonishing innovations in cargo operations over the last three decades. These innovations are described in this edition along with future innovations presently being researched.

The greatest impact on the waterfront over this time has been made by the innovation of *containerization*. When the first edition was published in 1956, the viability of containerization was just being tested by Mr. Malcolm McLean. He was interested in knowing if carrying truck vans by water from Houston to New York would be feasible and economical. He had two T-2 tankers covered with a platform that formed a raised deck upon which the truck vans could be secured for the voyage. The "test" proved that the idea was sound. The next year, 1957, the Pan Atlantic Steamship Company (later known as Sealand, S.S. Co.) converted the *S.S. Gateway City* from a breakbulk carrier to the world's first cellular containership. She was able to carry two hundred thirty 35-ft containers. Something over 60% of the world's deep-sea liner cargo carriers are now containerized. Obviously, the impact of the "Box" has been great. This second edition has been written to point up these changes along the waterfront.

It is not possible for us to acknowledge all of our intellectual debts within the space allotted to this preface. However, there is need to mention a few people who made major contributions. Mr. Brian Starer wrote Chapter 3, "Cargo Responsibility." As an admiralty lawyer practicing in New York, he provided insight and advice that will be extremely informative and valuable to the ship's officer. CDR Herman Allen, USMS, Associate Professor in the Department of Marine Transportation, U.S. Merchant Marine Academy, revised Chapter 2, "The Shipowner's Organization." Honorable mention is given to LCDR J. Hall and LT T. Keefe for their help regarding containers and bulk cargo, respectively. We hope that the many other friends and associates who assisted us in gathering, evaluating, and editing the material will not feel that we are ungrateful for their assistance.

CHARLES L. SAUERBIER
ROBERT J. MEURN

Arroyo Grande, California
Kings Point, New York
March 1985

CONTENTS

CHAPTER 1

THE CARRIAGE OF GOODS AT SEA

INTRODUCTION

The business of shipping is governed by the spirit and the political makeup of the period. With the growth of independent states as a result of the Renaissance, there arose conflicting nationalisms. By the 18th century every kind of restriction on and inconvenience for trade was in force. There was no way for commerce to break its fetters except by violence. Navigation Acts and monopolies resulted in illegal trading, smuggling, and privateering, and these frictions caused wars. The 19th century was slow to resolve these tangles. It became clear that agreements on such issues as shipowners' liabilities, common forms of chartering, and marine insurance were mutually desirable. The International Maritime Committee did much valuable work toward the end of the century in clarifying and simplifying the mass of confused regulations hampering the shipping business. The responsibilities and liabilities of the shipowner with respect to the handling, loading, and stowage of cargo are defined primarily in the *Carriage of Goods by Sea Act* of 1936 and the *Harter Act* of 1893. One of the principal purposes of the Harter Act was to prevent shipowners from evading their obligation to properly load, stow, care for, and deliver cargoes carried on their vessels.

With a view toward making uniform the rules governing the responsibilities and immunities of shipowners under a bill of lading, a code of rules known as the "Hague Rules" was drawn up in 1921 by the Maritime Law Committee of the International Law Association. These Hague Rules form the basis of the Carriage of Goods by Sea Acts of Great Britain and the United States. There can be little doubt that responsibility for proper stowage rests primarily on the shipowner, and on the master as his representative.

Because domestic markets are limited, nations must turn to world markets for commerce. Nations can draw upon gold reserves for only so long, and soon learn they must trade or fade. The sea-power historian Mahan taught that national sea power is a politico-military force composed of naval and commercial fleets capable of transiting the oceans of the world at will, served by overseas bases strategically located throughout the world, and exploiting worldwide markets and sources of raw materials to the fullest possible extent.

Trade, of course, requires movement of goods. The only practical method of transporting large volumes of goods across the oceans of the world is in merchant ships. A ton of cargo may be moved in a ship for only a fraction of the cost of air transport. If a nation is going to trade with the rest of the world, it must utilize merchant ships. A nation's vessels plying the shipping routes of the world have a stabilizing effect on freight rates charged to that nation's businessmen. Economically, the need for a strong, healthy merchant marine is tremendous.

INNOVATIONS IN THE CARRIAGE OF GOODS AT SEA

The most astonishing and whole-hearted acceptance of any principle of materials handling in the past two decades has been that of unitization. It has been proven that materials handling becomes more efficient as the size of the unit increases.

Over the past two decades, billions of dollars have been invested by the shipping industry in the production of marine containers, the construction and modification of ships to carry those containers, the development of marine terminals to handle efficiently both the ships and the containers, and, lastly, the manufacture of other equipment and systems to support the maritime intermodal triangle of container, ship, and marine terminal. This does not even take into account the expenditures of other transportation interests generally considered only land oriented: railroads, truckers, equipment-leasing companies, and so forth. After this very busy and costly era of development, it would seem that a plateau of attainment and rest might be in order. Such is certainly not the case!

Intermodal containerization has grown from approximately 30,000 twenty-foot-equivalent units (TEU) to current worldwide inventory estimates of $2\frac{1}{4}$ million TEU. Growth estimates through the end of the century dwarf even this incredible increase. As regards ships to carry these TEU, not only will the number and size be increased, but there is also the prospect of a supersized containership with a 6000 to 10,000 TEU capacity.

This would seem to imply that the era of the breakbulk ship has ended. Although the number of breakbulk vessels has dwindled, the need for their flexibility of service still exists. Therein lies the advantage of such vessels. Even though a containership may be far more productive and cost effective, it is so *only* in the carriage of containers. A breakbulk vessel can carry breakbulk cargo; dry and liquid bulk cargo; outsized cargo such as lumber, rails, and piping; refrigerated cargo; heavy lift cargo; and yes, even containers.

The concept of lower costs is more likely to change carrier attitudes toward breakbulk handling than is that of new equipment. Carriers tend to think in terms of unit costs and to disregard improvements in traditional handling methods without bothering to assess their net benefits. Effective breakbulk handling can require more intensive management than does containerization, but is rarely accorded it, due to a tradition of fragmented responsibility. With adequate thought and simple technology, breakbulk handling rates can almost match those of containerization, and for much lower investment and operating costs. Optimum results depend on proper handling at the source, which principally involves efficient slinging. Carrier resistance arises from expenditure on nonreusable slings, but this can be recouped by only a small productivity increase. Reduction of worker resistance to the use of slings depends on proper consultation and on convincing dock labor that it involves no increase in effort or reduction in safety. Modern vessels with open hatches and cranes are best in terms of offsetting productivity rating against daily hire, but many carriers still use old and unsuitable ships.[1]

The type of vessel required depends upon the trade route. Containerization cannot be the answer for a less developed country; the economic viability of this technique depends on the nation's being able to stuff into containers a major portion of its trade goods, to create a series of inland depots to receive the goods, to stuff and de-stuff containers, to redesign its railroad system and its road system, to abandon intermediate ports, and to redesign its major ports. This may take several generations. It is obvious that the trade route will determine the mode of transportation. In view of this, the intent of this first chapter is to discuss the major innovation in the carriage of goods at sea in the past 30 years, namely, containerization.

CONTAINERIZATION

The concept of containerization was not a new one when Malcolm P. McLean initiated the container revolution in April 1956 with the sailing of a converted T-2 tanker, the *Ideal X*. Malcolm McLean brought about containerization in much the same way that the forklift truck brought about palletization in 1935. The *Ideal X* was converted to allow for cargo movement by both truck and ship, thereby forming an integrated transportation system. In examining this innovation the remainder of this chapter will discuss the container, the ship, the terminal, and the lashing of containers aboard ship.

The Container

Containerization is considered the most successful solution to the problem of moving cargo in international trade. Containerization has achieved its primary purpose of minimizing the handling of cargo. It ensures the efficient, reliable, and rapid delivery of the undamaged goods with through-transportation that can utilize all modes of transport. The end result is intermodalism. Intermodalism can best be depicted by the example of a "house to house" or "door to door" container. A shipper will

---

[1] S. Wade, "The Forgotten Challenge of Improved Breakbulk Handling," *Fairplay International Shipping Weekly,* August 26, 1982.

stow (a term preferable to "stuff") a container, and the consignee will unload ("strip") the container. The cargo is only handled twice, which fulfills the objective of containerization.

To allow this intermodalism several aspects of transportation had to be studied and some had to be altered. Trailer manufacturers had to develop a chassis that the container could easily and swiftly be lifted from or placed on and locked into position for road transport. Larger road networks had to be developed in various countries, including areas throughout Europe. Interfacing with the railroads became important. Transportation of containers by rail necessitated altering various bridge and tunnel clearances. Lift requirements for piggyback containers had to be adapted for the Association of American Railroads. Through the efforts of the American National Standards Institute (ANSI) and, on a world scale, the International Organization for Standardization (ISO), containers have reached a certain degree of standardization. The term "ISO container" is a well-recognized one in the transportation industry.

The success of containership operation (containerships carried 84% of the cargo moved in 1981) is due to the following advantages:

a.   The amount of time a vessel spends in port is reduced. A containership can load and unload in a matter of hours, as compared with the older breakbulk ships, which take days to load and unload in port. This fast turnaround time in port greatly increases the number of voyages a containership can make per year as compared with a breakbulk ship. This means that a new fleet of containerships could replace an old breakbulk ship fleet at a ratio of one to three. In essence, one containership can do the job of three to four comparable breakbulk ships, depending on trade route and size. The following comparison illustrates this advantage.

HYPOTHETICAL VOYAGE COMPARISON BETWEEN A
CONTAINER VESSEL (CV) AND BREAKBULK SHIP (BB)

Assumptions

1.   Route New York to San Juan—round trip sea time for both ships, 6 days.
2.   Both ships carry 8000 tons of cargo discharging 8000 tons in New York and 8000 tons in San Juan.
3.   A 350-day year, 7-day week, 12-hour day for in-port cargo handling.
4.   Breakbulk ship works five hatches at one time. Container vessel works three hatches (crane availability, one hatch separation).

5. Cargo handling rate for a breakbulk ship: 18 tons/hatch/hour (nonpal-
letized).
6. Cargo handling rate for a container vessel: 370 tons/hatch/hour (20 containers at an average of 185 tons per container).

Cargo Handling Rate

Breakbulk—12 hours × 5 hatches × 18 tons/hatch = 1,080 tons/day
Container—12 hours × 3 hatches × 370 tons/hatch = 13,320 tons/day

Port Time

$$\text{Breakbulk} \frac{2 \text{ ports} \times 8000 \text{ tons}}{1,080} = 14.8 \text{ days}$$

$$\text{Container} \frac{2 \text{ ports} \times 8000 \text{ tons}}{13,320} = 1.2 \text{ days}$$

Days/Voyage	Sea Time	+ Port Time	= Voyage Time
Breakbulk	6 days	+ 14.8 days	= 20.8 days
Container	6 days	+ 1.2 days	= 7.2 days

Voyages/Year	350 ÷ Voyage Time	#Voyage/Year
Breakbulk	350 ÷ 20.8 days	16.8
Container	350 ÷ 7.2 days	48.6

Therefore, it takes approximately 3 BB ships to carry same amount of cargo as one container vessel

Annual Handling Costs

Ships × Port Time (in days annual) × Hours/day × Hatches × Labor Cost

$$\overset{(16.8 \times 14.8)}{\text{BB}} = 3 \times 248.6 \times 12 \times 5 \times \$750 = \$33,561,000$$

$$\overset{(48.6 \times 1.2)}{\text{CV}} = 1 \times 58.3 \times 12 \times 3 \times \$1300 = \$2,728,440$$

(includes Gantry Rental
and Hustler Rate)

The use of a container vessel fleet for a trade route with appropriate marine-materials-handling equipment can be seen to provide substantial cost savings.

b.   The container, being standard, can easily be interchanged among the various transport modes. This allows intermodalism.

c.   A decrease in transit time has a tendency to decrease inventory costs because a shorter lead time is required at reorder points.

d. A closed and sealed container provides protection against pilferage because what goods are in the container is not common knowledge.

e. The handling of goods themselves is dramatically reduced. This is especially true of the "house to house" container, with which the goods are handled once by the shipper and once by the consignee. (The container may be handled several times, but the goods are handled only twice.)

f. The container can serve as protection against the elements, thereby reducing the need for increased packaging.

Container operation is not without its indigenous problems, however. These include the following:

a. Equipment balance necessitates having the required number and types of containers located conveniently for the shipper's access. The container must be structurally sound, clean, and seaworthy. If it is not, the possibility of cargo damage increases.

b. The container must be correctly loaded. If it is not, the cargo, the container, and the handling equipment may be damaged. When "house to house" containers are used, responsibility for proper stowage shifts from the carrier to the shipper, who may have less expertise in cargo stowage.

c. Certain types of cargo may not be transportable by container simply because of the restrictive size or carrying capacity of the container.

d. Certain containerships are limited in their ports of call, due to poor capabilities for the loading or unloading of containers at some terminals.

Container Materials

The construction materials used in assembling containers are steel, plywood/fiberglass, and aluminum. *Steel* containers are strong but heavy and susceptible to corrosion. Steel keeps damage from puncturing and collisions with equipment to a minimum. *Plywood/fiberglass* containers generally have steel frames and smooth walls of ¾-in. plywood reinforced by fiberglass that do not reduce internal space. These walls are easy to maintain and when maintained properly will not corrode. Such containers are not comparable in strength to aluminum or steel types. *Aluminum* containers are made of wide, lightweight aluminum sheets and have one-piece roofs and sides. Their light weight is a considerable advantage in over-the-road transport. The interiors are generally lined with plywood for protection and insulation. Aluminum is more resistant to corrosion than is steel.

Basic Container Elements

Since the container is the package that is handled by the steamship, an understanding of the terms commonly used to talk about containers is a necessity (see Fig. 1-1). *Side rails* provide the strength to resist bending (compression and tension) when the container is lifted at the corners. The *bottom structure* bears the weight of the cargo resting on the floorboards (wood planks on plywood) and distributes this weight to the cross members (I or Z beam) understructure. In addition, the I or Z beams distribute weight to the bottom side and end rails. *Corner castings* are steel fittings that allow the insertion of lifting devices. They also provide a means for handling, stacking, and securing. *Corner posts* are the steel vertical strength members that make stacking possible. The stress is transmitted through the corner castings. The *sides* and *front* are the flat sidings attached to the side and front posts, which are bolted to the top and bottom rails. These aluminum posts may be inside or outside the siding on aluminum containers. Steel-container siding is made of corrugated steel sheets, which eliminates the need for posts. Plywood/fiberglass containers do not use posts for their fiberglass-reinforced plywood panels. The *roof* is usu-

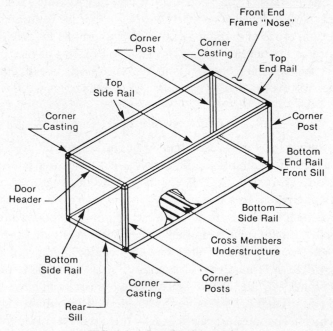

FIG. 1-1.   Container elements.

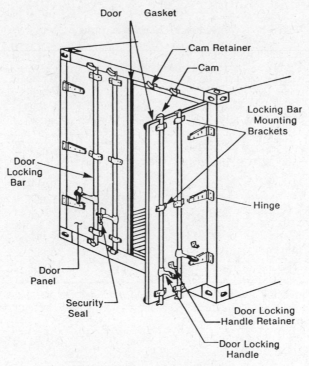

FIG. 1-2. Rear door components.

ally made of the same material as that of the side panel. It is desirable to have one-sheet construction to resist water entry. The *security seal* (see Fig. 1-2) is used in conjunction with the locking door handle. Once the door is closed and the seal is in place the door cannot be opened without the seal breaking. Such seals are number coded. The *door gasket* is attached to the door edges to help provide a waterproof seal.

The Container Choice

There are several different types and sizes of containers to meet the varying needs of shippers. A thorough knowledge of these container types allows the shipper to plan shipments effectively. Selection of the most suitable container can be done by conferring with a container-leasing company or the carrier or referring to *The Official Intermodal Equipment Register*. The *Register* contains, among other things, the dimensions and capacities of the various containers currently in use. Typical information for some of the more commonly used containers is found in Table 1-1.

TABLE 1-1

COMMONLY USED CONTAINERS

Exterior Dimensions W × H × L	Approximate Interior Dimensions W × H × L	Interior Approximate Cube Capacity (Cubic Feet)	Approximate Weight Capacity (lbs)[a]	Door Openings W × H
8' × 8' × 20'	92" × 87" × 230"	1,065	40,000	89" × 84"
8' × 8'6" × 20'	92" × 93" × 230"	1,138	40,000	89" × 89"
8' × 8'6" × 24'	92" × 94" × 282"	1,411	45,900	90' × 89"
8' × 8'6" × 35'	92" × 93" × 415"	2,054	50,000	90' × 90$\frac{7}{8}$"
8' × 8' × 40'	92" × 88" × 473"	2,216	50,000	89" × 84"
8' × 8'6" × 40'	92" × 93" × 473"	2,342	55,000	90" × 89"
8' × 8'6" × 40'	92" × 105" × 473"	2,644	55,000	90" × 102"
8' × 9'6½" × 45'	92" × 106½" × 534"	3,035	64,300	92" × 102½"

[a] It should be realized that there are weight and size limits on containers for state and interstate over-the-road transportation. This information may also be found in *The Official Intermodal Equipment Register*.

Container Types

The following are some of the different types of containers, along with possible commodity applications:

Flatracks (Fig. 1-3) are used for cargo that does not conform to the normal dry cargo container. The cargo may be awkward or too large; the flatrack allows such cargo to be carried on containerships. The ends are collapsible and can be interlocked for cost-saving in handling and storage. Examples of cargo not susceptible to water damage that may be carried in this container are boilers, generators, machinery (agricultural, construction), and transport vehicles (boats, buses, tanks, and trucks).

Ventilated containers (Fig. 1-4) provide ventilation and protection against water damage. They are similar to the basic dry cargo container and therefore may be used as such on the return leg of the voyage. Cargo carried in these containers includes fruits and vegetables, green coffee (bagged), and spices (bagged).

Half-height containers (Fig. 1-5) are normally only 4 ft high and are designed for cargoes of high density (i.e., those that are heavy and take up little space). They act as a space-saver, because two can fit into the space

FIG. 1-3. Flatrack type container.

FIG. 1-4. Ventilated type container.

11

FIG. 1-5.   Half-height type container.

occupied by one standard container. Cargo carried in these containers includes steel products such as pipes, rolls, and beams; drums; and ores.

Tank containers (Fig. 1-6) have many different configurations. Their function is to carry bulk liquids in small lot shipments. Shipping in tanks is an alternative to drum shipping and may save labor. Examples of cargo carried in this container include chemicals (hazardous and nonhazardous), potable liquids, liquid foods, gas, and helium.

The *open or canvas-top* container (Fig. 1-7) allows loading both through the top with an overhead crane and through the rear doors by conventional means. It is advisable that cargoes susceptible to water damage be kept in a weatherproof pack inside the container. Overheight cargoes may be carried in this container. Examples of cargo carried in such containers include machinery, lumber, and pumps.

The recently developed *high cube* containers (Fig. 1-8) are large-volume containers (9 ft 6 in. high) specifically for low-density (light, large) cargo. Such containers may present problems in foreign countries with low rail and bridge clearances. Examples of cargo carried in this container include household goods, stoves, refrigerators, washers, furniture, carpets, tobacco, and cargoes with high stowage factors.

FIG. 1-6. Tank type container.

FIG. 1-7. Open or canvas-top container.

13

FIG. 1-8. High cube container.

Bulk containers (Fig. 1-9) answer the need for reducing the expense of bagging dry bulk commodities that flow freely. Certain bulk products require additional protection such as polyethylene liners. Examples of cargo carried in these containers include grains, cereals, flour, malts, sugar, fertilizers, and coal.

Reefer (refrigerated) containers (Fig. 1-10) have precise temperature-control capabilities for situations in which refrigeration or freezing is necessary. They run on diesel, electrical, or liquid gas power, and can be used for ocean transport, storage, or over-the-road transport.

Other specialized containers include livestock, produce, insulated, tilt, side door, and automobile containers.

Stowing (Stuffing) the Container

For the sake of clarity, we shall use the term "stowing" to denote placement of the container aboard the ship. The term "stuffing" shall mean the

FIG. 1-9. Bulk type container.

FIG. 1-10. Refrigerated container mounted on a chassis.

15

placing or packaging of the goods within the container. This term (stuffing) is a misnomer because it suggests packing a container in a careless manner, which is not how stuffing should be done.

Prior to the advent of containerization, the shipper had to rely on the expertise of the stevedore to stow the cargo aboard ship properly so as to withstand the hazardous rigors of an ocean voyage. But with modern, house-to-house containerization, the shipper "stuffs" the container. Therefore, the responsibility of proper packaging has shifted from the carrier to the shipper. Because of this shift in responsibility, the shipper should be aware of all forces that the container (and the cargo contained within) will encounter from the start of its journey to the final destination. These forces are due to the motions of the road, railroad, and the ship. *Road motions* include acceleration and deceleration (braking); impact (coupling of the chassis with the truck and backing up against loading docks); and vibration, sway, and road shocks from poorly paved roads. *Railroad motions* include acceleration and deceleration, impact from coupling, vibration, and sway. *Ship motions* (see Figs. 1-11 and 1-12) include rolling, yawing, pitching, surging, sway, and heaving. In addition, there are wave-impact stresses such as panting and pounding. *Panting* is a stress due to unequal water pressures on the bow as it passes through successive waves. *Pounding* results from the bow lifting out of the water and then slamming down onto the crest of the next wave.

It is apparent that the ocean voyage presents the most hazardous situation for the container. To stuff for the worst conditions is to stuff for the oceanic leg of the integrated transportation system.

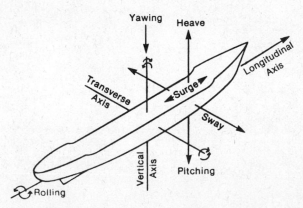

FIG. 1-11. Six motions of a ship in a seaway.

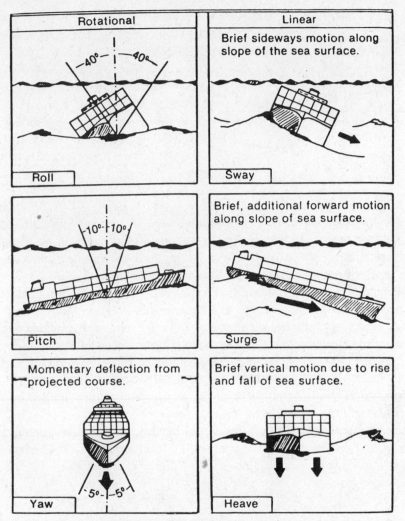

FIG. 1-12. Ship motions in a rough sea.

The principles of stowing a breakbulk ship (see Chapter 4) are applicable to the stuffing of containers. Containers must be loaded so as to prevent movement within them. The distribution of weight should be even throughout the length and breadth of the container, with heavy cargo on the bottom and light cargo on top. The permissible floor load capacity must be adhered to, and cargo within the container must be compatible; for example, wet hides should not be packed in a container with food-

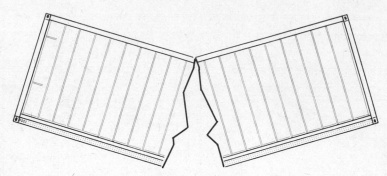

FIG. 1-13. Container too heavy in the middle.

stuffs; the hides will taint the foodstuffs. Also, wet cargo should never be stowed on top of dry cargo. Finally, all void spaces should be filled to prevent shifting of the cargo. Not adhering to these principles will cause the types of damage depicted in Figures 1-13 to 1-16.

In addition to the motions (forces) already mentioned, the container experiences *transfer forces* (forces experienced between modes) such as lifting, stacking, racking, and restraint. *Lifting* emanates from rapid acceleration and deceleration. *Stacking* results from forces applied at the corner castings (in the marshaling yard and aboard ship). *Racking* forces are side forces that tend to make the top and bottom of the container no longer square with its sides. *Restraint* results from forces applied by the various lashing systems used aboard ship.

As was stated above, the principles of stowing a breakbulk ship are similar to those of stuffing a container. The techniques involved in secur-

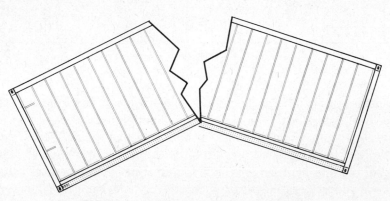

FIG. 1-14. Container too heavy at both ends.

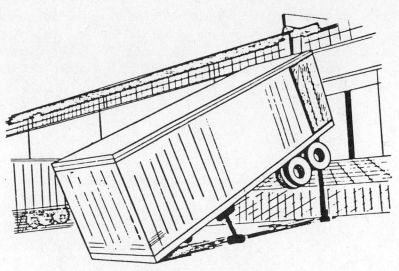

FIG. 1-15. Improper loading; container heavy at front end.

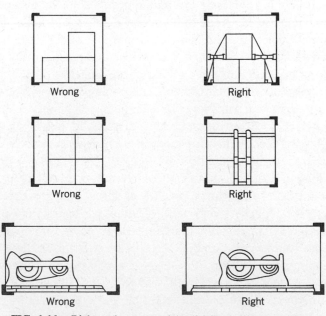

Wrong   Right

Wrong   Right

Wrong   Right

FIG. 1-16. Right and wrong weight distribution in a container.

ing cargo in a breakbulk ship are also similar to those used for securing cargo in a container. The use of dunnage, whether lumber, plywood, steel strapping, or fiberboard, for bracing and shoring is discussed in detail in Chapter 6. Adherence to these techniques will help to minimize damage to the cargo and the container.

The Ship

Container stowage aboard conventional breakbulk ships is difficult and time consuming. One limiting factor is the hatch opening, which usually does not extend for the full breadth of the vessel. Stowing in the wing spaces below deck requires additional horizontal motion. To alleviate this problem a ship of cellular construction was developed. The containership, with hatch openings extending for almost the full breadth of the vessel, incorporates vertical cell guides that allow confined positioning below deck and vertical stacking (see Figs. 1-17 and 1-18).

The Operation

The overriding consideration in performing cargo operations on containerships is keeping port time to a minimum. To achieve this end a cargo stowage plan and the sequence of container loading is arranged by shoreside personnel prior to the ship's arrival in port. The chief mate has only a brief period to look at the stowage plan, since cargo operations usually commence immediately on the docking of the ship.

Once the containership has docked, the lashings are removed from the on-deck containers and the shoreside gantry crane commences discharging. The hatch cover is then removed. The crane will then begin discharging cargo below deck, removing containers from one vertical cell only (usually six or seven containers) until that column or cell is empty. This allows loading to begin and to run in conjunction with unloading. Time is saved because every time the crane with its spreader moves it is carrying a container (see Figs. 1-19 and 1-20). The actual loading and unloading operation is quite simple, but planning where each container is to be located aboard ship can become quite complex. Because of the vast number of containers aboard ship, it is necessary to have a location identification system. A numerical system has evolved that utilizes the three dimensions involved in cargo stowage. This system calls for numbering containers longitudinally from forward to aft by hatch and bay number,

FIG. 1-17. Cell guides for vertical stacking.

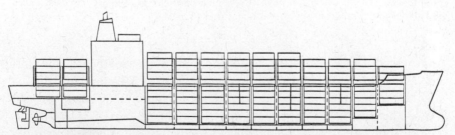

FIG. 1-18. Profile view of vertical stacking below and on deck. The hatch covers are designed for a heavier deck load capacity because of the weights incurred when containers are stacked on deck.

21

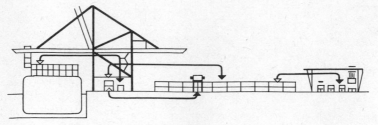

(a)

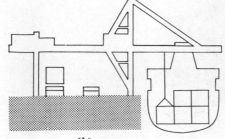

(b)

(c)

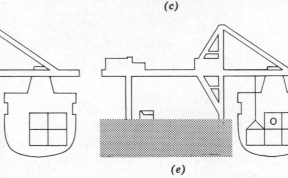

(d)   (e)

FIG. 1-19. (*a*) Crane sequence of operations in discharging an import container and loading an export container. (*b*) Crane engages container. (*c*) Import container lifted onto trailer waiting ashore. (*d*) Crane engages export container on trailer. (*e*) Export container loaded into ship's hold. Spreader then off-loads container marked "O."

transversely beginning at the centerline and moving outboard (the port side being designated by successive even numbers, the starboard side by successive odd numbers), and vertically beginning at the tank top and moving upward to the uppermost tier on deck (see Figs. 1-21 and 1-22). The bay concept arose to accommodate the fact that it is possible to stow two 20-ft containers in the same cell as one 40-ft container and thus it is necessary to differentiate between the two sizes on the cargo plan.

FIG. 1-20. Crane spreader that locks into the tops of containers for lifting aboard ship.

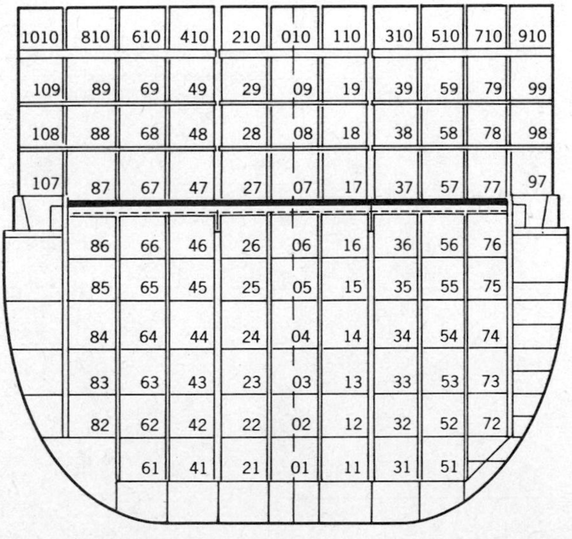

1010	810	610	410	210	010	110	310	510	710	910
109	89	69	49	29	09	19	39	59	79	99
108	88	68	48	28	08	18	38	58	78	98
107	87	67	47	27	07	17	37	57	77	97
	86	66	46	26	06	16	36	56	76	
	85	65	45	25	05	15	35	55	75	
	84	64	44	24	04	14	34	54	74	
	83	63	43	23	03	13	33	53	73	
	82	62	42	22	02	12	32	52	72	
		61	41	21	01	11	31	51		

FIG. 1-21. Transverse and vertical numbering system for containers aboard a container-ship.

23

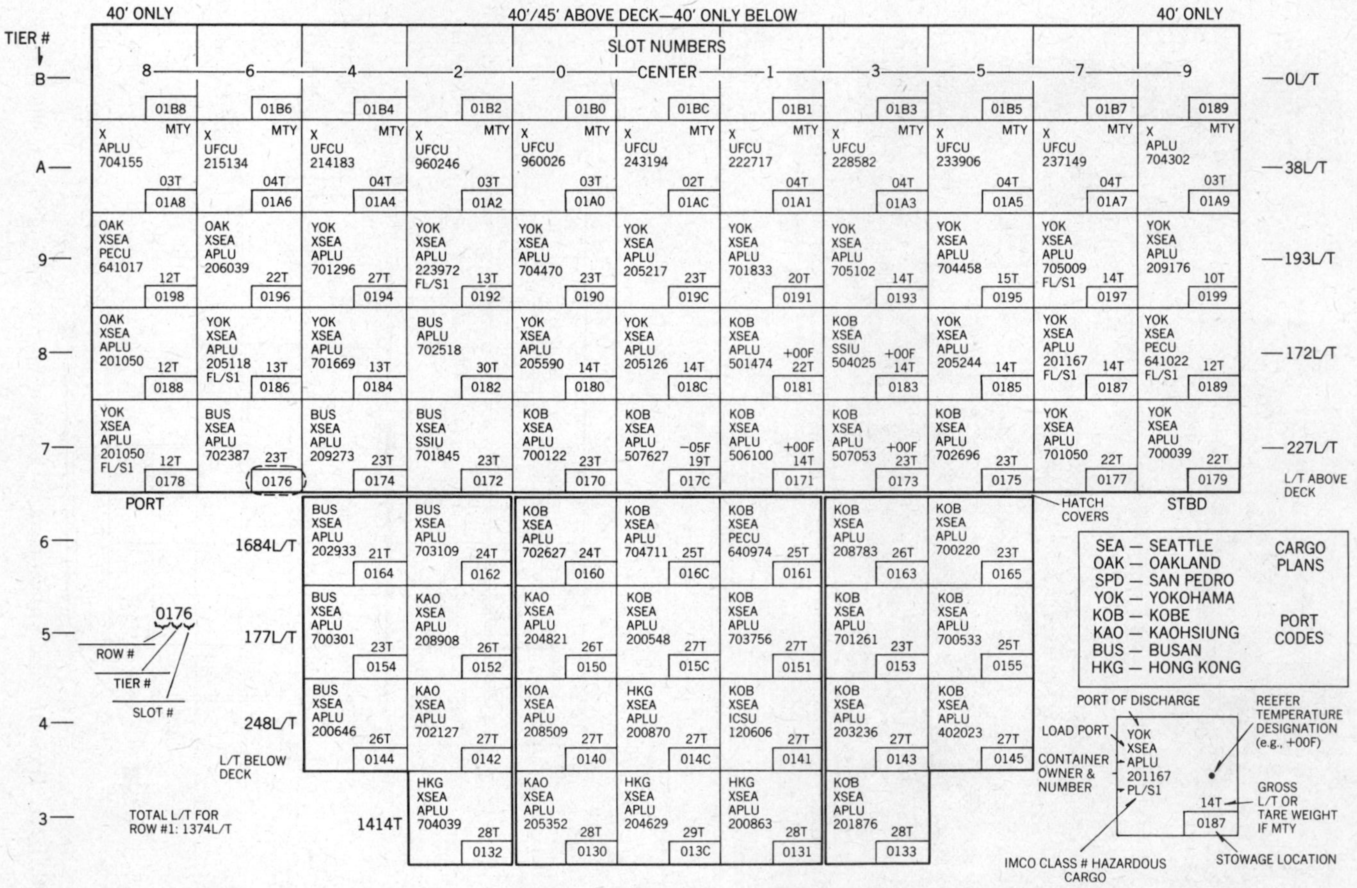

FIG. 1-22. Computer plan layout illustrating bay concept. Courtesy of M/N M. G. Heiserman, Class of 1986, United States Merchant Marine Academy (USMMA).

PORT CODES

SEA — SEATTLE
OAK — OAKLAND
SPD — SAN PEDRO
YOK — YOKOHAMA
KOB — KOBE
KAO — KAOHSIUNG
BUS — BUSAN
HKG — HONG KONG

CARGO PLANS

REEFER TEMPERATURE DESIGNATION (e.g., +00F)

GROSS L/T OR TARE WEIGHT IF MTY

STOWAGE LOCATION

PORT OF DISCHARGE
LOAD PORT
CONTAINER OWNER & NUMBER
IMCO CLASS # HAZARDOUS CARGO

Stowage Considerations

Stowage considerations that pertain to containerships include stability, trim and list, vessel stresses, stack height and weight limitations, port of discharge, and hazardous cargo regulations.

Considerations of *stability, trim,* and *list* dictate that it is necessary to stow the heaviest containers below deck, with the bottom tiers heavier than the ones on top; to build up the stacks of containers on deck that are closest to the ship's centerline (again with the heavier containers on the bottom); and to distribute weight evenly between the port and starboard sides to limit ballasting and to prevent a list. It should be noted that under certain listing conditions a containership cannot work cargo owing to the inability of the container to be inserted into the cell guides. Finally, the weight must be distributed longitudinally so as to give the desired trim.

Vessel stresses (shear stresses and bending moments) should not exceed the design limits of the vessel. To this end no more than two consecutive hatches, depending on design limits, should be empty at the same time, nor should they carry only empty containers.

Stack height is a concern when stowing a high cube (9 ft 6 in.) container under deck, since this type of container will occupy the space of two 8 ft 6 in. containers. This results in lost volume. Stowing of high cube containers on deck is also a problem, due to lashing system constraints.

Stack weight limitations must be considered in that the total weight of the stack should not exceed the deck load capacity of the tank top for under-deck stowage or the deck load capacity of the hatch cover for on-deck stowage. In addition, the strength of the bottom container must also be considered, since stacks may be as high as 11 containers under deck and as high as 5 containers on deck.

The use of more than one crane at the *port of discharge* will probably dictate a need for bay separation when stowing the cargo. Because of their size, container cranes cannot load or discharge adjacent bays (see Fig. 1-23).

Hazardous cargo regulations dictate certain stowage restrictions that are set by the Department of Transportation in the United States and the International Maritime Organization (IMO) for international shipping.

In addition to these considerations, there may be *special cargo stowage restrictions*. For example, refrigerated containers are generally loaded on the first tier on deck for accessibility to a power source. These containers should be stacked only one high to allow servicing at sea. First-tier on-deck stowage of containers containing wet animal hides or

FIG. 1-23. Adjacent container cranes. Courtesy of M/N J. Robben, Class of 1986, USMMA.

other odorous cargoes must also be considered. Aluminum containers must be stowed in a protected location away from possible damage by heavy seas. They should not be stowed on deck forward or in the outboard slots. During cold weather it is necessary to stow "Keep from Freezing" (KFF) cargo under deck. Open or canvas-top containers must be stowed under deck or at least under another container on deck. High cube containers should be stowed on deck, taking into consideration any lashing constraints. Flatracks and value cargo containers should be stowed below deck to limit access.

Some of the considerations and restrictions noted above could be included on a port detail sheet to assist proper stowage planning. Table 1-2 is a fictitious port detail sheet that includes some of this information.

The Containerized Terminal

At the breakbulk terminal the equipment, starting at the ship, consists of the following: First, there is the married fall rig or the ship's crane, with a capacity of about 3 tons unless it is doubled up for heavier lifts. Most

TABLE 1-2
PORT DETAIL SHEET

PORT OF CALL:

BERTHING	*Depth of Channel/Berth*	*Berth Use*	*Docking Side*	*Shift Starting/ Finishing Times*	*Holidays*
	Channel—40' Berth—38' Channel width —900'	Multiuse No first come first serve	Starboard side-to	0730–1530 1530–2330 2330–0730	

CRANES	*Number & Type of Cranes*	*Crane Moves/ Hour*	*Types of Spreader*	*Crane Reach Height/Across*	*Bay Separation Necessary for Crane*
	4-Paceco	18.0	Telescoping	5 High all bays 13 Across all bays	2 Bays for all cranes

STOWAGE	*Stowage Preference*	*Reefer Capacity*	*KFF Facility*	*Heavy Lift Capacity*	*Hazardous Cargo Restrictions*
		Terminal plug-in	Keep from freezing, limited	Mobile crane 150 tons	

PIER FACILITIES	*Fuel Available*	*Fresh Water Available*	*Customs*	*Terminal Equipment*	
	Yes, by barge	Yes, at berth	At terminal	Straddle Carriers Forklifts	

COST	*Cost/Box*
	Normal move $
	Restow $
	Bay shift

loads handled by a breakbulk ship do not exceed 1 ton. When the load reaches the pier apron it is either picked up by a forklift truck or landed on a dock trailer and hauled away by a small warehouse tractor as part of a three-trailer train. In the transit shed, protected from the elements, the load is stacked with a forklift truck. Some types of cargoes can be handled better with the Ross Carrier, which is a type of straddle truck much smaller than those seen today in use at containerized terminals. For heavy lifts and special cargo types, floating cranes or special heavy-lift cranes mounted on the ship are used. The equipment list is short and almost universal: ship's gear, forklift truck, tractor-trailer trains, and straddle truck. Whether the terminal is of high or low volume, the same basic set of equipment and methods is used. For high volumes the ship is simply worked longer hours and more hatches are worked. In the past, some breakbulk ships with double gear at four hatches and single gear at two hatches (six-hatch vessel), side ports in the upper and lower 'tweendecks, and overhead chutes and conveyors in operation would require as many as 23 gangs working the vessel. The entire operation required an army of men (as many as 400) and many machines.

Containerization has changed things. The circumstances and controls are not the same when working with the "box." Gone are the 1- to 3-ton lifts. The lift may be as much as 40 tons if the container is a 40 footer loaded with dense cargo. However, if so, the container must be lightened before being hauled over the road due to limits on total weight of cargo and equipment. Some of the containers will be empty because of the need to keep the supply of containers somewhat equal throughout the system. On the average, about 75% of the containers handled will be loaded and the average load will be about 14 tons. As a result the married fall rig cannot be used, and the ship's own heavy-lift gear, which use to be used for a few lifts, is just too slow to work a large number of containers and at any rate cannot reach over them. The dockside equipment has changed too; we will look now at the equipment found at container handling terminals.

Gantry Crane With Trolley

One of the most common types of cranes used to move containers to or from the ship is the gantry crane with a trolley controlling the hook. The trolley is positioned vertically over the container by running it out on the gantry tracks; the operator then carefully lowers a spreader with fittings that mesh with four corner slots on the container. One control locks the spreader onto the container, which can then be hoisted clear of the hold. On fully containerized vessels the containers are stacked one on top of another in vertical cells constructed with heavy metal guides at the four corners.

Several manufacturers make these gantry cranes, some of which are installed on the ship to make the ship self-sufficient. When Sea Land pioneered this system, the ships needed to carry the container handling cranes on board, as the terminals they visited were not equipped to handle the containers. The ability to handle the containers with ship's gear is still important on vessels that act as feeders from smaller, less well equipped ports to the larger major container-oriented ports. Large mobile cranes with slewing booms are also used to lift the containers on and off the ship, but these cranes cannot match the cycle times of the gantry-and-trolley type cranes.

Two types of gantry-and-trolley cranes have been developed. In one, known as the A-frame design, the boom that extends out from the dock over the ship is topped up when not in use and lowered to the horizontal position when a ship is positioned alongside the dock. The other type has a lower profile; its boom is always in the horizontal position but slides inboard and outboard. The uprights supporting the latter type of crane are spaced quite far apart, usually enabling the containers to be handled over a greater distance, sometimes as much as that from areas inland to the pier apron. Both types provide a boom that is long enough that the trolley can be positioned over any point on the vessel in line with the gantry tracks as the trolley travels athwartships.

The crane is mounted on rails on the dock and can be moved forward or aft by the operator. Thus, the hook can be positioned vertically over any point on the ship. The cycle time of the hook on these cranes is 2 to 4 min, depending on the design of the crane and the weight of the load being handled. Cranes with sliding or telescopic booms with wide spans between their uprights provide more space for landing containers on the pier apron. The cranes at Sea Land's Elizabeth, New Jersey, facility can land as many as seven containers abreast of one another. This capability reduces the dependence on the tractor's chassis meeting the hook cycle exactly. The short cycle time and average load of about 14 tons obviously enables high tonnages to be handled per ship hour. (A 2-min cycle time for the hook of a breakbulk married fall system is good, but the hook carrys a 1-ton rather than a 14-ton load.)

Equipment on the Terminal Side

Once the gantry or slewing crane has landed the container on the dock, various types of equipment are used to place the container in a marshaling area and from there onto a road chassis for the haul to the consignee. Three systems have evolved for this purpose over the brief history of containerization. These systems may be designated by the equipment that they use. The three common systems are (1) the Chassis System, (2) the Straddle Carrier System, and (3) the Straddle Gantry Crane System. Additional equipment other than that which we will discuss under these three names is also found in use at containerized terminals. These additional pieces of equipment will be described separately.

The Chassis System.   The chassis system (see Fig. 1-24) was the system first developed and used by Sea Land when that company pioneered containerization in 1956. Sea Land still uses it at large and busy terminals. The container from the ship is landed directly onto a road-haul chassis waiting on the pier apron. Either a tractor that can operate over the road or, more commonly, a yard tractor then pulls the chassis with its container load to a slot in a large marshaling area. Once the yard tractor takes the chassis from the pier apron to the designated slot in the marshaling area and drops it off with its container load, the yard tractor then picks up an empty chassis and returns to the pier apron to receive another container from the ship. The number of yard tractors working in the system will depend on the hook cycle of the ship. With a crane discharging a container every 2 min, about four tractor-chassis trains would be needed. The distance of the slots in the marshaling area from the pier apron is also a factor. The most important consideration is that enough tractors must be working in the cyclical system to make sure the hook on the ship does not have to wait.

FIG. 1-24. A 40-ft chassis.

One immediately evident advantage of this system is that no other handling of the container is necessary. The customer (consignee) can take the chassis with the container directly from the terminal to his facility. The consignee has a certain amount of time to unload the container and bring the chassis and container back to the terminal.

The chassis system requires a lot of land and a large inventory of chassis. For example, the Sea Land terminal at Elizabeth, New Jersey, covers 232 acres and provides space for more than 6600 chassis. Besides the marshaling space for the chassis with containers, there are supporting facilities such as a gateway control area with 20 lanes for trucks: 7 for entry, 7 for exit, and 6 that may be for either entry or exit, depending on the need. A freight consolidation station servicing trucks and railroad cars with less than container load (LCL) shipments is under a 306,000-ft^2 (over 7 acres) building. Another, 100,000-ft^2 complex is provided for repair and maintenance work on the many types of equipment that must be kept in good condition. There are also administration buildings. The pier is 4520 ft long and can accommodate six large container vessels. It is equipped with six gantry cranes of the low-profile extension boom type.

The Straddle Carrier System.   The straddle carrier (see Fig. 1-25) picks up one container from under the hook spot on the dock and transports it to a predesignated slot in the marshaling area. Some straddle carriers can stack the containers three high. However, there is a need to be able to get at any one of the stowed containers for customer servicing. This involves placing the container on a consignee's road-haul chassis with tractor. Because of this need, the stacking limit is one and a half high for 20-ft containers and one high for 40-ft containers. This means that for four slots, assuming 20-ft containers and a 45° parking pattern as in Fig. 1-26, two would be one high and two would be two high. With this pattern the straddle truck can shuffle the upper containers to get at any of the bottom containers. The 45° pattern makes it unnecessary for the straddle truck to maneuver down long lines of containers, which may damage containers and is slower.

The rate at which a straddle carrier can handle containers determines the number needed to support any given operation. One straddle carrier can handle about 12 containers/hour if receiving them from the ship's hook and taking them to a marshaling area. The containers are landed on the pier apron and a carrier with a spreader sling moves over the container and drops the spreader down on top of the container so that special fittings mesh with the container's corner fittings. The operator then locks

FIG. 1-25. The straddle carrier.

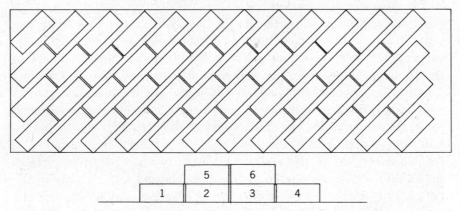

FIG. 1-26. *Top:* The 45° parking pattern that may be used for containers in the marshaling area handled by straddle carriers. In the 200- by 65-ft area 66 twenty-ft containers can be placed so that all are readily available. With allowances for aisles between blocks the number of containers stored per acre will be slightly less than if they were stored in parallel rows. *Bottom:* Diagram illustrating why stacking containers one and a half high is necessary for easy availability. Of the six 20-ft containers in the row, numbers 1, 5, 6, and 4 are easily available. To obtain number 2 or 3 the operator must place 5 or 6 on top of 1 or 4; he then can pick up 2 or 3 and pass over the remaining containers to place the desired container on a chassis or under the ship's hook. If the containers were stacked two high throughout, the bottom containers would still be accessible, but not without considerable time being taken to shuffle and move other containers.

the spreader onto the container, picks it up, and carries it away to the marshaling area.

Picking up a container and placing it on a chassis bed is a slower process, because the operator must carefully maneuver over the chassis bed and lower the container so it meshes with special fittings that will secure it to the chassis. The rate of loading onto chassis is about 8 containers/hour.

Once a loaded inbound container is at the terminal, it does not remain there very long. The ship operator will have contacted all consignees and notified them as to when their shipment would be ready for delivery. The consignee then has a limited time within which he must come and remove the container, take it to his own facility, unload it, and bring it back. About 48 hours is allowed for the consignee to come and pick up his shipment, and he has about the same time to unload it and bring it back to the terminal. If he keeps it beyond the specified time he is charged a daily rental fee. The outbound containers are delivered to the terminal so as to coincide with the ship's schedule. It is the responsibility of the freight department to coordinate these movements.

The Straddle Gantry Crane System (SGCS).   The straddle gantry crane (SGC) (Fig. 1-27) is not used to transport the container from or to the

FIG. 1-27.   The straddle gantry crane system.

pier apron. A tractor and chassis train is used for that purpose in this system. One lane under the SGC is reserved for the tractor and chassis. The SGC is used to pick up the container from the chassis and place it in a predesignated slot in the marshaling area.

If the container is being loaded onto a ship or an outgoing road-haul chassis, then the SGC picks up the container and places it on the chassis. Thus, the transport between the marshaling area and the ship is by the yard chassis; the SGC does the handling. As we have already seen, straddle carriers perform both the transport and the handling.

It may appear that the straddle carrier would be the more economical of the two systems, because of its flexibiltiy. Under some operating conditions, it may be. However, when the number of containers that must be handled is high, there may be some economy in switching to the use of the SGC and chassis trains to service the ship's hook or to load trucks. The SGC system utilizes the customer's equipment and driver to move containers between the yard entrance gate and the stack in the marshaling area. The savings resulting from its use are thus due to a reduction in the cost of needed equipment, especially when the terminal workload is such that there is a need for a high truck-loading rate. For example, the SGC can handle 15 truck-loading operations and 30 ship-loading operations per hour, whereas the productivity level for the straddle carrier is such that when the need for multiple units develops in order to keep up with delivery and receiving rates of containers passing through the terminal, then the SGC system, at some point, becomes more economical.

Other types of container handling equipment. Certain equipment at the container terminal has been designed for special tasks. Three examples are the *front-end loader* (Fig. 1-28), the *side loader* (Fig. 1-29), and the *shifter* (Fig. 1-30).

Some ships have their own container handling equipment (see Fig. 1-31). This equipment allows greater flexibility, but does take up space that can be utilized for the carriage of additional containers.

Fixed and Variable Costs of Terminal Operations

The primary objective of the container terminal is to provide good and complete service to its customers, the consignees and shippers. This service is the receipt and delivery of the containers with as little delay and difficulty as possible. The terminal is the interface between the ship and the land transport systems. Its efficiency can be measured by the *cost per*

FIG. 1-28. Front-end loader.

FIG. 1-29. Side loader.

35

FIG. 1-30. The shifter.

(a)

FIG. 1-31. Self-sustaining containerships. (*a*) The *Contender*, offering a maximum capacity of 1307 TEU. *Following page:* (*b*) A Tarros class vessel that combines a container and RO/RO vessel. (*c*) The *Strider*, one of the most versatile medium-sized self-sustaining containerships.

36

(b)

(c)

FIG. 1-31. (*Continued*)

37

container handled. If we know the total cost of an operation and the total number of containers handled per year we can determine the cost per container.

The total *cost* is composed of fixed and variable costs. The fixed costs for the terminal consist of the money invested in the land and in the equipment. Once the land and equipment have been procured, the annual costs remain fixed regardless of how many containers are handled. The variable costs are those that depend on the number of containers handled and the size of the installation being set up. These costs include labor, repair and maintenance, development costs (paving, buildings, etc.), and administration costs.

Estimating equipment needs and costs. Let us assume that we must load 76 containers on chassis per hour to meet an over-the-road delivery schedule, and that we also need to operate the ship's hook at the rate of 30 containers/hour. What equipment would be needed for each operation? Table 1-3 gives the required information.

According to this analysis the straddle carrier system would need 13 units, and the SGC system would need 6 units plus four yard-chassis trains. If we obtained the cost of the various needed components, we could determine the annual cost of the capital investment for equipment.

Estimating land needs and costs. The area needed for marshaling is less with the SGC than with the straddle carrier system. This is for two reasons: (1) the ability of the SGC to stack containers higher; and (2) the space between containers can be less with the SGC. With the straddle

TABLE 1-3

EQUIPMENT COMPARISON

Factor	Straddle Carrier	SGC
Ship handling rate	12	30
No. units needed to handle 30 containers/hour	2.5	1
Truck-chassis loading rate	8	15
No. units needed to handle 76 containers/hour	9.5	5
Yard-chassis trains needed	0	4
Standby units needed	1	0

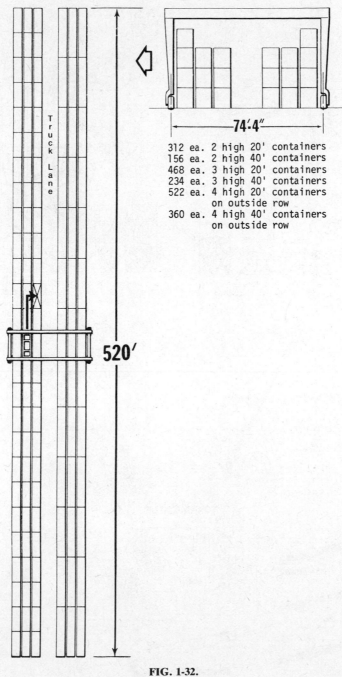

Truck Lane

520'

74'-4"

312 ea. 2 high 20' containers
156 ea. 2 high 40' containers
468 ea. 3 high 20' containers
234 ea. 3 high 40' containers
522 ea. 4 high 20' containers
      on outside row
360 ea. 4 high 40' containers
      on outside row

FIG. 1-32.

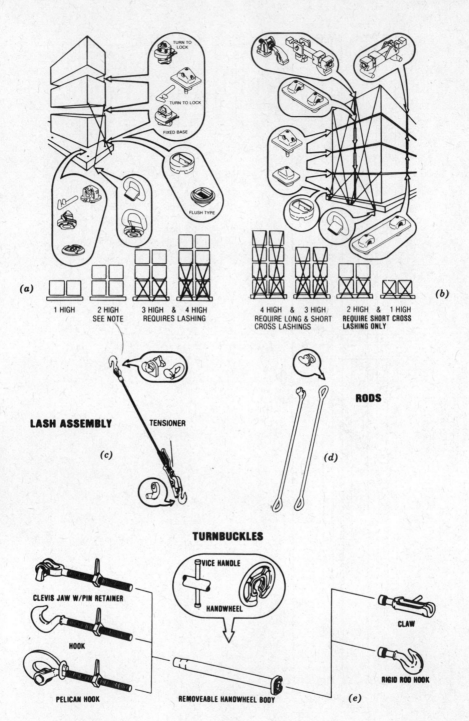

(a)

TURN TO LOCK

TURN TO LOCK

FIXED BASE

FLUSH TYPE

(b)

1 HIGH

2 HIGH
SEE NOTE

3 HIGH & 4 HIGH
REQUIRES LASHING

4 HIGH & 3 HIGH
REQUIRE LONG & SHORT
CROSS LASHINGS

2 HIGH & 1 HIGH
REQUIRE SHORT CROSS
LASHING ONLY

LASH ASSEMBLY

TENSIONER

(c)

RODS

(d)

TURNBUCKLES

VICE HANDLE

HANDWHEEL

CLEVIS JAW W/PIN RETAINER

HOOK

PELICAN HOOK

REMOVEABLE HANDWHEEL BODY

CLAW

RIGID ROD HOOK

(e)

40

carrier the space between rows of containers must be 4 ft to control damage to containers as the straddle carrier runs down the rows. With the SGC the space between rows can be reduced to 1 ft, and the lanes between blocks can be narrower.

The height of stacking with the straddle carrier is limited to two containers high, or for good customer service, which is important, to one and a half containers high. Three high is possible but not practical except for the storage of a few containers when selectivity is not important. The carrier can pass over a stack two high while carrying a container, but cannot pass over a stack three high. The SGC can stack three or four high and retain the ability to pick out of the stack any of the containers for delivery. Figure 1-32 illustrates this point. The plot illustrated is 600 ft long and about 94 ft wide, or 1.3 acres. There are 20-ft containers on the left half of the stack and 40-ft containers on the right. The layout shows that the straddle carrier system could stow 54 forty-ft and 108 twenty-ft containers, or 216 TEU. This amounts to 167 TEU/acre. With the SGC the same area could handle 130 forty-ft and 260 twenty-ft containers, or 520 TEU. This amounts to 403 TEU/acre, or a 240% improvement in land use.

To estimate the fixed cost of the needed land, we must know the level of productivity that must be met. For example, let us assume that the terminal must be capable of holding 2000 containers. The land needed for the straddle carrier system would be 2000/167 or about 12 acres. For the SGC system the land needed would be 2000/403 or about 5 acres. With these data plus the cost per acre of land we could determine the annual fixed cost of the land.

The Lashing of Containers Aboard Ship

Great strides have been made in lashing systems since the inception of containerization. New equipment is continually being introduced and today there are several systems available. Information on which system might be best for an individual ship can be obtained from the manufactur-

◄ **FIG. 1-33.** Lashing components. (*a*) Lock–Lash System. (*b*) Stack–Lash System. (*c*) The lash assembly shown is used on the Lock–Lash or Stack–Lash Systems. The strength of the lash is determined by the loads involved. End fittings should be selected to suit lashing arrangement and deck fittings. (*d*) Rod strengths should be determined by dynamic analysis using the ship's characteristics. End fittings and best arrangement can be selected to suit individual operating requirements. (*e*) Turnbuckles. Courtesy of Peck & Hale, Inc.

FIG. 1-34. Lashings with tensioners and turnbuckles. Note twist locks at the bottom of the figure. Courtesy of Peck & Hale, Inc.

42

FIG. 1-35. Use of rigid rods and turnbuckles. Courtesy of Peck & Hale, Inc.

43

ers of the various container securing systems. It is commonplace for the
manufacturer, such as Peck & Hale, to run a computerized dynamic
analysis of container accelerations, which determine the required lashings
and allowable weights.

Three common systems in use today are the locking, lashing, and
buttress systems. The *locking system* can be utilized with relatively

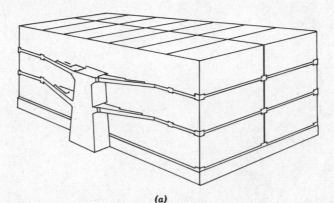

(*a*)

FIG. 1-36. The buttress system. (*a*) Buttress system in place. (*b*) Buttress towers aboard
the *Sea Land Freedom*.

(a)

(b)

FIG. 1-37. Removable stacking frames. (a) In place aboard ship. (b) On the pier.

lightly loaded containers stacked no more than two high. Locking is achieved by the use of stacking fittings such as twist-lock stackers and stacking cones (see Fig. 1-33*a*). The *lashing system* (Fig. 1-33*b*) uses wire rope (Fig. 1-33*c*), rigid rods (Fig. 1-33*d*), chains, or combinations thereof with various tensioning devices such as turnbuckles (Fig. 1-33*e*). Stacking fittings are used in conjunction with the lashing equipment for alignment of the container and additional restraint. Diagonal lashings are usually preferable to vertical lashings because they provide antiracking strength (see Figs. 1-34 and 1-35). The *buttress system* uses buttress towers that are permanently fixed to the deck at the fore and aft ends of each hatch. Removable stacking frames that allow several containers to be blocked and locked as one integrated unit fit into each of the buttresses. After a tier of containers has been loaded, the stacking frame is placed on top and the next tier is loaded on top of the stacking frame. The frame thus provides a rigid means of restraint. This system reduces container securing time but cannot accommodate containers of different heights except on the uppermost tier (see Figs. 1-36 and 1-37).

CONCLUSION

The largest containership built to date is the United States Lines' 4200-TEU vessel. This containership (see Fig. 1-38) represents the latest stage in this most remarkable innovation in the carriage of goods at sea. The

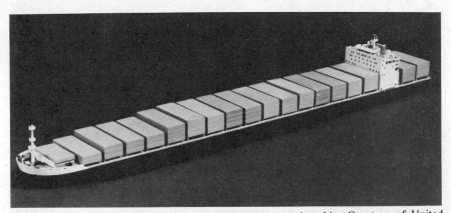

FIG. 1-38. United States Lines' newest 4200-TEU containership. Courtesy of United States Lines.

growth of containerization in the past 30 years to the point where, in 1981, 84% of all cargo moved was moved by containership is indicative of this success. The era of the breakbulk ship is not over, however. The need for its flexibility of service still exists. With improved unitization of loading and discharging and innovative marine materials handling equipment, the costs of a breakbulk operation can be significantly reduced.

CHAPTER 2

SHIPOWNER'S ORGANIZATION FOR CARGO PROCUREMENT, STOWAGE, CARE, AND DELIVERY

INTRODUCTION

This chapter is a brief account of the duties and responsibilities of ship operating company management personnel who carry out the functions related to procuring, receiving, loading, carrying, discharging, and delivering breakbulk general cargo. The organization of companies engaged in carrying only bulk cargoes is much simpler. Functions of cargo procurement, receipt, and delivery are nominal or eliminated entirely from the carriers' responsibility.

Figure 2-1 is an organization chart showing the titles and line of authority of the principal department heads of a linear company providing breakbulk or combination breakbulk and container service. Such companies are still found, especially in third-world country trade. (An organization chart for a containership company is shown in Fig. 2-6).

The president is usually the chief executive officer and a member of the board of directors. He must have a good understanding of the business of general cargo liner ship operations. In addition, he needs to know much about finance, economic geography, government policies toward shipping and trade, and conferences. Training in business management is also highly desirable in today's complex world. An executive vice president is often employed to handle many details for the president and to provide continuity of leadership during the president's absence.

The organization and areas of responsibility are quite different in liner companies operating only containerships from those in companies offering only breakbulk service. Some cargoes cannot be carried in standard containers, and there will always be a need for ships able to carry these goods, many of which are very specialized, such as refrigerated ships and heavy-lift vessels. The organization of such traditional companies will be described first. A later section of this chapter will deal with the structure

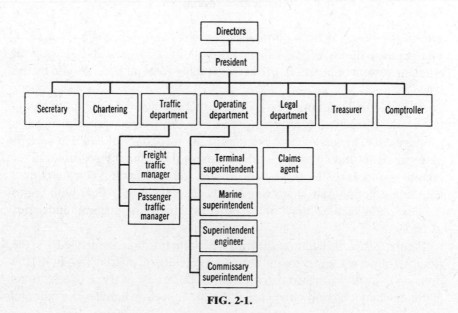

FIG. 2-1.

and added functions required of a truly intermodal containership liner company.

THE RATE CONFERENCE

A brief explanation of the term "conference" seems in order. Liner companies, in an effort to control competition, have developed semimonopolistic organizations known as conferences. A conference is in reality a form of trade association that seeks to produce agreements between members relative to freight rates, sailing schedules, and ports of call on a given trade route. Such conferences may be quite informally conducted or very formally organized. A strict and formal organization is necessary when a number of lines operate a cargo service over the same route. Under such conditions, competition is very intense with regard to efficiency and quality of service.

THE TRAFFIC DEPARTMENT— BREAKBULK/COMBINATON SERVICES

The Traffic Manager

The traffic department is normally headed by a *freight traffic manager*. The primary duties of this officer are soliciting cargo and managing an efficient system of processing the cargo to and from all ports served by the line.

Although solicitation of cargoes and management of the groups doing the clerical processing of such cargo are the freight traffic manager's primary duty, he has other, very important secondary duties. He participates in steamship company conferences as the company representative, conducts and analyzes surveys of routes and extensions, considers new ship construction and chartering needs insofar as they deal with cargo services, and keeps abreast of the flow, progress, and general tendencies of traffic.

Figure 2-2 is a diagram showing the organization controlled by the freight traffic manager. A universal characteristic of the freight traffic division is its division into two sections. The largest of these two sections is the one handling outbound cargo; the other section handles the inbound cargo.

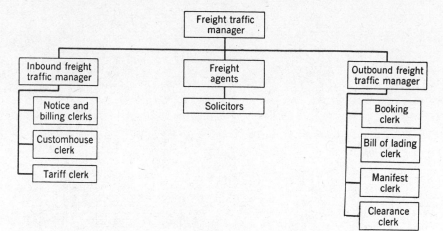

FIG. 2-2.

The Outbound Section

The *outbound freight traffic manager* supervises the staff as shown in the diagram. This staff is concerned with booking cargo, preparing the shipping documents necessary for the smooth flow of cargo, and paying the charges connected with transportation of cargo. The practical operation of this part of the freight department is best covered by presenting a narrative description of the functions of the most important desks together with appropriate explanatory comments (see Fig. 2-3).

The booking desk. The booking desk is the logical starting point for our description. A shipper wishing to use the services offered by a common carrier serving an overseas route would first contact the booking desk and give the booking clerk all necessary information concerning his shipment. This is done over the phone or via telex in many instances. The booking clerk has before him a schedule of all ships on the service for which he is booking cargo. Companies operating ships over several routes may have a desk for each service. The schedule lists all discharging ports and loading ports, the dates the ship is to be ready for loading at these ports, and the sailing dates.

The first step in booking for a given ship is to prepare data on the ship and enter them on the first page of what is to become the *engagement sheet*. The engagement sheet is the controlling document for booking a ship. The data entered on this sheet will include such things as the cargo deadweight capacity available; the cubic capacity available; information

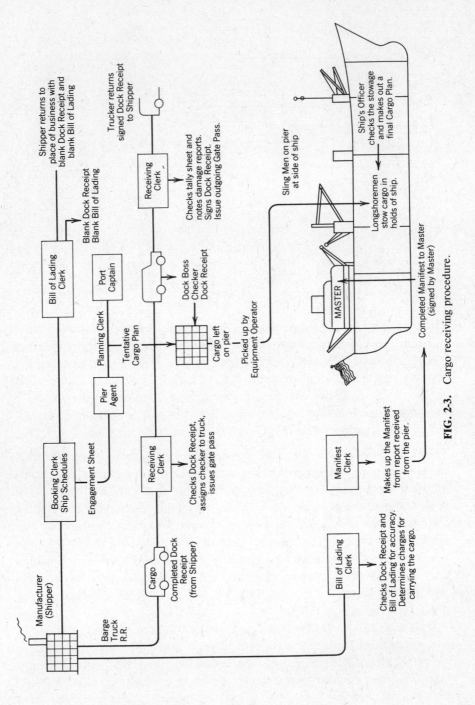

FIG. 2-3. Cargo receiving procedure.

Manufacturer (Shipper)

Shipper returns to place of business with blank Dock Receipt and blank Bill of Lading

Bill of Lading Clerk

Blank Dock Receipt Blank Bill of Lading

Booking Clerk Ship Schedules

Engagement Sheet

Pier Agent

Planning Clerk

Port Captain

Tentative Cargo Plan

Trucker returns signed Dock Receipt to Shipper

Receiving Clerk

Checks tally sheet and notes damage reports. Signs Dock Receipt. Issue outgoing Gate Pass.

Dock Boss Checker Dock Receipt

Cargo left on pier

Picked up by Equipment Operator

Barge Truck R.R.

Cargo

Completed Dock Receipt (from Shipper)

Receiving Clerk

Checks Dock Receipt, assigns checker to truck, issues gate pass

Bill of Lading Clerk

Checks Dock Receipt and Bill of Lading for accuracy. Determines charges for carrying the cargo.

Manifest Clerk

Makes up the Manifest from report received from the pier.

Sling Men on pier at side of ship

Ship's Officer checks the stowage and makes out a final Cargo Plan.

Longshoremen stow cargo in holds of ship.

MASTER

Completed Manifest to Master (signed by Master)

52

concerning deep tanks, heavy lifts, length limits, and refrigeration capacity; and similar information relating to the ship's ability to carry cargo.

Broken stowage.   The cubic capacity that the ship has available is based on an assumed value of 25% for the broken stowage. Broken stowage values will always vary because they are a function of several uncertain factors (see Chapter 4).

The engagement sheet.   The booking clerk maintains the engagement sheet day by day until all booking is completed. On the pages following the data page, all information regarding each shipment booked for a particular ship is recorded. The engagement sheet is a running account of the status of the proposed cargo to be laden on the ship. As such, it is obvious that it must contain such basic information as:

1. The weight of each package and the weight of the entire shipment.
2. The maximum measurements of the shipping units and the total cubic of the shipment.
3. The total number of packages in each shipment.
4. The name of the shipper.
5. The name of the consignee.
6. The descriptive or shipping name of the cargo. The name used must explain the characteristics of the cargo well enough that a decision can be made regarding its care and custody.

At the end of each day, the cargo booked is recapitulated, and the outbound freight traffic manager can tell at a glance how much of the total capacity of the ship is consumed. One page of the engagement sheet is devoted to the recapitulation of all bookings daily. The total booked at any moment can be determined easily. The total booked is compared with the total actually received at the terminal from information phoned into the freight department daily by the receiving clerk's office at the terminal. In some operations, the cargo is booked only a day or so before delivery to the terminal. In other operations, booking may occur several days to a few weeks before the ship is ready to receive the cargo. Copies of the engagement sheets are sent each day to the operations department for stowage planning.

Full and down condition.   The profit realized by the carrier will vary directly with the experience and knowledge of the freight traffic manager in arranging for and booking cargoes. It is during this first step that the success or failure of the venture may be decided. In times of heavy competition, when cargo is scarce and full loads are not available, some of the problems connected with the booking desk are solved by the circumstances. The objective, however, is always the same: to book a cargo that will use up all of the cargo deadweight of the ship and at the same time consume all of her available cubic capacity. When a ship is loaded so that these conditions are met she is said to be loaded *full and down* because her volume is "full" and she is "down" to her legal draft marks. Theoretically, when loaded in this fashion, she is carrying her greatest paying load.

In times of cargo shortage, there is little probability of overbooking a ship. Many ships will sail neither full nor down. However, when plenty of cargo is available, the ship may be overbooked as a regular policy of the company. Overbooking by about 10% of the ship's capacity, either deadweight or cubic, helps to protect the shipowner from the nondelivery of cargo to the terminal in time to be loaded on the ship. There may be cases in which cargo is booked and delivered to the terminal, but still cannot be loaded, as when there is a failure to obtain an export license. If cargo is booked and the ship becomes loaded before that cargo has been placed on board, the decision as to which cargo will be left behind must be made by consultation between the traffic and operations departments. The likelihood of such "shutouts" should be evident by the time 80% of the ship's capacity has been used. The value of the cargoes still ashore, the identity and importance of the shippers, and the possible effect on future business with regard to each customer must be considered carefully. Close cooperation and coordination within the carrier's organization is absolutely essential to arriving at the most reasonable solution.

After booking the cargo, the shipper is provided with blank copies of the *dock receipts* and *bills of lading*. The shipper must fill out these forms with all descriptive data concerning the cargo, such as the name of shipper and consignee, the volumes and weights of the packages; a descriptive name for the commodity, the type of container, the export license, and the export declaration numbers.

The dock receipt.   The dock receipt is a shipping paper that fully describes a given shipment. When the shipment has been duly received by the carrier at its terminal, the signature of an authorized agent on the dock receipt is the shipper's evidence that the cargo described thereon has been delivered to the custody of the carrier. Some notations may be made on the face of the dock receipt if the cargo received is not found to be precisely as described on it. This is very important to protect the carrier against possible future claims. The signed dock receipts are sent to the traffic department to be matched up with the bills of lading. Any exceptions noted on the dock receipt must also be recorded on the bill of lading.

The bill of lading.   The bill of lading is the most important of all the shipping papers used in negotiations between shipper and carrier. It contains a complete description of the shipment, and it also sets forth the provisions under which the cargo is shipped. It is the contract between the shipper and the shipowner for the transportation services involved. Its provisions are based on either the *Harter Act of 1893* or the *Carriage of Goods by Sea Act of 1936,* depending on whether the carrier is operating on domestic or foreign trade routes, respectively. Under certain conditions, the bill of lading is negotiable and the shipper can use it as the basis for a draft.

There are two major types of bills of lading. Under the *order* bill of lading, the shipper has not yet been paid for the cargo being shipped. Therefore, before the cargo can be obtained by the consignee at the port of discharge, the shipper must release the cargo by presenting the consignee with an endorsed bill of lading. The carrier is authorized to release the cargo to the consignee only if he presents the endorsed bill of lading. This arrangement is for the financial protection of the shipper. The other type of bill of lading is the *straight bill.* The straight bill of lading is drawn directly in the name of the consignee and needs no endorsement by the shipper before delivery is authorized at the port of discharge. A little more discussion of the immunities and responsibilities of the shipper and shipowner will be found in Chapter 3; these factors are controlled by the provisions of the bill of lading used on ocean carriers.

The Terminal Superintendent and Staff

We now shift to Fig. 2-4, which is a diagram of the organization under the *terminal superintendent.* The terminal superintendent supervises the operation of the various piers or berths making up the terminal facilities. He is in charge of the cargoes received, delivered, loaded, and discharged at the terminal. The terminal superintendent may have several assistants to manage individual piers or wharf areas, or specific ships. The assistant superintendents are supplied copies of the engagement sheets by the freight department. From these sheets the assistant superintendents pre-

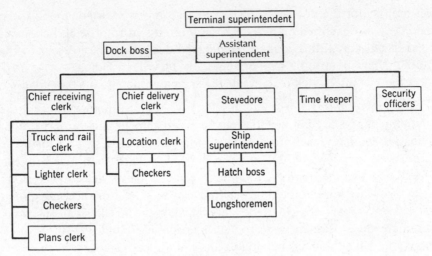

FIG. 2-4.

pare tentative stowage plans for the cargoes. These plans and the problems involved in laying out the ships will be discussed in greater detail in Chapter 5. Cargo is brought to the piers by truck, by railroad, or on lighters or barges. The assignment of appropriate space on the terminal for these cargoes is an important function of the assistant superintendents, who work closely with the dock boss in this regard.

Receiving procedure at the terminal.   The *chief receiving clerk* supervises the staff in charge of receiving and checking the cargo at the terminal. The receiving office is equipped with windows that are manned by clerks prepared to process the papers of specific connecting carrier types. If the amount of traffic warrants it, there will be a window for each of the three connecting carriers mentioned above (i.e., trucks, railroads, and lighters or barges). If not, the number may be reduced so that one window may take care of two types and so on. In smaller ports one office may handle both receiving and delivery. The clerk manning the window processes the dock receipt, which should be filled out by the shipper before it is given to the trucker (or other connecting carrier representative). The routine at each window is similar. The trucker presents his copies of the dock receipt to the clerk manning the truck window. The clerk checks the copies over to see that they are properly made out. The clerk also checks the export license number and the export declaration number. The receiving clerk may also have a copy of the engagement sheet, and the shipment will be checked off on that. If there are any discrepancies, or if he receives special instructions from the traffic department, the clerk may receive the shipment but stamp all copies of the dock receipt with the words "hold on dock." This hold-on-dock order will prevent the cargo from being loaded until it is properly cleared.

After checking all copies of the dock receipt, the receiving clerk enters the trucker's truck license number on the dock receipt, assigns a number to the dock receipt, makes out a gate pass, and finally assigns a checker to the truck. The checker accompanies the truck to the point on the terminal where the cargo is to be unloaded. This point is designated by the *dock boss*. The cargo will rest there until it is picked up for movement to the terminal apron. At the apron it is attached to the ship's cargo hook and hoisted on board. The terminal position where the cargo is held prior to being moved to the terminal apron is often referred to as the *pickup point*. At some terminals, the truck proceeds to an unloading platform, where the cargo is unloaded directly onto a pallet. The pallet will be taken to the pickup point from the unloading platform. This last arrangement obviates the need for the trucks to go into the actual warehouse area. Control over pilferage is much better if trucks can be kept out of the warehouse area.

As the truck is unloaded, the checker checks each package. He checks the dimensions, marks, and general characteristics of the cargo to see that they are as stated on the dock receipt. If there are any damages, shortages, or discrepancies, the checker makes appropriate notations on the dock receipt. These notations are, in effect, corrections to the data listed on the dock receipt by the shipper. Later, these modifications or, as they are more accurately called, "exceptions," are used to correct the bills of lading so that the carrier is not held responsible for shortages and damages that occurred before the cargo was received into his custody.

When the truck has discharged the cargo appearing on the dock receipt and it has been checked, the checker makes out a *tally slip*. The tally slip contains information concerning the entire shipment, such as the total number of packages, consignee marks, shipper's name, ship on which the shipment is to be loaded, weight, and measurements. One copy of this tally slip is secured to the shipment.

The trucker returns to the truck window of the receiving clerk's office with a copy of the tally slip and the dock receipt. The receiving clerk now signs the dock receipt and presents a copy to the trucker. This copy of the dock receipt is the shipper's receipt for cargo delivered into the custody of the shipowner. Later, this dock receipt is presented by the shipper to the bill of lading clerk in the freight department when he picks up the bill of lading for the shipment.

Additional Clerical Work in the Freight Department

We now leave the cargo at the terminal and return to the freight department. The shipper appears with his signed copy of the dock receipt and copies of the bill of lading. The bill of lading clerk checks all the data appearing on the dock receipt against those on the bill of lading. The bill of lading is then passed to a desk where the tariff is assigned, and is given a number. The assignment of the tariff rate is very important. If the carrier is a member of a conference, it must be done in accordance with the rules set down by that conference. The clerk doing this has several books listing all the commodities that might be carried as cargo on any of the various services. If no listing appears, there is a general cargo rate that will apply. The general cargo rate might be listed as follows: 0.80/1.50. This means that the cargo is shipped for $0.80/ft^3 or $1.50/100 lb. The choice of which rate the shipper is charged is at the shipowner's option. The choice is always made so that the greatest income is obtained. Additional surcharges are also added as applicable. These may be for heavy lifts, extra lengths, currency adjustment factors, bunkers, or port congestion. A freight invoice is made up, after which the shipper pays the invoice and picks up the bill of lading. One copy of the dock receipt will have been forwarded to the freight department from the receiving clerk's office, and the bill of lading data will have been checked before the shipper or his agent appears. The bill of lading is then signed by an authorized person on behalf of the ship's master to signify the shipowner's agreement to the terms of the contract represented by the bill of lading provisions.

Copies of the bill of lading go to the *manifest clerk* or to an electronic data processing specialist who will use computer equipment to make up the manifest. The manifest clerk makes out the manifest from these bills of lading. The manifest is a complete list of the cargo loaded on the ship. It gives the descriptive names of the cargo, marks, destinations, shippers, and consignees, weights, cubics, number of packages; and other data. Preparation of the manifest requires speed and accuracy. It must be used to clear the ship, yet it cannot be filled out until the loading of the ship starts. Mistakes on the manifest may leave the shipowner open to fines.

The Inbound Section

The above has been concerned with outbound cargo. Inbound cargo demands much less detailed work, and hence a smaller staff is needed to handle it. The *inbound freight traffic manager* is concerned primarily with obtaining rapid and correct customs clearances for all cargo and sending out *arrival notices* to all consignees to insure prompt removal from the terminals. The consignee is given a copy of the bill of lading, which he presents at the customhouse. If all is in order, the customs office issues the consignee a release that authorizes the customs office at the terminal to release the goods. Following the issuance of the customs' release order, the consignee returns the bill of lading to the inbound freight office, where he receives a company release order for the shipment.

When the cargo is discharged from the ship, it enters the control area of the *chief delivery clerk,* who is under the supervision of the terminal superintendent or assistant superintendent. The delivery clerk is charged with receiving the cargo from the ship and then releasing it to the consignee or his agents according to a set legal procedure. When a trucker arrives at the terminal to pick up cargo, he should have in his possession a *release order.* There are several types of releases. The release may be from the consignee of the cargo; in this case it is called a *paid delivery* release. *In-transit releases* are issued when the cargo is to be delivered to another carrier for transportation to the hinterland, generally with the understanding that all customs and duty fees will be paid there. An entire shipment or parts of a shipment may be taken to the U.S. Appraiser's warehouse so that its value may be appraised to check on the duty required or for other reasons. When cargo is being picked up for delivery to the appraiser's warehouse, customs issues an *appraiser's release order.* This allows the cargo to be picked up and transported to the warehouse. *Re-export releases* are issued when the cargo is to be picked up to delivery to another terminal where it will be loaded on another ship for re-export. If a free zone exists in the port, *releases to the free zone* will also be used.

Regardless of the type of release, the delivery routine is the same. The chief delivery clerk will have had a *delivery book* prepared from a copy of the ship's manifest some time before the consignees start calling for their cargo. In the delivery book, all the cargo is listed alphabetically by consignee. Each entry contains all the descriptive data concerning the cargo that would appear on the bill of lading. Another book, which may be called a *location book,* is kept by a clerk of the same office. The location book shows the location of each shipment on the pier after it has been discharged. The dock boss, under directions of the assistant superintendent, designates the stowage location of all shipments discharged from the ship.

On arrival at the terminal, the trucker proceeds to the delivery clerk's office and has his release papers checked. He then proceeds to the customs office on the dock and gets clearance permits. He returns to the delivery clerk's office, where a checker is assigned to his truck. The clerk keeping the location book directs the checker and trucker to the point where the cargo is. The checker checks the cargo onto the truck and makes out three copies of a tally slip. The trucker returns to the delivery office with one copy of the tally slip. A clerk checks the data on the tally slip against those appearing in the delivery book. If they are the same and all is in order, the delivery clerk signs the tally slip and issues the trucker a gate pass, or the tally slip may be used as a gate pass. The trucker also signs the delivery book, indicating his approval of the checks made. When a given shipment has been completely delivered to the consignee or his agents, the word "completed" is stamped over the entry concerning the shipment in the delivery book.

Cargo left at the terminal beyond the *free time period* may be removed by the customs office to a customs warehouse. The free time period is generally 5 days, beginning the day after all the cargo has been discharged. However, in some cases, such as with coffee, 10 days are permitted. Demurrage will be charged against the consignee for every day the cargo is left at the terminal beyond the free time period.

The Claims Agent

The *cargo claims agent* may be under the supervision of either the traffic manager or the terminal superintendent or be attached to the legal department. The routine in the claims office is similar regardless of the office having supervisory control. Claims can be minimized by alert checking when the cargo is received at the terminal and by alertness on the part of deck officers when the cargo is being loaded onto the ship. All information regarding damaged and recoopered cargoes is routed to the claims office on standard company forms. Such information is filed according to ship and shipper upon receipt. The information may never be referred to again unless the company is contacted by a claimant. If a request for a settlement is received, the data are checked and a quick settlement is attempted.

The majority of cases are settled without litigation. At times, however, the need for litigation arises, and the case goes to court. In such cases, the provisions of the bill of lading are used by both parties concerned as a basis for settlement. These provisions will be discussed in Chapter 3.

The most frequent type of claim is for shortages, such as when 500 cases are checked onto the ship and only 494 are checked off. The missing packages may never have been loaded, or their contents might have been pilfered during the time that the packages were in the hands of the carrier. In such cases, the ship is liable and settlement is quick. The claims agent will review the facts of the case and make certain that the claimant is not claiming more than the actual loss. In the case of cargo damage from any cause, an effort may be made to settle on a percentage basis if there is reason to believe that insufficiency of packaging or some cause beyond the responsibility of the carrier contributed to the damage of the shipment in question. When such cases reach the courts, the shipper generally maintains that the cause of the damage is attributable to either improper stowage or improper care and custody. The carrier will maintain that the damage was attributable to one of several other causes, frequently mismanagement on the part of the ship's complement.

Variation of Systems

The foregoing has been an attempt to outline briefly the organization and the mechanics involved in procuring cargo for the common carrier and getting it to the marine terminal through which it must pass as it leaves the land carrier and comes under the jurisdiction of the water carrier. It must be pointed out that the description given of the process is not representative of any one company. The exact procedure varies from company to company depending on the size of the organization and the trade routes served by it.

The Stevedore

Before concluding this orientation discussion, it is necessary to cover the functions of the *stevedore* and the *ship's officers* as they apply to cargo operations. The *stevedore,* who may be a company employee or a contractor, is in charge of the actual loading or discharging procedure. He may assist the terminal or assistant superintendent in *laying out* the ship. Laying out is a phrase used to describe the preparation of the tentative stowage plan. The stevedore's primary duty, however, is to execute this plan. The stevedore will generally work with more than one vessel at the terminal; hence, he needs assistants. His immediate assistants are known as *ship superintendents* or *walking bosses*. (The term "walking boss" is used primarily on the west coast of the United States.) For every hold or cargo compartment working cargo, there will be a *gang* of longshoremen. The number of longshoremen in each gang is controlled strictly by union regulations. One man in the gang, known as the *gang boss,* acts as the foreman. The gang boss is the intermediary between the longshoremen and the walking boss. Generally eight men are assigned to the hold and six men to the dock; there are two winch drivers, a hatch tender, and some drivers of lift trucks or other equipment. The gang generally runs to about 16 or 20 men, but there is no universal standard.

Most contract stevedores nowadays supply the steamship operator with complete terminal operations, which include, of course, the functions of receiving and delivering as described above. Such work is done in close liaison with the steamship company freight and operations departments.

Ship's Officers

Leaving the dock and going aboard the ship with the cargo, we come in contact with that part of the shipowner's organization that will be responsible for the cargo during its time on the ship. This group consists of the

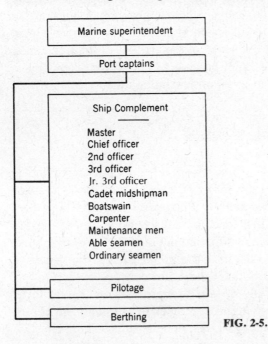

FIG. 2-5.

master and his deck officers, supervised by the *marine superintendent* (see Fig. 2-5). The master of the ship is the responsible ship's officer who acts as the shipowner's agent in many instances during normal operation.

In these days of shiploads of hundreds of general cargo items and rapid operating techniques, it is impossible for the master of the ship to carry out the duties and assume the responsibilities that many laymen attribute to him. Large modern steamship companies have complete staffs to do the work of procuring cargoes, checking them on board, and processing the many legal shipping papers necessary to operate the vessel. However, the master still stands as the shipowner's expert representative with respect to the handling and stowing of the cargoes booked for carriage. As such, he should instruct and organize his officers so that they will understand precisely what is desired in the stowage of the cargo on hand and should have the spirit and depth of knowledge to insist intelligently on what is best. Briefly stated, the master and deck officers have the responsibility of seeing that there are never grounds for a shipper of cargo to claim improper care and custody as a cause of cargo damage aboard ship.

Briefing of Officers Concerning Cargo Operations

The chief officer should be briefed by the master and the walking boss or the pier superintendent regarding the tentative plans for stowing the vessel's cargo. The second and third officers should be present during this meeting also. This briefing is frequently missing in practical operations in spite of the obvious need for responsible officers to be informed of any operation over which they will be expected to have some degree of control. Company operations managers should insist that such essential briefings always take place.

Duties While Loading

Once the loading of the cargo commences, all of the ship's officers must concern themselves with checking on the stowage of all items that come on board the ship. They must know what is and what is not correct stowage, and they must insist that all stowage be done correctly. What constitutes correct stowage is a question that will be answered in detail in Chapter 5.

The ship's officers also make certain that the ship's cargo handling equipment is correctly rigged, and they keep a record of where all cargo is stowed on the ship. Keeping this record should be the primary duty of one of the ship's officers. From the data taken by observation, a final-as-stowed cargo plan should be prepared. In most large steamship companies, a *plans clerk* is attached to the office of the chief receiving clerk. The plans clerk is responsible for preparing a final stowage plan. In practice, however, the interests of the shipowner are served best by the officers of the ship keeping such a plan. Plans clerks are not as closely associated with the actual stowage as are the ship's officers, nor are they by nature as interested. Serious mistakes may appear on the plans prepared by clerks simply because of the mechanics used in their preparation. Preparing a plan of the stowage will increase the ship's officers' knowledge of the stowage on the ship that they are going to take to sea. The firsthand manner of acquiring the information will necessitate their being in the ship's holds during the loading of the cargo and circulating from compartment to compartment. Knowing their ship and knowing the cargo on board their ship is a fundamental principle of safety and of efficient merchant ship operation. Any philosophy or circumstance of operation that tends to minimize the importance of the ship's officers' knowledge of their ship and her cargo or fails to encourage the acquisition of such knowledge is to the serious detriment of the ship, company, or merchant fleet so affected.

Cargo Exceptions and Damage Reports

As mentioned above in discussing the receiving clerk's duties, the checker checks the cargo as it is loaded from the connecting carrier to the dock to see that it is as represented upon the dock receipt. This is but one check that the shipowner's organization makes against accepting cargo that is not in good condition on receipt. The ship's officers must always be making checks also. They must visually check cargo while it is being loaded. Obviously, they cannot check all cargo that comes on board, but they can spot check almost all shipments. Any discrepancies should be reported by them on a *cargo exception form* or *cargo damage report form*. One form may be used, but the context of the remarks will vary depending on whether the report is in fact an exception or a cargo damage report.

The cargo exception is a listing of cargo defects that should be excepted on the bill of lading. In other words, the cargo is received as stated in the terms of the bill of lading except for the points mentioned in the exception report. The cargo damage report is simply a statement of damage to cargo giving all details.

As mentioned above, most cargo exceptions will be noted first by the checker who visually checks the cargo when it is received at the terminal. It is obvious, however, that this is not going to preclude some packages or shipments being missed by the checker in one detail or another. The officers of the ship are the next responsible members of the shipowner's organization to come into contact with the cargo. The damage or condition being excepted may have been caused while the cargo was on the terminal; if so, the claims department should know this. The most important period for the ship's officers to be on the alert for exceptions is when loading cargoes in foreign ports where the shipowner's agent may not have a trained or experienced dock force available to check cargo on receipt from the shipper. In these instances, the ship's officers' check may well be the only check. The neglect of proper exceptions could mean the loss of a great deal of money for the shipowner. Exceptions should be noted on the mate's receipts. These exceptions will then be noted on the bill of lading, thus protecting the shipowner.

The cargo damage report is a statement of fact regarding the condition of cargo when discovered in the ship in a damaged condition. Generally there are a number of such reports to be made when the cargo is discharged. During the loading of cargo, longshoremen frequently damage the cargo. In such case, a report is made out and a representative of the stevedore will sign it, acknowledging the facts. If damaged cargo is found in the ship during loading and the cause cannot be determined, a report should be made. Some companies require that the damaged cargo be sent back to the dock and recoopered before being stowed in the ship. When this is done, careful checking is required so that the bill of lading will state correctly the condition of the cargo.

The cargo exception and the damaged cargo report are similar in context. Many companies do not require the ship's officers to differentiate between them; they require a report of the damage or exception on a standard form. Such reports can be made more meaningful, however, if the ship's officer is aware of the differences between the two reports. Both types of reports should record six essential items:

1. Consignee marks.
2. Commodity type.
3. Container type and number involved.
4. Stowage location when discovered.
5. Extent of the damage.
6. Remarks concerning the facts of the discovery.

The principal difference between the cargo exception and the damaged cargo report will be in item 6. As an example, the remarks on a cargo exception concerning damaged steel rods would read as follows:

> Steel rods arrived alongside the ship in a gondola car with rods bent and considerable rust in evidence.

The officer making such a notation could make the report more useful by giving a count of the number of bundles involved or any other relevant extension of the remarks.

The cargo damage report would lack any reference to the fact that cargo had just been received from the connecting carrier. A cargo damage report on the same type of cargo would read along the following lines:

> Steel rods bent by longshoremen in breaking them out of stowage. Top tiers were very rusty, probably caused by heavy condensation during the voyage.

THE TRAFFIC DEPARTMENT— CONTAINERSHIP MANAGEMENT ORGANIZATION

The carriage of general cargo in containers suitable for intermodal transport by ship, barge, truck, or railroad started in 1956. Breakbulk operators provide only the marine transport of cargo from port to port; the containership operators have extended their services much further. These innovative carriers' responsibilities may begin with the delivery of cargo containers to the shippers' premises, which could be hundreds of miles from the ship's port of call. When the shipper has filled the containers, the containership company may then arrange for all phases of transport to the ultimate destination, providing these additional services involves the ocean carrier in problems of container ownership, inland transport, and equipment location control. Inland cities become "ports" and depots must be established at major locations for the storage and repair of containers.

The development of containership services has made necessary substantial changes in management organization. To manage and promote the greatly expanded services of the carrier effectively, the organization structure had to be revised as shown in Fig. 2-6. New capital expenditures

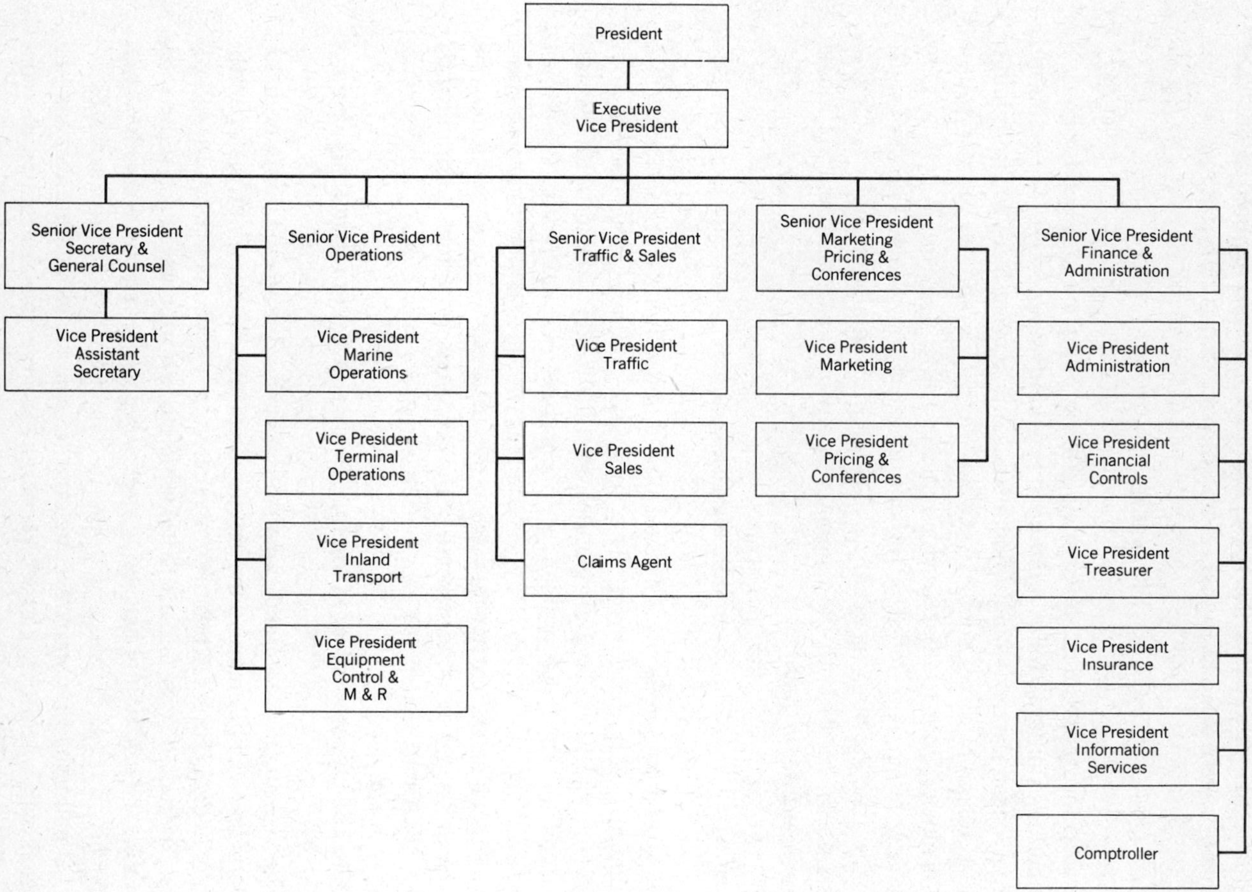

FIG. 2-6.

President

Executive Vice President

Senior Vice President Secretary & General Counsel
- Vice President Assistant Secretary

Senior Vice President Operations
- Vice President Marine Operations
- Vice President Terminal Operations
- Vice President Inland Transport
- Vice President Equipment Control & M & R

Senior Vice President Traffic & Sales
- Vice President Traffic
- Vice President Sales
- Claims Agent

Senior Vice President Marketing Pricing & Conferences
- Vice President Marketing
- Vice President Pricing & Conferences

Senior Vice President Finance & Administration
- Vice President Administration
- Vice President Financial Controls
- Vice President Treasurer
- Vice President Insurance
- Vice President Information Services
- Comptroller

approaching the cost of the vessels themselves were required. Significant changes were needed to market and direct the new services.

The Traffic Department

The traffic department is responsible for booking cargoes and for clerical functions similar to those required in breakbulk operations. The extension from ocean carrier to multimodal transport service has brought about a need for more direct contact between the carrier and the shipper and/or consignee. Door-to-door service requires traffic department personnel to have a broad knowledge of the entire transport system. The staff must coordinate the booking of cargoes and empty container requirements with the personnel of both the equipment control department and the inland transport department.

The Sales Department

In some companies the solicitation of cargoes is the responsibility of a separate sales department. The complexity of providing door-to-door service and increasing competition require freight solicitors to have an extensive knowledge of inland transport services and tariffs. They must be aware of the shippers' and consignees' needs and alert to any possible way in which the carrier may assist the customer. The company must be so organized that it can respond to the customers' needs as determined by the solicitor.

The Department of Marketing, Pricing, and Conference Affairs

The previously mentioned expansion of carrier involvement into door-to-door service has enlarged the geographical area within which the operator may actively look for business. Research into market opportunities, analysis of trade trends, and the development of information to assist the sales staff is the function of marketing specialists. Increased emphasis has been placed on these activities with the introduction of container services.

Containership companies belong to the same conferences as the breakbulk operators. Carriers generally charge the same rates for containerized cargo as for breakbulk transport, but special rates have been established for some full container loads. *Through rates,* sometimes

known as *point-to-point rates,* have also been included in container tariffs
for movement of full container loads between selected inland locations.

The Operations Department

Four interdependent operating functions are involved in a container transport system.

The *marine operations* department is concerned with the operation of
the ships. The principal differences between breakbulk and containership
vessel operations are in the handling and stowage of the cargo. Containerships are discharged and loaded in hours instead of days. This has created
a need for special attention to the requirements of services for the vessel
itself. The marine department must arrange to take care of all the ships'
needs during the very limited time that it is scheduled to be in port.

The *terminal operations* department is responsible for the careful and
efficient handling of cargo at a reasonable cost. Some liner companies
own or lease the terminal facilities at major ports of call. Arrangements
for berthing and for the handling of cargoes by terminal operators at other
ports must be made by this department. Costs and performance are periodically reviewed for each terminal to assure that appropriate standards
are maintained.

The *inland transport* department arranges for all movement of equipment beyond the marine terminals. The cost of these movements is an
area of major expense for the ocean carrier. The inland transport organizations to which the equipment is assigned for transport must all give
proof of reliability and show evidence of having suitable insurance. Negotiations with these carriers regarding liability and the rates for all movements are the responsibility of this department. The cost and quality of
the services that the inland transport department controls are of great
importance to the overall success of the entire venture.

The *equipment control* department represents another new area of
operations not required in breakbulk services. Containership operators
provide the containers free of charge to the shippers. No fee is assessed
for the placement of the empty unit at the shipper's door, nor for removal
of the empty unit from the consignees' premises. The shipper or consignee has a limited time within which to unload or load (empty or stuff)
the container. About 48 hours of free time may be allowed; beyond the
allowed limit a daily rental charge is assessed. Major containership opera-

tors may have as many as 90,000 containers and 30,000 chassis in service. The purchase, maintenance and repair, safety certification, and location control of all this equipment are tasks requiring good organization and planning. Traffic flow is seldom equal in both directions; thus a surplus of empty equipment must be dealt with at minimum cost to the company. The positioning of empty containers for ready availability to the shippers at the desired loading time and the repositioning of empties after discharge must be carefully controlled.

Operation of the Complete Containership System

The following example illustrates the manner in which the full containership system serves a customer.

The *marketing* department learns of the sale of a product manufactured in the midwestern United States to a consignee in central Europe. This information is passed on to the sales department, which has a solicitor call on the prospective shipper. The shipper agrees to use the solicitor's line and specifies a need for five containers. Production schedules will assure the filling of one container every 2 days. The containers are then to be hauled to the seaport for loading on the first ship available. The booking clerk is advised of these requirements by the solicitor or by the shipper's traffic manager. This clerk notes that two or three containers may arrive at the port for loading on one ship and the balance will arrive later for loading on a ship sailing the following week. These bookings are noted on the ship's booking sheets in a manner similar to that in the breakbulk operation. In this case, however, the ship's capacity is specified in terms of container spaces available aboard each ship, often given in TEUs. The ship's TEU capacity may be subdivided into spaces or "slots" under deck and spaces on deck. Further details may also include the number of locations where electric power is available for operation of the refrigeration machinery of reefer containers.

The booking clerk advises the equipment control department of the new booking, giving specific details of the type of equipment needed and the dates and locations at which it is to be available. The control department must then evaluate where suitable equipment is likely to be located closest to the loading site at the dates specified. If appropriate containers formerly holding import cargo are likely to be empty in the general area of the shipper, the problem of supply is simplified and the costs of positioning empty containers are minimized. More likely it will be necessary to move some empty containers from an inland "port" at which a reasonable supply of equipment is regularly maintained.

The equipment control department advises the inland transport department of the booking, the shipper's requirements, and what equipment is to be made available to the customer. The transport department then informs the booking clerk of the details regarding the supply of empty containers as well as what arrangements are planned for transport of the full containers from the shipper to the port of loading. A growing number of shippers in the United States make their own arrangements for transport of the loaded containers. The booking and transport information is also relayed to the terminal department so that it may plan to receive the equipment on the designated dates for loading on assigned ships. The marine operations department is notified concerning the bookings in order that it may plan the ship's stowage. Booking information includes the types of containers, the sizes of the containers to be used, and loading/discharging ports. Considerations such as hazardous cargoes, goods susceptible to heat or cold, and other special requirements are also recorded.

The marine terminals department receives the loaded containers in somewhat the same manner as in breakbulk shipments. Although the ocean carrier may have arranged for the inland transport, the actual responsibility for protection and safe delivery of the container and its cargo is in the hands of the land carrier.

The procedures at the marine terminal are typically as follows. The truck driver arrives at the terminal entrance gate, which has a number of booths similar to highway toll booths. These booths are staffed by receiving clerks as described for breakbulk operations. The truck driver need not leave his vehicle to hand over the dock receipt and bill of lading for the container load. The clerk notes the time of delivery and compares the container identification numbers on the papers with the numbers of the containers being received. In many cases the truck and chassis are standing on a scale and the receiving clerk records the gross weight as well as equipment weight on his documents.

The doors of the loaded containers are always firmly secured with tamperproof seals. The seals are individually numbered, and these numbers are also shown on the dock receipt. The seal number and the fact that it was intact are noted by the *equipment inspector*. Any broken seal or wrong seal number would alert this individual to the need to have the container opened by terminal security personnel and its contents checked before responsibility for the cargo could be accepted. Normally the ocean carrier accepts the shippers' declaration of the contents of containers.

When the clerk has completed his tasks the driver is told to proceed into the terminal to deliver the equipment at an assigned location. After doing so, he returns to the gate to retrieve the signed dock receipt and proceeds on his way.

While the receiving clerk is processing the delivery documents at the gate, an equipment inspector checks on the condition of the container and chassis and fills out an "equipment interchange receipt" (EIR) or "trailer interchange receipt" (TIR). This document acknowledges the return of the equipment from the land carrier to the terminal. Any discrepancy or damages noted on the EIR or TIR will be compared with a similar report filled out when the trucker took the equipment. Liability for damages is the responsibility of the party having custody of the equipment. The equipment control division compares EIRs and charges the responsible parties for damages that occurred while equipment was in their possession.

The traffic department and operations departments are regularly advised of all containers received at the terminal. Stowage plans are made up based on the bookings and actual receipt of cargo at the marine terminal. An evaluation must also be made about which containers booked may reasonably be expected to arrive prior to the cut-off time.

The ports for which the containers are destined are informed by modern rapid communications as to the numbers of inland destinations of the units they are to receive on each ship. They may thus plan before the ship sails for the discharge and forwarding of containers to consignees. The same requirements for efficient utilization and fast turnaround of the containers exists in every area. In our example, the overseas organization must at this first notification already begin to make plans regarding inland transport arrangements and provide for the best possible utilization of equipment.

The problems of marketing, traffic, sales, pricing, and operations exist at all locations serviced by a containership line. Thus the organization of a successful liner company must include elements of all these departments wherever it does business.

Shippers of small lots or less than container load (LCL) cargo may have their shipments consolidated with other cargoes in a container destined for the same port. This consolidation is done at the marine terminals, with receiving operations performed in the same manner as for breakbulk cargo. Inbound LCL cargo is handled in a like manner. Transit sheds are thus required at container terminals for this important business. Consolidation may also be done by the ocean carrier at inland depots in major cities.

New organizations known as "non-vessel-owning common carriers" or NVOCCs have also established facilities convenient to many shippers/consignees at which LCL cargo can be handled. NVOCCs relieve their customers of much paperwork and have depots at locations that are usually more accessible and have less traffic congestion than the marine terminals. The fully loaded containers are booked with containership companies by the NVOCC just as by any other shipper. Special rates are applicable to this traffic, since the NVOCCs often use their own or leased containers and carry many varied commodities. A cooperating organization of the shipping NVOCC at the container's destination will unload the container and forward the individual LCL shipments as required.

On the east and Gulf coasts of the United States the longshoremen's union has negotiated a special provision in its contract with stevedore companies regarding containerized cargo. The so-called 50-mile rule requires that all containers carrying cargo of two or more shippers/consignees coming from or going to points within 50 miles of a port must be loaded or unloaded by longshoremen's union employees. This has preserved some of the work of handling smaller shipments at the marine terminals. Shippers have found the services of the NVOCCs to be most convenient, however, with the result that the amount of LCL cargo handled at marine terminals has declined.

It is obvious that successful container liner services are very complex, requiring many people, an efficient organization, good planning, large amounts of capital, and suitable equipment. The electronic computer is an indispensable tool in these operations. It provides all the departments mentioned with rapid and accurate information concerning bookings, equipment location and availability, individual container movement data, ships movements, and so forth. Computers are programmed to prepare ships' stowage plans, to assign terminal locations for storage of individual containers, to construct terminal loading sequence plans, to design inland transport routings, and many other applications.

The importance of this phase of the operations has given rise to the establishment of a separate *department of information services*. The vice president in charge reports to the senior vice president for finance and administration. Other departments within the organization use the same equipment for traffic analysis, financial reports, payroll preparation, and other uses too numerous to mention. The present level of operating efficiency and the low rates charged for container services would not be possible without the electronic data processing equipment now available.

THE MARINE TERMINAL

Although a short chapter could easily be devoted to the design of marine terminals, we will limit our discussion to a brief survey of the principal factors.

Function of the Marine Terminal

The principal function of a general cargo marine terminal is to act as an interchange point for cargo moving from land routes to the seagoing carrier. The *transit shed,* which is the only superstructure of any consequence at the terminal, gets its name from the temporary nature of the storage facilities that it is supposed to offer.

A marine terminal cannot be said to be modern or efficient unless it can accommodate with expediency the latest developments in the water and land transportation systems that it serves and also provide for full use of all modern materials handling equipment. Today's terminals must be as efficient as the ships they serve in order to permit fast cargo handling and speedy departure of the vessels.

Two Basic Types of Construction

The two basic types of marine terminal construction are the *pier* and the *wharf.* The pier is a long, relatively narrow structure that juts out into the harbor waters, roughly at right angles to the shoreline. This type of construction is desirable in areas where water frontage is at a premium and where there is sufficient clearance in the stream for the pier. This type of construction was common in American ports. In many ports, piers have been demolished or transformed into terminals of the wharf type or marginal bulkhead terminals by filling in the waterways between adjacent piers.

The wharf, or quay as it is called in many foreign ports, is constructed along the shoreline roughly parallel to it. It is used where there is no shortage of water frontage or where the waterway adjacent to the terminal site is too restricted to permit pier construction. The wharf is well adapted to use at river ports and at new terminals located in previously underutilized port areas. This type of construction predominates most southern ports; Portland, Oregon; Oakland, California; and Port Elizabeth, New Jersey.

Basic Marine Terminal Requirements

Because of the great increase in the use of the large semitruck and trailer as a connecting carrier, the modern marine terminal must be designed to accommodate large numbers of these carriers daily without congestion. In other words, space must be planned for expeditious ingress and egress roadways and loading and discharging areas beyond the space necessary for the temporary stowage of the cargo moving over the terminal. The pier type terminal offers a greater problem here than does the quay or wharf. Modern terminals generally provide loading platforms with heights equal to those of the truck platforms and railroad cars and are designed so as to obviate the necessity of these vehicles entering the transit shed storage area.

Everything possible should be done to provide adequate traffic control of the waiting lines of trucks coming to the terminal to pick up or deliver cargo. The greater the efficiency within the terminal, the less the need for shoreside space and control.

Current Terminal Developments

The design of terminal facilities in any port will be influenced by current and prospective trade patterns. Important factors to be considered include the types of cargo to be loaded and discharged, container traffic anticipated, RO/RO traffic needs, and the amount of money available for development. Flexibility of operations must be provided for, since marine terminals are often planned for 50 years of service.

Specialized Cargo Facilities

A few commodities are transported along regular routes in sufficient quantities to justify specialized facilities designed for the handling of each commodity alone. Bulk liquids, such as petroleum products, can be pumped into and out of the ship's tanks; bulk ore and grain can be gravitated into the ship's hold through chutes and unloaded by means of huge clamshell type buckets and cranes or, in the case of grain, sucked out with vacuum pumps. Large areas of hard-surfaced level land are required for the storage of containers waiting to be loaded aboard ship, as is space to accommodate all those being discharged. Decisions must be made as to whether containers will be stored on chassis for ready access or stacked two, three, or even six units high.

The availability and/or cost of land, plus the costs of land development, will influence decisions regarding terminal size, location, and specific use. These considerations are generally evaluated by port management officials. Port administration, an activity closely associated with marine cargo operations, is being recognized increasingly as important to the welfare of the local community as well as to shipping interests.

CHAPTER 3

CARGO RESPONSIBILITY

"The carrier shall be bound . . . to make the holds . . . in which goods are carried, fit and safe for their reception, carriage and preservation."

—Carriage of Goods by Sea Act
Title I, Section 3
Approved April 6, 1936

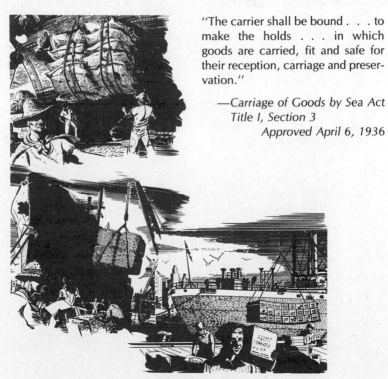

Drawing courtesy of Cargocaire Marine Systems.

INTRODUCTION

A competent ship's officer should have a basic familiarity with the laws governing carriage of goods by sea. By understanding where the responsibility for cargo lies throughout its carriage, a ship's officer will be better prepared to avoid cargo loss or damage. In an effort to explain the necessity of and reasons behind some of the routine duties required of the ship's officers and crew members, we will provide here an overview of the laws governing cargo handling. In addition, we will give suggestions on ways to meet the requirements of the various laws governing cargo handling and thereby reduce shipping costs and strengthen the carrier's position in case of litigation over cargo damage.

THE HARTER ACT OF 1893

Under the general maritime law, the common carrier had historically been regarded as the insurer of all cargo carried. Prior to the passage of the Harter Act of 1893, the shipowner was held responsible for cargo damage unless he could prove that such damage was due to one of the following causes:

1. An act of God.
2. An act of public enemies.
3. Inherent vice.
4. Fire.

Because the English allowed shipping companies to protect themselves by disclaimers contained in the bills of lading, the American shipping companies were at a disadvantage. Thus, in 1893, Congress passed the Harter Act, which gave some protection to our shipping companies.

With the passage of the Harter Act, shipowners obtained six additional exceptions under which they would be not liable for cargo loss or damage:

1. Errors in navigation or mismanagement of the ship.
2. Perils of the sea.
3. Insufficiency of the packaging.
4. Seizure under legal process.
5. Act or omission of shipper.
6. Saving or attempting to save life or property at sea.

Thus, the shipowner enjoyed 10 immunities, counting the four previously enjoyed.

Congress also demanded that the shipowner fulfill three definite responsibilities in order to enjoy the 10 immunities:

1. He must properly stow and care for the cargo.
2. He must exercise due diligence to properly equip, man, and provision the ship.
3. He must exercise due diligence to make the vessel seaworthy in all respects.

The Harter Act has no provision relieving the carrier of responsibility for damage or loss caused by fire; however, shipowners are covered by the Fire Statute of 1851. This statute protects shipowners by excluding liability of the shipowner and vessel for fire unless the fire occurred as a result of the shipowner's or its agents' neglect. Employees such as the master and crew members are not agents of the shipowner under this statute; the term "agents" in the statute refers to shoreside management. The problem with the Fire Statute of 1851 is that it applies only to bareboat charterers, shipowners, and the vessel. Time and voyage charterers are left exposed and liable for damages and losses caused by fire.

The Harter Act applies only to vessels operating in the domestic trade. For the shipowner to enjoy the benefits of the six additional exceptions to liability, the Harter Act must be incorporated as governing the relations between the parties in order to apply. The Harter Act does not apply to tug-tow relations. The shipowner is responsible for the cargo from yard to yard. Thus, it behooves the ship's officers to insure that the cargo is properly cared for while on the dock prior to loading and after discharge from the vessel.

Besides directing the shipowner to issue a bill of lading, the Harter Act also prohibits the shipowner from avoiding the responsibilities specified above. However, the carrier can accept more responsibilities by so contracting with the shipper.

The carrier has the burden of proving freedom from negligence if he is to enjoy the benefits of the Harter Act. If the cargo is damaged by two causes (attributable to the shipowner and/or the cargo owner) and the parties to an action are unable to establish who caused what, the shipowner must pay for all the losses or damages suffered by the cargo interests.

With respect to tug-tow relationships, the barge is not considered cargo of the tug. However, if there is cargo on the barge and the bill of lading states that the Harter Act applies with respect to the cargo, then the barge owner is responsible for the cargo on the barge so carried.

Even if the cargo is carried aboard a foreign vessel, if it is being shipped in the domestic trade of the United States, the Harter Act applies. Thus, in *Knott v. Botany Worsted Mills,* a foreign vessel engaged in domestic U.S. trade was held responsible for water damage to wool. The water came from drainage off an adjacently stowed cargo of sugar. The shipowner was held to be negligent in its care and custody of the cargo. The failure to trim the vessel properly was construed as negligence in care and custody.

Section 3 of the Act, referring to errors in navigation and management of the vessel, was interpreted interestingly by the courts in *May v. Hamburg.* The vessel was stranded in a German river, and the court ruled that the shipowner used due diligence in making the vessel seaworthy and that the stranding was caused by an error in navigation; therefore, no liability for loss or damage could be assigned to this stranding. However, the shipowner sent a managing agent to examine the ship for necessary repairs. While en route to the repair yard, the vessel was involved in a second accident in which it was again stranded. When the court examined the case, they decided that when the shipowner sent its agent to Germany, it was as if a new voyage had commenced. Thus, the shipowner had to prove that it exercised due diligence to make the vessel seaworthy in all respects for *that* voyage. Since it was unable to prove that due diligence was exercised after the first stranding, the shipowner was liable because the damage was not from an error in navigation. The most significant element of the *Hamburg* ruling is that the damage did not necessarily have to be causally related to unseaworthiness. If the shipowner cannot prove that it exercised due diligence to make the vessel seaworthy in all respects, it loses the six exceptions to liability granted by the Harter Act.

The importance of determining the cause of damage to cargo is illustrated by numerous water damage cases. Whenever sea water gets into the hold, in order to enjoy the benefits of the exceptions of the Harter Act, the shipowner must prove that it was not negligent and that the cause comes under one of the exceptions of Harter. In the case of the *Folmina,* a classic case on the common occurrence of salt water damage to cargo, the court was unable to determine how this sea water got into the hold. The shipowner claimed salt water damage to the cargo was a peril of the sea and disclaimed liability. However, the court ruled that the doubt as to the cause of the entrance of the sea water must be resolved against the carrier.

Even when there is evidence that the cargo damage was caused partially by an excepted peril and partially by the carrier's negligence, the carrier has the burden of proving what portion was caused by the excepted peril. In *Schnell v. The Vallescura*, the cargo of onions, which required ventilation, was damaged by decay. The vessel encountered heavy weather en route and was therefore unable to ventilate the cargo properly for part of the voyage. However, the court also noted that the vessel neglected to ventilate the cargo at night during fair weather. Because the vessel was unable to prove what portion of the damages were attributable to the excepted peril of the heavy weather, the Supreme Court ruled that the vessel must bear the total cost of damage to the cargo.

As is obvious from these cases, it is important that the ship's officers obtain and log all facts surrounding any damage to cargo.

THE CARRIAGE OF GOODS BY SEA ACT OF 1936

In an effort to provide more protection to merchant shipowners and more certainty in establishing the relationship of the ship to its cargo, Congress passed the *Carriage of Goods By Sea Act* (COGSA) in 1936. COGSA differs from the Harter Act in that it has its own fire clause protecting time and voyage charterers. It is not necessary to insert clauses in the bill of lading to obtain the benefits of COGSA, and the shipper must establish a causal connection between damage to cargo and unseaworthiness of the vessel to recover for such damage. COGSA applies to vessels carrying goods between an American port and a foreign port. However, it may apply to vessels in the domestic trade if the bill of lading so states. In such a situation, COGSA supersedes the Harter Act.

Under COGSA, the carrier is responsible for the cargo from tackle to tackle. Thus, as soon as the cargo is loaded aboard the ship, the carrier's responsibility commences, and as soon as the cargo is discharged from the vessel, the carrier's responsibility ends. Recent court decisions, however, have tended to extend the responsibility to arise as soon as the carrier loses or obtains custody of the cargo (i.e., when it enters or exits the yard).

COGSA may be incorporated into a charter party by specific reference. The charterer would thus be able to benefit from the provisions of the act.

By inserting a so-called Himalaya clause[1] into the bill of lading, the carrier can extend the protection of COGSA to stevedores employed to handle the cargo. As long as the stevedores are assisting in the loading and discharge of the cargo and the language of the Himalaya clause is clear and unambiguous, the courts have permitted COGSA to be extended to protect the stevedores.

COGSA also sets up some statutes of limitation. If cargo is damaged, the claimants have 1 year from the date of delivery to proceed against the carrier; if the cargo is not delivered on the date specified in the bill of lading, the time period is 1 year from the time the cargo should have been delivered. In addition, a claimant has 3 days from the time of delivery within which to make a complaint of latent defects in cargo.

The additional immunities granted to the carrier under COGSA are:

1. Act of war.
2. Quarantine restrictions.
3. Strikes or lockouts.
4. Riots and civil commotion.
5. Insufficiency or inadequacy of marks.
6. Latent defects not discoverable by due diligence.
7. Any other cause arising without the actual fault and privity of the carrier.

The additional responsibility set upon the carrier by COGSA is the necessity to make all cargo spaces fit and safe for stowing the cargo.

Summary of Immunities

To sum up all of the immunities of the carrier under COGSA, neither the carrier nor the ship shall be responsible for loss or damage to cargo arising from any of the following:

1. Error in navigation or mismanagement of the ship.
2. Fire or loss caused by the fault or privity of the carrier.

[1] *Himalaya* is a vessel; the name of the clause stems from a case involving that vessel. The clause extends all limitations and exonerations of liability of a carrier to independent contractors. These independent contractors include stevedores and longshoremen doing work normally done by the carrier.

3. Perils of the sea.
4. Act of God.
5. Act of war.
6. Act of public enemies.
7. Arrest or seizure under legal process.
8. Quarantine restrictions.
9. Act or omission of the shipper.
10. Strikes or lockouts.
11. Riots or civil commotion.
12. Saving or attempting to save life or property at sea.
13. Inherent defect, quality, or vice of the goods.
14. Insufficiency of packaging.
15. Insufficiency or inadequacy of marks.
16. Latent defects not discoverable by due diligence.
17. Any other cause arising without the actual fault and privity of the carrier.

The Carriers' Responsibility

Just as in the Harter Act, the carrier, to enjoy the immunities of COGSA, must fulfill the following conditions:

1. He must exercise due diligence to make the ship seaworthy.
2. He must exercise due diligence to man, provision, and equip the ship properly.
3. He must properly care for and stow the cargo.
4. He must make all cargo spaces fit and safe for stowing the cargo.

Sample Cases

In an effort to clarify the real meaning of the exemptions and responsibilities delineated by COGSA, we shall summarize below cases wherein the courts have interpreted their meanings. Through understanding how the courts have interpreted the provisions of COGSA, a ship's officer should be better able to grapple with the various problems encountered in cargo handling.

Improper Stowage Versus Insufficiency of Packaging

In the case of *Martin v. S.S. Ensley City*, a ship a Basra, Iraq, loaded 732 cases and 319 bags of licorice extract from open lighters. The licorice had been shipped by rail from Turkey and had been on the dock waiting for the ship for about 1 month. The cases were marked with the words: "Stow away from the boilers."

When loading commenced, the temperature of the atmosphere was about 115°F. The master of the ship noted that some of the cases were broken and in some instances the licorice was oozing out. He had all the damaged containers recoopered and he endorsed the bill of lading to the effect that the cases were badly damaged and recoopered but that the ship could not be responsible for the loss of contents or a short delivery.

The entire cargo of licorice was stowed in the forward part of the no. 4 'tweendecks, directly aft of the engine room. The cargo extended from the deck of the compartment to within 12 in. of the overhead. The cases were stowed about eight high, with the bags on top. The ship proceeded to an East African port and loaded chrome ore. Some of this chrome ore was placed just aft of the licorice, with a temporary wooden bulkhead erected between the two cargoes. During the voyage, no temperatures were taken of the hold spaces, nor were any inspections made. In Baltimore, where the licorice was to be discharged, it was found that the extract had broken out of most of the cases and bags in a glutinous state and therefore had become a hard mass. Many of the cases and bags were stuck together. Wood, nails, and other foreign material were imbedded in the licorice. This made it impossible to remove the cargo from the ship case by case or bag by bag. It had to be broken apart with picks before it could be removed from the hold. Recovery was sought for the expense of removing it, melting it down to remove the foreign material, and making it marketable.

In court, the shipper argued that the damage was due to improper stowage. He claimed that the licorice should have been stowed in one of the lower holds, preferably in the no. 1 lower hold, where the temperature was less than in the no. 4 'tweendecks, and also that dunnage should have been used between tiers to allow better air circulation.

The carrier claimed that the licorice extract was not in good condition when loaded and that its stowage was entirely proper. It also claimed that any change in the condition of the licorice during the voyage was due to inherent vice and to the insufficiency of the packaging.

It was brought out in court that the recognized place for stowing such cargo was below the waterline of the ship, where temperatures are lower than in the upper deck. The carrier introduced as evidence the temperatures of another ship that had carried licorice extract successfully in this upper deck, but the court rejected these records because it was purely a matter of speculation whether the temperatures on the first ship were the same as those on the second. In fact, there were two things that made it seem likely that the temperature on the second ship was probably higher. First, chrome ore in bulk will heat up when damp. Such cargo was stowed adjacent to the licorice, and although there was no evidence that it did or did not heat up, the court could not accept the submitted temperatures with so much conjecture about them. Secondly, the ventilators in the after part of the 'tweendecks where the licorice was stowed led into a special locker, which prevented full circulation of the air in that space. This was not the case on the other ship.

It may have been true that the temperatures on the two ships were the same, but the ship's officers failed to take any temperature measurements and log them; therefore, the carrier had no proof of proper temperatures.

The court held in favor of the shipper, saying that he had proved, as he must, that the cargo was in a marketable condition when delivered to the ship, but that on arrival in Baltimore it was not marketable. The court also concluded that the proximate cause of the change was not insufficiency of packaging or inherent defect, but improper stowage.

Unseaworthiness as a Cause of Loss Under COGSA

In *Riverstone Meat Company v. Lancashire Shipping Company (The Muncaster Castle)*, cargo was damaged by sea water due to an improperly replaced valve inspection cover. The shipowner argued that he exercised due diligence to make the vessel seaworthy in that he delegated the tasks to a reputable shipyard. The court disagreed, stating that the warranty of seaworthiness and duty of due diligence are nondelegable. Thus, the shipowner was held liable for all cargo damage caused by the sea water.

Errors in Navigation and Management Under COGSA

In *Mississippi Shipping Company v. Zander and Company (The DelSud)*, the court created the concept of voyage by stages. *The DelSud,* while in Santos, Brazil, struck the concrete pier, resulting in a hole in the side of the vessel. This occurred while the ship was being turned around for the purpose of leaving. The court ruled that the voyage had already commenced. The vessel then sailed to Rio de Janeiro without anyone noticing the hole, and from there to Curaçao and New Orleans. In New Orleans, extensive salt water damage to the cargo was discovered. The cargo claimants from the port of Santos and ports of loading prior to Santos argued that the vessel should have been made seaworthy on departure from Rio. The court ruled that the failure to make repairs at an intermediate port, such as Rio or Curaçao, was an error of management for which the vessel and carrier were not liable. As long as the vessel owner exercised due diligence to make the vessel seaworthy prior to departure from the loading port for the respective cargo, any subsequent unseaworthiness would be seen as an error in management or navigation. Thus, for cargo loaded at Santos, the voyage commenced with respect to that cargo at Santos, and the shipowner had a duty to exercise due diligence to make the vessel seaworthy for that cargo prior to departure from that port.

It is important to note that if a shipowner or his representative comes aboard the vessel to take full command of a situation involving repairs, the courts look at this situation as destroying the continuity of the voyage. The shore staff is then required to exercise due diligence in inspecting the vessel and making it seaworthy lest the shipowner forfeit his immunity. Applying the ruling of the court, the shipper of cargo loaded in Curaçao, if that cargo was damaged by the leakage caused in Santos, would be able to recover from the shipowner for his failure to exercise due diligence to make the vessel seaworthy at the port of loading.

Although the concept of voyage by stages is somewhat beneficial to the shipowners, it would be prudent for the ship's officers to inspect the vessel thoroughly whenever there is a suspicion of damage to vessel or cargo.

The Fire Exception of COGSA

COGSA specifically excepts liability for damage caused by fire unless the fire was caused by the actual fault or privity of the carrier. The Act also obviates the need for repeal of the Fire Statute. The Fire Statute holds that the owner was not liable to the shipper for losses caused by fire on board "unless such fire is caused by the design or neglect of such owner." As previously noted, the COGSA exception covers any carrier, whereas the Fire Statute does not cover time and voyage charterers. The courts have interpreted actual fault or privity to have the same meaning as design or neglect. As it stands, it is a nearly impossible burden for the cargo owner to prove that the fire was caused through the actual fault, privity, design, or neglect of the shipowner or shoreside personnel.

Cargo damage claimants have been successful in contending that unseaworthiness of firefighting equipment should make the vessel owner liable for damage, notwithstanding the fire exemption.

Despite the Supreme Court's ruling that due diligence to provide a seaworthy ship is not a prerequisite to enjoying the benefits of the Fire Statute or COGSA's fire exemption, two appellate courts have ruled otherwise. These courts have been harshly criticized by the majority of the appellate courts, which claim that the design and neglect test is the exclusive one to be applied in a fire damage situation.

"Neglect," as thus used, means negligence, not the breach of a nondelegable duty. For the shipper to recover he must prove that the carrier caused the damage, by proving either that a negligent act of the carrier caused the fire or that such an act prevented the fire's extinguishment. Thus, it is incumbent upon the ship's crew to maintain the firefighting equipment in top condition.

Jurisdictional Clauses in Bills of Lading Under COGSA

Often bills of lading contain forum selection clauses indicating, for instance, that New York is the place to litigate any disputes with respect to the contract of carriage. While the courts like to uphold reasonable forum selection clauses, they do not want to lessen liability in the process. In *Indussa Corporation v. The Ranborg,* the bill of lading stipulated that Dutch law would apply to any disputes between shipowner and cargo interests. In addition, the jurisdiction clause required that any dispute arising under the bill of lading would be decided in the country where the carrier has his principal place of business. In this case, the company was based in Netherlands. The U.S. court held that any bill of lading outlawing or prohibiting American courts from deciding causes otherwise properly before them is against the Congressional intent of COGSA. The court implied that, especially with a small claim, the mere fact of letting the trial proceed in Holland would effectively lessen the carrier's liability (which is expressly prohibited by COGSA).

Damages Under COGSA

If the value of the cargo has not been declared somewhere on the bill of lading, the carrier and ship are not liable for any loss or damage to the cargo exceeding $500 per package or, in the case of goods not shipped in packages, per customary freight unit. It is important for the parties involved in any dispute over cargo damage or loss to define clearly what the customary freight unit or package is. Although a ship's officer should exercise special care for all of the cargo, if he sees a bill of lading listing a large number of packages, he should take extra precautions to make sure that the cargo is properly cared for. This, of course, is because the $500 package limitation is multiplied by the numerous packages listed in the bill of lading could result in excessive liability for the vessel in the unfortunate event of damage or loss.

The growing trend is for the courts to consider the bill of lading as manifesting the intent of the parties. Thus, if under the heading "number of packages" the bill of lading lists one container, some courts are likely to rule that the intermodal container is the package. Although this has been disputed among the various courts, if the container is supplied, stuffed, and delivered by the shipper to the carrier in a sealed condition, most courts are likely to rule that the container is the package for the purposes of COGSA. These courts reason that the carrier has no knowledge of the contents of the container unless they are delineated in the bill of lading. Thus, if the bill of lading reads as previously stated, the carrier's lack of knowledge of the contents of the container will not disqualify him from claiming the $500 per package limitation. Even if elsewhere on the bill of lading there is a description of, for instance, the number of containers on a pallet, if under the heading "number of packages" the pallet is listed as the package, the court will not rule the former description to be binding on the carrier. In brief, the courts generally treat the number of packages delineated in the "number of packages" section of the bill of lading as binding upon the parties for the purposes of the package limitation. Thus, pallets, containers, and pails, as well as loose cases and drums, can all be ruled to be COGSA packages, depending on their positions on the bill of lading.

It is important for ship's officers to realize the vessel's exposure to great liability in the event that a court rules their conduct to be an unreasonable deviation from normal practice. For shipboard personnel, avoidance of subjecting cargo to greater risks can be accomplished in many ways. By making sure that cargoes are properly stowed and segregated according to the applicable regulations, by insuring that high-value cargo is stowed in secure areas of the vessel, and by abstaining from welding or burning near cargo, washing down decks near cargo, and the like, the ship's officers will be able to avoid any claims for deviation.

ADVANTAGES OF UNIFORM ALLOCATION OF CARGO RESPONSIBILITY

With the trend toward greater uniformity in laws applying to cargo, assignment of liability is becoming more predictable. Unification of law makes it easier for local parties involved in international transportation to be certain of foreign law so that adequate planning and protection may be assured. To the extent that cargo interests and carriers are able to predict which law will govern a potential dispute, they are able to avoid disputes and arrive at settlements without resorting to litigation.

Except where national laws apply, existing modal transportation conventions provide a significant amount of uniformity of laws for modal carriage. For example, the Warsaw Convention unifies the laws of international air transport, the Hague Rules provide a uniform regime for maritime carriage, and the CIM and CMR Conventions (European Train and Road Transportation Conventions) provide uniformity on a regional basis for rail and truck transport, respectively. For domestic rail and truck carriage, the Interstate Commerce Act mandates substantial uniformity among the 50 states. In fact, this national law goes one step further; it regulates multimodal transportation among surface modes.

Although a significant amount of legal uniformity has been attained, there are several gaps among the various governing laws and treaties that cause problems to shippers and carriers alike. One example of lack of uniformity confronting multimodal cargo interests and carriers is provided by a comparison of the limits on liability in the various conventions and national laws. Whereas the national laws tend to provide compensation up to the full value of the goods, the liability limits under the international conventions range between $13.63/lb under the CIM Convention and $500/package under the Hague Rules (COGSA). The variety of liability limits makes it difficult for banks, insurers, carriers, and cargo interests to indemnify the risk involved in carriage affected by more than one convention. The only time liability is uniform is when the carriage of cargo is governed by one convention. In addition, there is a substantial lack of uniformity regarding the defenses against liability in the various conventions and national laws. A variety of defenses exists among the various regimes, ranging from the 17 defenses under the Hague Rules (COGSA) to the strict liability regime of the U.S. Interstate Commerce Act. Furthermore, a great disparity exists among the laws that apply to international multimodal carriage, because there is no single international law that governs the interfacing among the various modal conventions. It is sometimes impossible to ascertain or predict the applicable law. Thus, if the carriage mode in which loss or damage occurred is not clearly established, the interested parties must fend for themselves. Concealed damage to cargo that is shipped by several modes of transport under various conventions leaves the interested parties stranded in the gaps. If they cannot prove that a particular convention applies, a legal stalemate results. Thus, the claimants must pay higher insurance premiums to cover the risks incurred in these "gaps." While uniformity may be coming soon, the keeping of accurate records of cargo condition on ships will insure that the point of damage is determined.

CARGO CLAIMS AND MINIMIZING LOSSES

We will now summarize the possible causes of damage that may bring the carrier under the exception clauses of the Harter Act and COGSA, thereby exempting him from liability.

Causes beyond the vessel's or cargo's control are typically excepted. Such causes include:

1. Fire.
2. Act of God.
3. Perils of the sea.
4. Quarantine restrictions.
5. Government restraints.
6. Strikes or civil commotions.

Several claims attributable to the ship permit the vessel to escape liability. These claims are:

1. Errors in navigation or mismanagement of the ship by the crew.
2. Unseaworthiness despite due diligence in caring for and stowing the cargo.
3. Latent defects; that is, those not apparent to usual inspection (e.g., metal fatigue).
4. Assisting vessels in distress.
5. Deviation for the purpose of saving life and property.

Claims arising due to problems originating within the cargo itself are excepted:

1. Inherent vice or poor condition prior to loading.
2. Fault or omission of the shipper or cargo owner (in shipping instructions, etc.).
3. Insufficiency of marking.
4. Insufficiency of packaging.

Any claims arising outside the normal course of seagoing transportation or scope of the contract of transport are also excepted claims permitting the carrier to escape liability:

1. Flood or fire at port warehouse.
2. Negligence of third party for whom the carrier is not responsible (e.g., shipper's freight forwarder).

Finally, time limitations on claims may free the carrier from liability.

Good Practices for Loss Prevention

When handling outward-bound cargo, it is important that the shipboard personnel check all markings. They should make sure that the marks on the bill of lading and the manifest are in concord. Any errors may cause fines to be imposed by customs authorities. The officers, prior to sailing, should survey the pier for cargo accidentally left behind. In addition, the tightness of all lashings should be checked; if the ship is carrying containers, all interbox connectors should be locked; all watertight doors and holds should be sealed; and, lastly, accurate bilge soundings should be obtained. It is prudent when multifarious cargo is being carried for the chief mate to ascertain that all of the containers or parcels aboard are listed in the vessel's stowage plan and are stowed in the exact locations indicated in the plan.

Special procedures should be used when loading incomplete or damaged cargo. The mate on watch should either reject the cargo or make sure that the bill of lading is amended accordingly. When cargo is insufficiently packaged, the mate should, if possible, give the shipper time to correct the defect or, if not, correct it at loading. The chief mate should mark the bill of lading "insufficient packaging." It is important that this

statement be placed on the bill of lading, as it will assist shoreside personnel in repudiating any claims. If anyone tries to issue a letter of indemnity
against a clean bill of lading, and the shipboard personnel consent, they
may be liable for fraud. In addition, if caught in this act, the carrier would
be deprived of all defenses and limitations contained in the bill of lading
under the various conventions.

Regarding stowage, ship's officers should make sure that for the cargo
stowed on deck there have been no "underdeck" bills of lading issued. It
is important that the ship's deck department be aware of the various
Federal regulations governing the proper stowage and separation of various cargoes. A daily inspection of the cargo areas over the course of the
voyage will help prevent false claims against the vessel. If the mate makes
entries in a log as to the condition of the cargo and the cargo spaces, and
observes how the cargo is riding as the vessel proceeds in a seaway,
possible losses and damages may be avoided, as well as spurious claims.

During discharge of general cargo, a check should be made for heavily
damaged cargo; if found, it should be safely stored and recoopered. If the
cargo is badly damaged or wet, it may be wise to call in a surveyor.
Consolidating cargo that is consigned to one recipient on the pier will
facilitate rapid, efficient delivery. High-value cargo should be tallied from
the vessel. This will help to minimize theft. Local stevedores should be
instructed to check the containers and parcels that are damaged or missing their seals. This should be done as soon as such items are discharged.
In addition, shipboard reports should be attached to the stevedore reports
to ascertain where the damage or loss occurred.

It is important that the vessel have the proper documentation for the
cargo and that no cargo be released without receipt of the original bill of
lading. Carriers should require that the person issuing the bill of lading
sign it and write his title on the front. This will assist the claims personnel
in tracing the facts surrounding the shipment, including the details on the
bill of lading. If the consignee or claimed consignee offers a letter of
indemnity with a bank guarantee in order to effectuate release of the
cargo, the carrier personnel or company staff in the respective port should
obtain clearance from the home office prior to releasing the cargo without
receiving the original bill of lading.

When the cargo is lost or damaged, it is incumbent upon the ship's
officers to write a report describing the facts surrounding the loss or
damage as soon as possible. The sooner the report is written, the more

accurate it will be. The information is more accessible if the report is completed quickly; thus, good relations are enhanced with both the shipper and the home office. Last, prompt reporting saves both the shipper and the company money.

It is important that the ship's officers give accurate and detailed descriptions of all surrounding circumstances, no matter how seemingly insignificant. If possible, photographs should be obtained and log entries made regarding the chronology of the events surrounding the loss or damage to the cargo.

In addition to the log entries on cargo damages, times of starting and finishing unloading, and so forth, the officers should make entries on cargo gear inspections, malfunctioning of any cargo gear or hydraulic equipment, and, of course, bilge levels.

Because the ship's officer makes a countless number of log book entries during his career, there is always the danger that he will become careless about the manner in which they are made. The distribution of responsibility for the cargo should make it evident that the log book entries can affect the shipowner's position materially.

When a hold is inspected to determine its fitness for receiving cargo, the determination and some details concerning it should be noted carefully in the log. The timing of fueling and of bilge soundings reflect due diligence, or the lack of it, so definitively that the prosaic task of sounding bilges must never be overlooked, nor, what is just as important, should the entries in the log. Monthly inspections of all cargo gear should be entered in the log book simultaneously with entries in the chief officer's cargo gear register. All efforts to ventilate a hold properly or to determine or control conditions in a hold before or during the voyage should be indicated clearly in the log. If this is not done, the officer is jeopardizing his employer's position in the eyes of the court in case of litigation, as well as his own position in the eyes of his employer. All efforts to prevent the elements from entering the hold during loading operations or from reaching cargo in lighters alongside the ship should be logged. Every time a lashing is checked or set up or added to deck cargo, this fact should be entered in the log.

Whenever there is any doubt as to whether an action to protect cargo is warranted, the decision should be in favor of doing it. But it is just as important to log the fact that the action was taken. All such special entries should be signed and the time of the action indicated. Entries should never be erased, and it is a poor policy to add entries long after the action has or should have taken place in an effort to deceive a surveyor or counsel. Postscripts to the log can be uncovered easily, leaving the responsible officer in an embarrassing position. When honest mistakes are made in writing an entry, officers should delete them by drawing a single neat line through the error, leaving it clearly discernible.

CHAPTER 4

PRINCIPLES OF STOWAGE

The fundamental objectives when cargo is stowed in a ship are (1) to protect the ship, (2) to protect the cargo, (3) to obtain the maximum use of the available cubic of the ship, (4) to provide for rapid and systematic discharging and loading, and (5) to provide for the safety of the crew and longshoremen at all times. These objectives may be referred to as the five principles of stowage.

(1) PROTECTING THE SHIP

The problem involved in meeting this first objective is the correct distribution of the cargo weight. The distribution must be correct vertically, longitudinally, and transversely. The weights must not be concentrated on any deck in such a fashion that the structure's supporting strength is exceeded.

Vertical Distribution of Weight

Vertical distribution affects the stability of the ship. If too much weight is in the upper decks of the ship, the ship will have a small amount of stability and be in a condition known as *tender*. If too much weight is concentrated in the lower holds, the ship will have an excess of stability and be in a condition known as *stiff*. The tender ship has a long, slow, easy roll. The stiff ship has a fast, whiplike roll that makes her especially uncomfortable in a heavy sea and is often the cause of cargo shifting transversely.

If a ship is excessively stiff, she will roll with such violent motion that damage to the ship can be caused by heavy wracking stresses on the hull. Some of the defects caused by these heavy wracking stresses will be apparent immediately after they occur, such as cracked porthole or window glass in the superstructure, standard compasses being whipped from their pedestals, and topmasts or antennas being shaken loose. Some defects having their basic cause in these wracking stresses will never be attributed to the improper vertical distribution of the cargo. These defects, which may not be discovered until the ship undergoes a thorough inspection in a dry dock, include cracked seams, hull weaps, and leaking tanks.

Longitudinal Distribution of Weight

Longitudinal distribution affects the trim of the ship and the hogging and sagging bending stresses that the ship's hull must withstand. The trim is the difference between the drafts of the ship fore and aft. The ship should be loaded so that she has an even trim or a trim slightly by the stern, 6 in. to 1 ft. A trim by the head, if only a few inches, does not affect the speed of the ship. Trimming by the head is usually avoided, however, because if the ship is deeply loaded and in a heavy seaway, there is more possibility that damaging green seas will be shipped on the foredeck. Hogging occurs when the ship has too much of the total weight concentrated in the ends and the hull bends as shown in Fig. 4-1a. Hogging results in compression stresses being imposed on the keel and tension stresses on the shear strake and main deck. Sagging occurs when too much weight is concen-

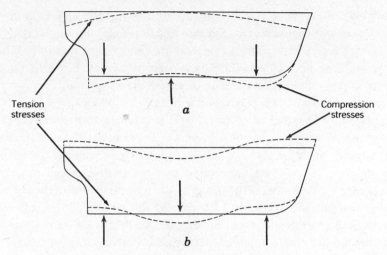

FIG. 4-1. (*a*) Hogging: too much weight in ends. (*b*) Sagging: too much weight amidships.

trated amidships and the hull bends as shown in Fig. 4-1*b*. Sagging results in tension stresses being imposed on the keel and compression stresses on the sheer strake and main deck.

If a ship is hogged or sagged when lying in still water, these excessive stresses will be accentuated when the ship is in a seaway, and the result may be a cracked deck or hull plate. Excessive hogging or sagging as a result of improper longitudinal distribution of cargo or ballast has been the cause of many ships breaking in two. But for every ship that has cracked into two pieces, there are hundreds of cases in which the shipowner has had to pay for the repair of cracks, large and small, in hull and deck plating. Many of the resulting expenses and delays could have been avoided by more intelligent longitudinal distribution of the cargo, voyage after voyage.

It must be understood clearly that a ship can be trimmed well while she is hogged or sagged dangerously. Good trim is simply a matter of correct balance of trimming moments. These can be precisely as desired and the concentration can still be such that great damage will be done, as explained above.

Transverse Distribution of Weight

Transverse distribution offers no problem. When stowing the cargo, the only necessity is to make sure that the weight is equal on both sides of the ship's centerline. This is accomplished by starting all loading on the centerline and stowing outboard, loading equal amounts in the wings, or, in the case of heavy lifts, putting them on the centerline if possible. If heavy lifts cannot be put on the centerline, then by careful planning an equal weight must be placed on the side opposite the heavy lift. This last problem may arise when forced to load a heavy lift on deck.

Although it is rarely of practical use, it is interesting to know that the period of the ship's roll can be affected by the concentration of the weight transversely. If the mass is concentrated inboard along the centerline of the ship, the roll will be very rapid and the period of the roll will be decreased. Conversely, if the weights are concentrated outboard in the wings, the roll will be slowed down and the period of the roll will be increased. Theoretically, then, this is a device that could be used to give a stiff ship a more comfortable roll. If a ship is unavoidably loaded stiff, the roll can be slowed down by concentrating weights outboard. Such refinements are seldom practicable, although they are possible.

Importance of Weight Distribution

Improper weight distribution due to improperly loaded cargoes has cost many shipowners much time and money in the past and in all probability is still costing time and money. Ships have been lost at sea due to the shifting of cargoes or ballast, caused primarily by violent rolling in a heavy seaway. For every ship lost, there are probably thousands of cases of minor but nevertheless costly damages. This causes needless trouble for all principals concerned. The solution is simple. All responsible officers must know what is meant by correct vertical distribution of weights, and furthermore, they must know how to accomplish said distribution with positive and scientific certainty.

The problem of proper longitudinal distribution is equal if not greater in importance than that of vertical distribution. According to the findings of a board of investigation established by the Secretary of the Navy in April 1946 to inquire into the design and methods of construction of welded steel merchant vessels, 970 ships reported some type of structural failure during the years 1943 to 1946. The total number of structural fractures reported by these ships was 4720. One hundred twenty-seven of these ships sustained fractures that weakened the main hull structure so that the vessel was lost or placed in a serious condition. Twenty-four ships sustained complete fractures of the main deck; one ship sustained a complete fracture of the bottom. Eleven of these ships broke in two, and of these eleven, seven were lost completely.[1]

Part of the answer to this problem of structural failures was the installation of crack arresters, modification of cargo hatch corners (where many of the fractures originated), elimination of square sheer-strake cutouts for accommodation ladders, and general elimination or modification of structural discontinuities. Another part of the answer was the use of expansion joints. The rest of the answer was to give more attention to loading and ballasting of the ships. Part of the answer to the problem, then, is virtually the same as given above for vertical distribution, namely, that responsible officers must know what is meant by correct longitudinal distribution of weight and how to accomplish it when loading ships.

The point to be made here is that improper longitudinal distribution of weights does contribute to structural failure of ships. These failures are of such a nature that in many cases the shipowner's organization may not be aware of the cost in time and money that improper distribution is causing. This is because minor failures might be attributed wholly to heavy seas or the age of the ship, whereas these factors only contribute to the failure and the principal blame should go to the improper longitudinal distribution of cargo and ballast. The operations department of every carrier should insist that the longitudinal distribution of weights be calculated, recorded, and attested to by the master of every ship that sails. Furthermore, some tabular or mechanical means for calculation of the bending moments of the hull should be employed. This should be in addition to an on-board computer to monitor stress points. A statistician should correlate the cost of repairs with the bending moments found in the ships.

To facilitate a complete discussion of the first fundamental objective of cargo stowage, the basic principles of stability, trim, and longitudinal stresses will be covered in this chapter. It is recommended, however, that every ship's officer, stevedore, or member of a shipping organization who has responsibility for the loading of ships consult the references mentioned in the footnotes for a more detailed explanation of the governing principles.[2,3]

---

[1] Report of a Board of Investigation, *The Design and Methods of Construction of Welded Steel Merchant Vessels,* Government Printing Office, Washington, D.C., 1947, pp. 1–3.

[2] J. La Dage and L. Van Gemert, *Stability and Trim for the Ship's Officer,* D. Van Nostrand Co., New York, 1947, Chaps. 1–3.

[3] J. La Dage, *Modern Ships,* Cornell Maritime Press, Cambridge, Maryland, 1953, Chap. 6.

The Ship's Stability

Stability is the tendency of a ship to return to an upright position when inclined from the vertical by an outside force. This discussion shall be concerned only with *initial stability,* which can be defined as the tendency of a ship to right itself when inclined less than 15°. Stability at large angles and damage stability are topics that every well-informed ship's officer should understand, but the discussion of them would be too lengthy for our coverage here and would not aid the discussion of the first objective of good stowage.

Three Important Points

For a study of initial stability we constantly refer to a diagram of a ship's midship section (see Fig. 4-2) and the relationship between three points on it. These points are known as (1) the center of buoyancy (B), (2) the center of gravity (G), and (3) the metacenter (M).

Archimedes' principle of buoyancy is the basis of any explanation regarding the stability of a floating body. This principle states that a body submerged wholly or partially in a fluid is buoyed up by a force equal to the weight of the fluid displaced. On a ship, the total of the upward forces of buoyancy is considered to be concentrated at point B. Actually, there are hydrostatic forces acting over the entire submerged surface of the ship's hull, but the total upward force is the force of buoyancy and acts through point B.

This force of buoyancy acting through B is opposed by an equal force

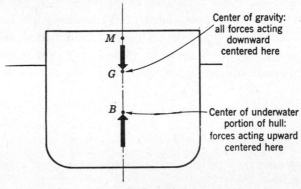

FIG. 4-2.

acting downward through point G. When a body is floating, the forces through B and G are equal. The force through B will never exceed the force through G. The force through G may exceed the force through B, in which case the body will sink. When the forces through G and B are equal, the body will float. If the floating body is at rest, that is, not revolving transversely or longitudinally (rolling or pitching), then the forces through B and G are exactly opposed: They are acting along the same vertical line. The force acting through G is the sum of the weights of all elements making up the ship's structure. For convenience, we consider them as acting through this single point G.

Point B will always be at the exact center of the underwater portion of the ship's hull. Point G will always be at the exact center of the entire mass of which the ship is composed. This mass includes every part that lies above or below the waterline, everything in the ship or on the ship. The addition of a weight, the removal of a weight, or the shifting of a weight will change the position of G. The only thing that will change the position of B is a change in the shape of the underwater portion of the hull. The underwater portion of the hull changes shape, of course, whenever the ship rolls or pitches; thus B moves about as the ship works in a seaway.

The Vertical Position of G

First we will discuss the location of G. The first step is to examine the problem of determining the *position of the center of gravity of a system of weights*. We use a physical law that may be explained mathematically as follows: *If a number of weights are part of a system of weights and each weight is multiplied by its distance from a reference line or surface, then the sum of all these products will equal the sum of all the weights times the distance of the center of gravity of the system from the reference line or surface.* This is a very important concept and the reader should not proceed until he is certain that he understands it. The rule as stated above applies to masses, but it can be made applicable to volumes as well. It can be used to find the position of a center of gravity in three dimensions; however, our problem in stability is somewhat simplified by the fact that our G will always be considered to lie on the ship's centerline, and we will be concerned only with the height of G above the keel of the ship. Let us consider, then, the relatively simple problem of finding the position of the center of gravity of a system of weights stretched out along a line.

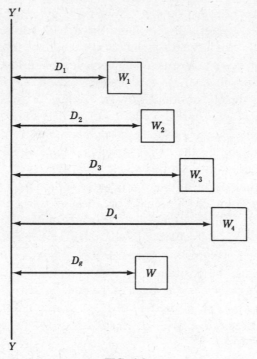

FIG. 4-3.

Figure 4-3 shows a system of four weights (W_1, W_2, W_3, W_4) with their respective distances (D_1, D_2, D_3, D_4) from a reference line yy'. The above law says that the following is true:

$$(W_1 \cdot D_1) + (W_2 \cdot D_2) + (W_3 \cdot D_3) + (W_4 \cdot D_4) = W \cdot D_g \qquad (1)$$

where D_1, D_2, D_3, D_4 = the distances of the weights from the line yy'.
$\quad\quad W_1$, W_2, W_3, W_4 = the weights.
$\quad\quad\quad\quad\quad\quad W$ = the sum of all the weights in the system.
$\quad\quad\quad\quad\quad\quad D_g$ = the distance of the center of gravity of the weight system from the reference line yy'.

Before proceeding further, it must be pointed out that the product of a weight and a distance ($W \times D$) in each case is a *moment*. A moment is defined as the product of a force acting through a distance. The units used on ships are long tons for W and feet for D; hence the units of moments are foot-tons. Knowing this, we can restate the important law given above as follows: *The sum of all the individual moments in a weight system is equal to the moment caused by the total weight of the system being concentrated at the distance of G from the reference line.*

Figure 4-3 illustrates the system of weights as being positioned relative to a vertical line outward in a horizontal direction. This was done because in everyday life it is the way that weight systems are generally visualized. The seesaw is a good practical example. However, the law holds for finding the position of *G* upward in a vertical direction with reference to a horizontal line. This latter method is the way we consider the problem when working to locate *G* on a ship. A numerical example pertaining to the vertical position of *G* follows. In this example, we will shift also to the terms used on board ship.

NUMERICAL EXAMPLE

In Fig. 4-4 the reference line is the keel of the ship. The lower hold has 3000 tons in it, with the center of this weight 8 ft above the keel; the lower 'tweendecks has 2000 tons centered 22 ft above the keel; the upper 'tweendecks has 1000 tons 30 ft above the keel; the deck load of cargo amounts to 500 tons 38 ft above the keel. There are 2500 tons of fuel, water, and stores on board, with the center of this weight 10 ft above the keel. The light ship structure has a weight of 4000 tons and the vertical height of its center of gravity is 25 ft above the keel. The question we want to answer with all these data is: How far does the center of gravity of all these weights (*G*) lie above the keel? Referring

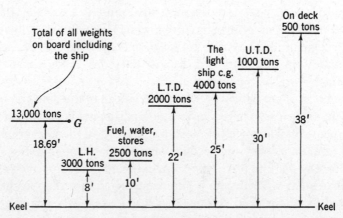

FIG. 4-4.

to the general statement of the problem, we proceed to multiply each weight by its distance above the reference line. We then add all of these moments. We also add all the weights involved to obtain the total amount of weight in the system, which is in this case the displacement of the ship. To obtain the answer we seek, we divide the sum of all the moments by the ship's displacement.

25 ·	4,000 =	100,000 ft-tons	Light Ship Moment
38 ·	500 =	19,000 ft-tons	On Deck Moment
8 ·	3,000 =	24,000 ft-tons	L/H Moment
22 ·	2,000 =	44,000 ft-tons	LTD Moment
30 ·	1,000 =	30,000 ft-tons	UTD Moment
10 ·	2,500 =	25,000 ft-tons	Fuel, Water, Stores
x ·	13,000 =	242,000 ft-tons	

Distance of G above keel   Total weight   Sum of all moments

Dividing 242,000 by 13,000, we obtain the value of x.

$$x = 18.69 \text{ ft}$$

Thus we see that G lies 18.69 ft above the keel of the ship.

The 13,000 tons is the weight of the ship and all she contains under the conditions given in the problem. This figure is the *displacement* of the ship, indicated in formulas by the Greek letter delta (Δ). The weight of the cargo and the fuel, water, and stores, 9000 tons, is known as the *deadweight lifting capacity* of the ship if this weight puts the ship down to her maximum legal draft. The *cargo deadweight* is the 6500 tons of cargo. The weight of the ship with nothing on board, 4000 tons, is known as the *light ship displacement*.

The solution of a practical problem requires all of the data given above. The weight of the cargo in the various compartments is obtained from the stowage plan. The distances of these weights above the keel are obtained partially from data concerning each compartment given on a ship's *capacity plan* (see p. 221) and partially from estimates made by the officer. If the compartment is filled with a homogeneous cargo, the weight can be considered to be centered vertically. If it is loaded with heavy goods on the bottom and lighter goods on top, assume the center of the weight to be about one-third of the height of the compartment above the deck of the

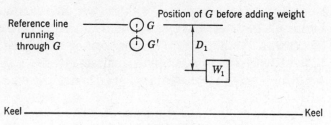

FIG. 4-5.

compartment. To find the center of all the fuel, water, and stores weights is a smaller problem solved in the same manner as the general problem. The weights of the fuel and water and the tanks in which they are contained must be obtained from the ship's chief engineer. The weights of stores and their positions with reference to the keel must be estimated as closely as possible. The weight of the light ship and the position of its center of gravity must be obtained from the ship's capacity plan or from stability data supplied for the ship by the builders. These data are determined by means of the inclining experiment.[4]

Shifts in Position of G

The position of G is affected by addition, subtraction, or shifts of weights in the system. It is important that the reader be capable of calculating these shifts in the location of G. When considering stowage problems, it is often necessary to estimate the shift in G owing to a change in the stowage plan or to the consumption of large masses on the ship, such as the burning of fuel oil from the double bottom tanks.

The first problem that we will examine is a shift in G owing to an addition or subtraction of weight in the system. Later, we will examine the shift in G caused by movement of a mass within the system. We refer to the basic law and consider a system comprised of the original total weight and the single added new weight. The reference line will be a horizontal line passing through the known position of G. The distance between our reference line and our biggest weight in our two-weight system is zero, and hence the moment is also zero. This is important. One other moment remains. This is the product of the additional weight and its distance from the old position of G (see Fig. 4-5). The sum of these

[4] J. La Dage and L. Van Gemert, *Stability and Trim for the Ship's Officer,* D. Van Nostrand Co., New York, 1947, Chap. 4.

products is equal to the value of the last product. We divide the sum of the moments by the old displacement plus the newly added weight. The quotient will be equal to the shift of G. The new position is referred to as G' ("G prime") and the distance of the shift is indicated by GG'. The shift is always toward the position of the added weight. Do not forget the formula for GG'.

$$GG' = \frac{D_1 W_1}{\Delta + W_1} \tag{2}$$

where W_1 = added weight.
D_1 = distance of added weight from old center of gravity.
Δ = displacement of ship before adding W_1.

When subtracting a weight, the same reasoning applies. The shift of G is away from the weight removed and the value of the weight system is the displacement less W_1. Equation 2 for GG' becomes:

$$GG' = \frac{W_1 D_1}{\Delta - W_1} \tag{3}$$

The shift of G (GG') caused by a shift of a part of the weight system, such as moving 100 tons up or down on the ship, is found using Eq. 4.

$$GG' = \frac{W_1 D_1}{\Delta} \tag{4}$$

where GG' = shift of G (always in the direction that the weight is shifted).
W_1 = the weight that is shifted.
D_1 = distance that W_1 is shifted.
Δ = the ship's displacement.

NUMERICAL EXAMPLE

Given: A ship displacing 15,000 tons with a *KG* of 30 ft; 2000 tons of fuel oil is burned from the double bottoms; this mass had a *KG* of 2 ft. (*Note:* The distance of G above the keel is generally referred to as *KG*. This notation will be used hereafter.)

Required: The ship's new *KG*.

Solution: This is a case of removing a weight from the weight system; hence we use Eq. 3. W_1 equals 2000 tons. D_1 equals 28 ft. Hence, the GG' equation becomes:

$$GG' = \frac{W_1 D_1}{\Delta - W_1} = \frac{2000 \times 28}{13,000} = 4.3 \text{ ft}$$

Since G has moved up 4.3 ft, the new KG is 34.3 ft.

Another method of solution uses the moments involved, as follows:

W	KG	Moment
15,000 ×	30 =	450,000
−2,000 ×	2 =	−4,000
13,000		446,000

$$\text{New } KG = \frac{446,000}{13,000} = 34.3 \text{ ft}$$

One more type of problem remains to be clarified with respect to the position of G and its shift due to the movement of weight on the ship. We have considered the solution for the shift in G due to the addition or removal of a single weight. There may be need to calculate the final position of G after loading several weights and discharging several weights. In this problem, it is necessary to consider each addition and removal separately and to separate all the movements into those that will move G up and those that will move G down. A net moment is obtained by adding these two sums. The net moment is divided by the *final* displacement to obtain the shift in G.

NUMERICAL EXAMPLE

Given: A ship with a KG of 20 ft. The displacement is 10,000 tons.

Two weights, W_1 and W_2, are loaded. $W_1 = 500$ tons and is placed 23 ft above the keel. $W_2 = 500$ tons and is placed 5 ft above the keel.

Two weights are discharged. $W_3 = 1000$ tons removed from 12 ft above the keel. $W_4 = 1000$ tons removed from 40 ft above the keel.

Required: The problem is to determine the ship's KG after the above operation. We actually would have four problems here if we solved each separately by Eqs. 2 and 3. It is easier to combine the data to form one problem. The procedure is to make two columns: In one column, list all the moments resulting in an increase of KG; in the other, list all those resulting in a decrease.

Solution: Before adding and removing the weights, the ship had a KG of 20 ft. Carefully check the effect of each weight on the KG. W_1 is added 3 ft above the initial G, which we are going to use as a datum level; hence, the movement of $3 \times W_1$ will increase our ship's KG. W_2 is added 15 ft below our datum level, and the moment will decrease the KG. W_3 is removed from 8 ft below G, and G will move away from a removed weight; thus, this moment will increase KG. W_4 is removed from 20 ft above G; thus, the moment will decrease KG. Next we calculate each of these moments, giving a plus sign to those that increase the KG and a minus sign to those that decrease the KG. Add the moments. Next divide by the final displacement of the ship. The quotient is the distance in feet that G has moved up or down. If the net moment has a minus sign, the movement is down and KG has been decreased. If the sign is plus, the movement of G is up and KG has been increased. The solution of the above problem is as follows:

Increase of KG (+)		Decrease of KG (−)	
$500 \times 3 =$	1,500 ft-tons	$500 \times 15 =$	7,500 ft-tons
$1000 \times 8 =$	8,000 ft-tons	$1000 \times 20 =$	20,000 ft-tons
$=$	+9,500 ft tons	$=$	−27,500 ft-tons

$$+9,500$$
$$-27,500$$
$$-18,000 \text{ ft-tons (net change in moments)}$$

A total of 2000 tons was discharged and 1000 tons were loaded; hence, the final displacement is 1000 tons less than the initial 10,000 tons, that is, 9000 tons.

Final step to solve for GG':

$$GG' = \frac{-18,000}{9,000} = -2 \text{ ft}$$

Therefore, the new KG is $20 - 2 = 18$ ft.

Determining the Position of B

The position of B can be calculated in a manner similar to that used to calculate G, except that volumes are used instead of weights. The calculations are complicated by the fact that these volumes are bounded by curved surfaces, but they are not exceptionally difficult. The reader should refer to a standard reference if he wishes to learn exactly how these calculations are made. For the practical determination of stability, there is no need to make the calculations involving the location of B. The reader needs only to be aware of the fact that B moves about as the vessel is inclined and of why it moves.

Movement of B

The definition of B must be recalled before we attempt an explanation of how and why B moves about when the ship is inclined. Point B is the geometrical center of the *underwater* portion of the hull. Looking at Fig. 4-6, we note that B is on the centerline of the ship. If the ship is upright and floating, G is either directly above or below B on a line perpendicular to the waterline. G is illustrated as above B on the diagram of Fig. 4-2. If the ship is inclined as shown in Fig. 4-6, a wedge of volume is removed from the underwater portion of the hull on the side away from the inclination and transferred to the side toward which the ship is inclined. What will happen to B? Obviously, B will move in the direction of the submerged side. This can be appreciated intuitively, for the center of the

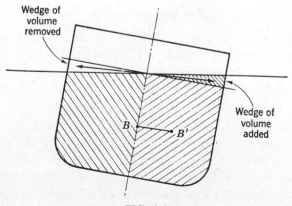

FIG. 4-6.

underwater portion now lies toward this side. The amount of the shift of B is not too important in our explanation of factors involved in stability, but a brief exposition of the method for calculating it is presented below.

Calculating the Shift of B

The technique for calculating a shift in B involves the same principles used with respect to finding the center of a system of weights. The only difference is that in place of weight values, volume values are used. Hence, the equations used to calculate GG' can be used to calculate BB'. In the case of the movement of B because of a shift of the volume wedge from one side to the other, Eq. 4 can be used, with the units changed as necessary:

$$BB' = \frac{v \cdot d}{V} \tag{5}$$

where BB' = the distance that B shifts.
v = volume of the wedge.
d = distance that the center of the volume wedge moved.
V = volume of the entire underwater portion of the ship.

The locus of the point B as the ship is inclined to larger and larger angles is an ellipse. This fact is of significance when studying stability at large angles of inclination.

The Metacenter (M)

The third and last point to be examined is the metacenter, M. To locate the metacenter, the naval architect takes the curvature of a circle that has the same curvature as a very small segment of the elliptical path of B, and draws radii of this circle. They intersect at the center of the circle; this point is M. M moves about as the ship is inclined. This fact can be appreciated because the curvature of an ellipse is at a minimum at the extremities of the minor axis and at a maximum at the extremities of the major axis. Therefore, the radii will be changing continually, and the point M will follow. For small angles of inclination, M will fall on the centerline of the ship. As the angle of inclination approaches $10°$, M moves upward and away from the side toward which the ship is inclined. When considering *initial* stability, it is assumed that M is always directly on the centerline. The distance from B to M is notated as BM and known as the metacentric radius. The force of buoyancy acting upward through B always passes through M, which will always be directly over B in a vertical direction normal to the waterline.

Distance from Keel to B and M

The distance from the keel to point B is known as KB and is calculated by the naval architect. BM is also calculated. The sum of KB and BM gives KM, a value of great importance in the calculation of initial stability. The value of KM varies with the ship's draft. For the ship's officer to know his ship's KM, he must refer to a set of hydrostatic curves or have a table giving the KM of the ship for the various drafts. The draft referred to is the mean draft, of course. This means that KM varies with the displacement. For the ship's stability to be calculated, the KM must be known. All ships should have a deadweight scale with the KM given on it or a table of KMs by draft or deadweight. Hydrostatic curves may be used if available, but they are more difficult to read and cumbersome to use.

Determination of Stability

The three points that must be understood clearly for an appreciation of ship stability have now been examined. The reader may already realize that a ship increases her stability with a lowering of G. The distance between G and M is referred to as the ship's GM. When a ship is inclined, initially B moves away from the centerline, G remains stationary, and for small angles we will assume that M remains stationary. Thus, our points would be located as seen in Fig. 4-7a if G was initially below M. The angle of inclination has been exaggerated in this diagram so that the elements of the figure can be seen more easily. Note that the forces through B and through G are no longer opposed to each other. They are forces acting through a distance; thus, they create a moment. The distance through which these forces act is the line GZ. GZ is perpendicular to the line BM, the metacentric radius. The moment tends to revolve the ship in a direction opposite to the inclination. In other words, it tends to push the ship back into an upright position. This moment is called the righting moment.

Stable, Neutral, and Unstable Equilibrium

When G is below M, the ship possesses *stable equilibrium* and tends to return to an upright position if inclined by an outside force. The tendency

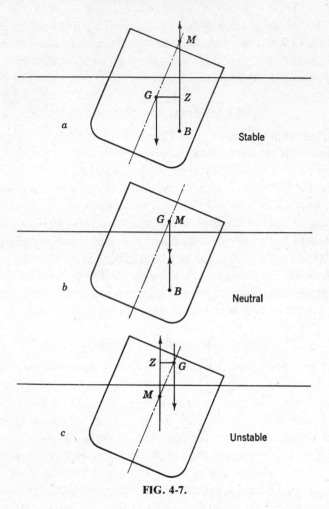

FIG. 4-7.

to right herself depends on the ship's displacement and the value of GZ, because the righting moment is the product of displacement and GZ.

If points G, B, and M are located as shown in Fig. 4-7b, the point G has moved up until the distance GZ has become zero and there is no righting moment. In this position, GM is also zero. GM and GZ vary directly. The ship is in a condition of *neutral equilibrium* when GM and GZ both equal zero. When a ship possesses neutral equilibrium, she will remain in whatever position she is placed, within certain limitations.

If G is moved up beyond M, the points and forces involved will act as shown in Fig. 4-7c. In this condition, the ship possesses *unstable equilibrium*. The ship will not remain in an upright position, but will assume a list either to the port or to the starboard. The angle of the list will depend on the distance of G above M. The greater the distance, the greater the angle of list.

Listing Due to Negative GM

The ship will not capsize just because she has unstable equilibrium, unless G is too far above M. When G is above but close to M, the ship will assume a small list. This list may be increased to a dangerous magnitude if G is moved up as a result of burning fuel oil from the ship's double bottom tanks. The reason why the ship assumes a small list and does not continue over is explained by the fact that the metacenter, M, moves up as the angle of inclination varies between 0 and 20°. Thus, M moves above G if G is sufficiently close to M, for G is stationary. When M is above G, the ship attains stable equilibrium, but she will have a permanent list to the port or starboard. The list will change from one side to the other when a force, such as centrifugal force during a radical course change or from strong winds or seas, is applied to the hull.

The ship's officer should be able to recognize a ship with negative GM by her behavior in a seaway. A ship with negative GM will have a list, but the list may change from port to starboard and back again with such forces on the hull as wind and sea or centrifugal force when applying heavy rudder. The ship will always have a long, slow, sluggish roll.

Correcting a List Due to a Negative GM

A common, and often serious, mistake made in the field of stability is that of not recognizing a list caused by negative GM or unstable equilibrium and therefore taking corrective measures that are improper.

If a ship has a list of 5 or 6° caused by off-center weights, that is, by G being to port or starboard of the centerline, the obvious correction is to move weights toward the high side. However, if a list is caused by negative GM, the movement of weights to the high side would cause the vessel to right itself partially and then suddenly take a much greater list to the opposite side. The sudden rush toward the opposite side would take place after about one-half of the previous angle of list had been removed. The list to the opposite side would be greater because it would be the result of two poor conditions: (1) negative GM; and (2) off-center weights.

The only way to correct a list due to negative *GM* is to add, remove, or shift weights so that *G will be moved down*.

Lumber carriers are frequently victims of negative *GM* conditions. To prevent or correct a list due to this condition, the double bottom tanks should be kept as full as possible. After a long voyage, the double bottom tanks may be so empty that the ship will have an excessive list on arrival in port. In such a case, the lumber deck load should be carefully removed in layers. It would be a serious blunder to attempt to right the ship by discharging from the low side first. The list can be removed by pumping salt water into the empty fuel oil double bottom tanks, but this is generally avoided if possible. Once salt water gets into the fuel oil system of the ship, the engineers may have endless trouble with it.

GM *as a Measure of Stability*

In the triangle *GZM* of Fig. 4-8*b*, the angle at vector *Z* is a right angle. From elementary trigonometry we have:

$$\sin \theta = \frac{GZ}{GM} \tag{6}$$

from which we get:

$$GM = \frac{GZ}{\sin \theta} = GZ \cdot \csc \theta \tag{7}$$

From Eq. 6, we see that *GM* varies directly as *GZ*. It has already been pointed out that *GZ* is one of two factors that determine the tendency of the ship to right herself. Because *GM* and *GZ* vary directly, *GM* is a measure of stability also.

Calculating GM

All of the points that we have examined are illustrated in Fig. 4-8*a*. They are arranged to indicate a positive *GM*. To calculate *GM*, data to calculate *KG* are required. In addition, a table of *KM*s for all the drafts from light to loaded is required. Then, *KM* − *KG* = *GM*. Once we know the *GM* of the ship, we have some knowledge of her stability and we know how she will react to a heavy seaway. With experience, such information gives the ship's officer a basis for making correct and prompt decisions concerning stability.

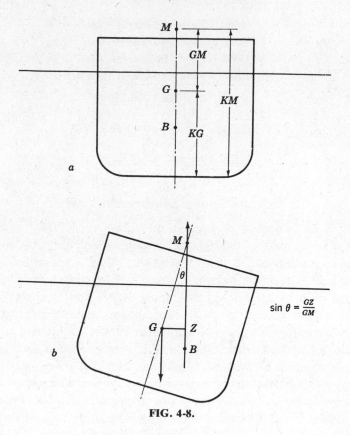

$$\sin \theta = \frac{GZ}{GM}$$

FIG. 4-8.

NUMERICAL EXAMPLE

Given: A ship with a light ship displacement of 4500 tons. Light ship *KG* = 25 ft. The following weights are on board: fuel, water, stores = 2000 tons with *KG* of 20 ft; lower hold cargo = 2000 tons with *KG* of 10 ft; 'tweendeck cargo = 1000 tons with *KG* of 30 ft. *KM* for a displacement of 9500 tons = 24.5 ft.

Required: The ship's *GM*.

Solution: We know the *KM*, so we only have to calculate *KG*. *KM* will be obtained from a table based on displacements or drafts. To calculate *KG*, we find the vertical moments of all the weights (including light ship), add these moments, and divide by the displacement. The result is the *KG*. Then *KM* − *KG* = *GM*.

$$4500 \cdot 25 = 112{,}500 \text{ ft-tons}$$
$$2000 \cdot 20 = 40{,}000 \text{ ft-tons}$$
$$2000 \cdot 10 = 20{,}000 \text{ ft-tons}$$
$$\underline{1000 \cdot 30 = 30{,}000 \text{ ft-tons}}$$
$$9500 202{,}500 \text{ ft-tons}$$

$$202{,}500 \div 9500 = 21.3 \text{ ft} = KG$$

$$24.5 - 21.3 = 3.2 \text{ ft} = GM$$

Light and Loaded Displacement GMs

Stability is the tendency of a vessel to right herself. This tendency is measured in the units of the righting moment, which are foot-tons. Considering a constant *GZ*, it can be seen that when a vessel is light, say displacing 6000 tons, the righting moment will be considerably less than when the vessel is loaded, say displacing 14,000 tons. The righting moment should be kept at a nearly constant value; hence, the *GM* that produces a comfortable and safe ship in the light condition will produce a very stiff ship in the loaded condition. The average *loaded* merchant ship is safe and comfortable with a *GM* between 2 and 3 ft. The *GM* of the ship with only 1000 or 1500 tons of cargo or ballast on board should be 4 to 5 ft. The important point here for the ship's officer to remember is that when the ship is only partially loaded the *GM* must be larger than when the ship is fully loaded.

A light ship may have a *GM* of 8 to 12 ft. With some fuel, water, and stores on board, this will generally be reduced 2 or 3 ft. The resulting *GM* will generally be too large for a ship even in the partially loaded or ballasted condition. At light drafts, the *KM* is reduced rapidly as the displacement increases; thus, adding weight so that the position of *G* remains fixed will reduce *GM* by dropping the metacenter down closer to the center of gravity.

Distribution of Partial Loads

It is neither feasible nor necessary to specify the precise distribution that should be given partial loads. The important thing is to be aware that if only 500 to 1000 tons are being carried, the load should be carried in the 'tweendeck areas. If the cargo amounts to 2000 to 3000 tons, the metacenter may drop so rapidly that some cargo will be required in the lower holds for a safe and comfortable *GM*. Each ship and her partial load will present a particular problem that should be solved on the basis of the facts available. The above is a discussion of the general case, and is not intended to lay down any precise rules or imply what should be done in a particular case. However, it can be stated safely that if a ship capable of lifting 7500 to 10,000 tons cargo deadweight is lifting only about 1000 tons, none of that weight should be in the lower holds.

When stowing a few hundred tons of ballast or cargo in the upper deck areas to reduce the *GM* of a ship going to sea with a partial load, the ship's officers should take every precaution possible to see that such ballast or cargo is heavily shored, braced, or tommed into position. In heavy weather, the ship may still roll rapidly and the cargo will be subjected to strong centrifugal forces. Ships are still likely to be lost at sea without a trace due to negligence of responsible officers in these matters. In January 1948, the *S.S. Samkey,* a British steamer, was lost with all hands while on a voyage to Cuba because of the sudden shifting of ballast carried in the 'tweendecks. The shifting, of course, was caused by heavy weather. The ballast was of a type known as Thames ballast, which consists of stones of various sizes mixed with sand, with an angle of repose of about 35°. A stiff ship in a heavy sea will roll to much greater angles than will a tender ship, and ballast with an angle of repose of only 35° would require heavy shifting boards. On this ship, inadequate shifting boards had been installed.

Estimating Vertical Distribution

It is often stated with reference to the stowage of cargo that, as a general guide for proper vertical distribution of cargo when carrying full loads, about one-third of the weight should be in the 'tweendecks and about two-thirds should be in the lower holds. **This rule should not be used.** Such distribution will often give a ship an excessive *GM*. The rule also leaves much doubt as to what should be done when the ship has three decks or four decks. The only safe procedure is to calculate the correct vertical distribution by the methods outlined above. However, if for any reason it is necessary to estimate the desired vertical distribution of a full load and data for calculating the amounts are not available, the safest procedure is to place in each compartment a proportion of the total weight equal to the ratio of the individual compartment's bale cubic to that of the ship's total bale cubic. This can be expressed by the following formula:

$$\frac{v}{V} \cdot T = t \tag{8}$$

where V = total bale capacity of the ship.

v = bale capacity of the compartment or series of compartments at equal distances from the keel.

T = total tonnage to be loaded.

t = number of tons of the total load that should be placed in the compartment or compartments in question.

This method cannot be used when taking on partial loads or under ballasted conditions, as is evident from the above discussion.

NUMERICAL EXAMPLE

Given: On a vessel, the bale cubics are as follows:

$$
\begin{aligned}
\text{U.T.D.} &= 145{,}865 \text{ ft}^3 \\
\text{L.T.D.} &= 134{,}025 \text{ ft}^3 \\
\underline{\text{L.H.}} &\underline{= 164{,}550 \text{ ft}^3} \\
\text{Total} &= 444{,}440 \text{ ft}^3
\end{aligned}
$$

The cargo deadweight is 8000 tons. No other information is available.

Required: The vertical distribution of the 8000 tons.

Solution: Simply multiply 8000 by the ratio of each compartment's cubic to the cubic of the entire ship.

$$\frac{145{,}865}{444{,}440} \cdot 8000 = 2640 = \text{tons in U.T.D. area}$$

$$\frac{134{,}025}{444{,}440} \cdot 8000 = 2440 = \text{tons in L.T.D. area}$$

$$\frac{164{,}550}{444{,}440} \cdot 8000 = 2920 = \text{tons in the L.H. area}$$

Note that the weight is divided up to equal about one-third for each level in this particular case. The resulting *GM* would depend on the tankage of the ship, which can cause the *GM* to vary widely. Tankage is the distribution of the weights in the numerous tanks on the ship. Free surface effect would also have to be given consideration, as in any practical calculation of *GM*.[5]

Trim Calculations

Definition of Trim

Trim is defined as the difference in drafts forward and aft on the ship. The trim of a ship is a function of the moments developed by weights acting forward of the ship's tipping center as opposed to those acting aft of the tipping center. With a ship on an even keel, the points mentioned in the explanation of stability would be arranged as shown in Fig. 4-9. These are the same points except that they are analyzed on the profile of the ship instead of the midship section. The center of gravity and the center of buoyancy are in the same place vertically as before. *M* is much higher and is identified as the longitudinal metacenter. The *GM* is a few hundred feet in length and is designated the longitudinal *GM*. It is important that the reader review the explanations of how *G, B,* and *M* are located and why *G* and *B* move.

When a ship is on an even keel and at rest, the points *G* and *B* are vertically one over the other. If they are not, the ship will revolve slightly, causing *B* to move until it comes under *G*. Hence, if we take the ship on an even keel and place a weight aft of point *B* or *G*, point *G* will be moved aft. The situation then would be as illustrated in Fig. 4-9*b*. This condition cannot exist, because the forces acting upward through *B* and downward through *G* are now acting through a distance, and this produces a moment. This moment will tend to revolve the ship counterclockwise as viewed in this diagram. As the ship trims by the stern, a wedge of volume is shifted from the forward part of the vessel to the after part. This causes *B* to shift aft and eventually come directly under *G* again, as shown in Fig. 4-9*c*. When this occurs, the ship will come to rest with a change of drafts forward and aft. The ship will have changed trim.

Moments Producing a Change of Trim

The change of trim can be expressed as a function of two completely different moments. The reader should be able to see that *GG'* multiplied

---

[5] J. La Dage and L. Van Gemert, *Stability and Trim for the Ship's Officer,* D. Van Nostrand Co., New York, 1947, Chap. 6.

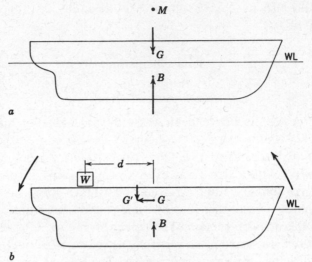

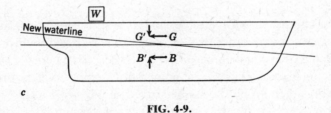

Adding weight W will cause G to shift to G'.

FIG. 4-9.

by the displacement is one moment that causes the ship to revolve. The displacement used in this calculation is the displacement in the final condition. Another moment, which would be equal to the first one mentioned, is the product of W times D, where W is the individual weight being considered at a distance D from the tipping center of the ship. Either of these two moments may be used to calculate the change in trim, but the methods used in each solution differ slightly.

As the vessel revolves in changing trim, it does so about the center of flotation, which is the geometrical center of the water plane at which the ship is floating. This point is also known as the tipping center and is notated as T.C. The tipping center changes its position as the displacement increases, generally moving aft.

Change of Trim Values

When given a set of drafts for two different conditions, the change of trim between the two conditions is calculated by first noting the trim in each case. Next, if the trim is in the same direction in both cases, the change in trim is found by subtracting the smaller from the larger.

NUMERICAL EXAMPLE

Given: A ship with the following initial and final drafts:

	Fwd., ft and in.	Mean, ft and in.	Aft, ft and in.	Trim, in.
Initial:	20 06	20 08½	20 11	5 aft
Final:	26 00	27 00	28 00	24 aft

Required: The change in trim.

Solution: By inspection it can be seen that 24 less 5 in. gives a *change of trim* of 1 ft 7 in. by the stern.

If two trims are in opposite directions, the change of trim is found by adding the two trims.

NUMERICAL EXAMPLE

Given: A ship with the following initial and final drafts.

	Fwd., ft and in.	Mean, ft and in.	Aft, ft and in.	Trim, in.
Initial:	20 00	19 06	19 00	12 fwd.
Final:	20 00	20 04	20 08	8 aft

Required: The change in trim.

Solution: Inasmuch as the trim is by the head in the initial condition and by the stern in the final condition, the change in trim is found by adding the trim. Thus, the *change in trim* is 1 ft 8 in. by the stern.

Moment to Change Trim 1 in. (MT1)

As stated above, the change of trim can be expressed as a function of either of two moments. When calculating the effect of loading, discharging, or shifting partial loads, that is, when the change of displacement is small, the moment obtained by multiplying the weight involved by the distance through which it acts is most commonly used.

When loading or discharging a weight, the moment is obtained by determining the distance between the center of flotation and the position of the weight and multiplying by the value of the weight. The change in trim is determined by dividing the above moment by a value known as the *moment to trim the ship 1 in*. This value is notated as $MT1$ and is found on the hydrostatic curves of the ship and on the deadweight scale. This value increases with displacement. If the exact position of the center of flotation is not known, the moment may be calculated about the midship section of the ship. If this is done, however, an error will be introduced. The center of flotation will generally be well forward of the midship section at light drafts. At loaded drafts, it may be slightly aft or slightly forward. The equation for the change of trim is:

$$\frac{W \times D}{MT1} = \text{Change in trim} \tag{9}$$

When shifting a weight, the moment is obtained by multiplying the weight by the distance that the weight is moved.

A typical trim problem is solved by first entering on the deadweight scale the ship's existing mean draft and finding the corresponding $MT1$. One then determines the trimming moment and divides by the $MT1$. The result will be the change in trim in inches. If the ship's tipping center is exactly at the midpoint longitudinally, the *change in draft* readings forward and aft are obtained by dividing the total *change in trim* by 2 and applying it to the old draft readings. If the ship's tipping center is not precisely at the midpoint, the change of *draft* (not trim) forward can be found by multiplying the change in *trim* by the ratio of the distance of the tipping center from the forward perpendicular to the total length on the waterline. In the same manner, the change in draft aft can be found. The sum of the changes in draft should equal the total change in trim. In practice, the error produced by assuming that the tipping center is at the midship section, whether it is or not, is less than the inaccuracies that enter into the actual reading of the drafts. Hence, the extreme precision obtained by multiplying by the ratio mentioned above is unnecessary.

NUMERICAL EXAMPLE

A trim problem involving a shift of weight.

Given: A ship with an *MT*1 of 1200 pumps 200 tons of fuel oil from a forward tank to an after tank, a distance of 60 ft. The drafts before the shift were: fwd., 23 ft 08 in.; aft, 23 ft 04 in.; mean, 23 ft 06 in.

Required: The total change in trim and the final drafts.

Solution:

$$\frac{200 \times 60}{1200} = 10 \text{ in. total change in trim by the stern}$$

Assuming the tipping center is amidships, the draft aft is increased 5 in. and the draft forward is decreased 5 in. Therefore, the new drafts would be: fwd., 23 ft 03 in.; aft, 23 ft 09 in.; mean, 23 ft 06 in.

In the above problem, if the ship was 480 ft on the waterline and the tipping center was 10 ft forward of the midship section, the exact change in drafts would be:

$$\frac{\text{Distance of tipping center from fwd.}}{\text{Length on the waterline}} \times \text{Change in trim}$$

$$= \text{Change in draft fwd.}$$

$$(230/480) \times 10 = 4.79 \text{ in.}$$

$$\frac{\text{Distance of tipping center from aft}}{\text{Length on the waterline}} \times \text{Change in trim}$$

$$= \text{Change in draft aft}$$

$$(250/480) \times 10 = 5.21 \text{ in.}$$

From this problem, it is seen that the error produced by assuming the T.C. to be amidships, when it was actually 10 ft forward, amounts to less than a quarter of an inch.

NUMERICAL EXAMPLE

A trim problem involving the addition of weight.

Given: A ship with an *MT*1 of 1000. 150 tons are loaded 100 ft aft of the tipping center. The drafts before loading were: fwd., 19 ft 02 in.; aft, 19 ft 04 in.; mean, 19 ft 03 in. Tons per inch immersion (T.P.I.) is 50.

Required: The total change in trim and the final drafts.

Solution: The *MT*1 and the T.P.I. are found on the deadweight scale. T.P.I. is defined as the number of tons required to increase the mean draft 1 in. It is obvious that to determine the final drafts, this figure would have to be known unless the deadweight scale was on hand to refer to. In practice the deadweight scale would be available, but when working problems the T.P.I. is usually stated.

The initial step in calculating the final drafts is to increase all the given drafts by an amount equal to the *mean sinkage*. The mean sinkage is obtained by dividing the number of tons loaded by the T.P.I. In this problem it amounts to 3 in. The reasoning used in the solution is that the weight is loaded directly over the tipping center first, then shifted to the actual position. Thus, the mean sinkage is added to all the given drafts first; then with these new drafts the change of trim is applied exactly as in the simple shift of weight problem already solved. Therefore, applying the mean sinkage to the given drafts we obtain: fwd., 19 ft 05 in.; aft, 19 ft 07 in.; mean, 19 ft 06 in.

Calculating the change of trim:

$$\frac{150 \times 100}{1000} = 15 \text{ in.}$$

The change forward is $-7\frac{1}{2}$ in., and aft it is $+7\frac{1}{2}$ in. The final drafts would be: fwd., 18 ft 09½ in.; aft, 20 ft. 02½ in.; mean, 19 ft 06 in.

Calculating the T.P.I.

The T.P.I. is found by taking the area of the water plane for the draft in question and dividing by 420. That is:

$$\frac{\text{Area of water plane}}{420} = \text{T.P.I.}$$

The derivation is simple. The number of tons required to sink a ship 1 ft is equal to the weight of sea water displaced by the volume of a 1-ft layer of the hull. Thirty-five cubic feet of sea water weighs 1 ton. Therefore, to find the tons required to immerse a ship 1 ft, we simply divide the volume of a 1-ft slice by 35. The volume of a 1-ft slice is equal to the area of the water plane multiplied by 1 ft. This is expressed by the equation:

$$\frac{\text{Area of the water plane} \times 1}{35} = \text{Tons per foot of immersion}$$

But we want to know the tons per inch, and an inch is exactly one-twelfth of a foot. Letting A.W.P. stand for area of the water plane, we have:

$$\frac{\text{A.W.P.}}{12 \times 35} = \text{T.P.I.}$$

Which is the same as:

$$\frac{\text{A.W.P.}}{420} = \text{T.P.I.}$$

Calculating MT1

When asked to calculate the *MT*1, the problem is to find the value of $W \times D$ (the trimming moment) that will produce a change of *trim* of 1 in. or, what amounts to the same thing, a change of *draft* at either end of $\frac{1}{2}$ in. One-half inch can be expressed also as $\frac{1}{24}$ ft. We will say then that our problem is to find an equation for $W \times D$ that will produce a change of *trim* of $\frac{1}{12}$ ft or a change of draft forward and aft of $\frac{1}{24}$ ft. When we do this, we will designate such a $W \times D$ as *MT*1. Referring to Fig. 4-10, if the weight, W, is moved aft some distance D, then G will move aft to some point G'. A moment will be set up causing the ship to revolve counter-

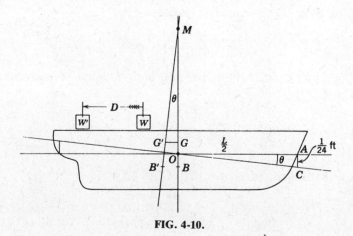

FIG. 4-10.

clockwise as viewed in the diagram. This will cause B to move aft until it is again under the center of gravity; that is, B will move to point G' or directly below this point and become B'.

Starting with the ship on an even keel, the right triangle MGG' has been formed. Also, right triangle OAC has been formed. AC equals $\frac{1}{24}$ ft. OA equals one-half the length of the ship, which we notate as $L/2$. The angles GMG' and COA are both equal to theta (θ).

From trigonometry:

$$\tan \theta = \frac{GG'}{GM} \tag{10}$$

$$GG' = \tan \theta \times GM \tag{11}$$

Also:

$$\tan \theta = \frac{1/24}{L/2}$$

$$\tan \theta = \frac{1}{12L} \tag{12}$$

From Eq. 4, p. 106:

$$GG' = \frac{W \times D}{\Delta} \tag{13}$$

Solving for $W \times D$ (our trimming moment):

$$W \times D = GG' \times \Delta \tag{14}$$

Setting the right-hand sides of Eqs. 10 and 12 equal to each other and solving for GG':

$$GG' = \frac{GM}{12L} \tag{15}$$

Substituting the right hand member of Eq. 15 in 14, we have:

$$W \times D = \frac{GM \times \Delta}{12L}$$

Because $W \times D$ produced a change of trim of 1 in. and by definition $MT1$ is the notation used for this particular moment, we substitute $MT1$ for $W \times D$ and obtain:

$$MT1 = \frac{GM \times \Delta}{12L} \qquad (16)$$

Changing Draft at One End of Vessel Only

The application of the principles of trim calculation to determine where a given weight should be loaded on a ship so that no change in draft will occur at one end of the ship may prove useful to the ship's officer. If the weight, W, is known and the T.P.I. is known, the mean sinkage is obtained by dividing W by T.P.I. Let us suppose that we wish to load W so that no change will occur in the present draft aft. It is immediately clear that we must load the weight at some point forward of the tipping center. It must be loaded so that the change of *draft* aft is equal to the mean sinkage due to loading the weight. The change of draft is considered equal to $\frac{1}{2}$ the change of trim; therefore, the change of trim obtained as a result of loading this weight must be equal to twice the mean sinkage. Now we have two formulas that we combine and then solve for the distance, D, that the weight must be from the tipping center.

$$\frac{W}{\text{T.P.I.}} = \text{Mean sinkage}$$

However, we have said that mean sinkage must be removed from the draft aft in order not to have any change aft. In other words, the change of trim must be equal to twice the mean sinkage or $2W/\text{T.P.I.}$ From Eq. 9, we have:

$$\frac{W \times D}{MT1} = \text{Change of trim}$$

Knowing that our change of trim must be equal to $2W/\text{T.P.I.}$, we substitute in the above equation:

$$\frac{W \times D}{MT1} = \frac{2W}{\text{T.P.I.}}$$

Solving for *D*:

$$D = \frac{2 \times MT1}{\text{T.P.I.}} \tag{17}$$

NUMERICAL EXAMPLE

Given: A ship with the following drafts must load 200 tons: fwd., 19 ft 00 in.; aft, 19 ft 06 in.; mean, 19 ft 03 in. *MT*1 is 1400. T.P.I. is 50.

Required: Where shall the weight be loaded so that there will be no change aft?

Solution: Using Eq. 17:

$$D = \frac{2 \times 1400}{50}$$

$D = 56$ ft forward of the tipping center

If this problem is stated without mentioning the weight involved, the solution is the same with the exception that the weight may be found by knowing the mean sinkage required and the T.P.I. For example, supposing the problem had asked for the position and the amount of the weight needed to give the ship an even trim with a mean draft equal to the present maximum draft. The present mean draft is 19 ft 03 in., and the problem states that it must be 19 ft 06 in. when the weight is loaded. Therefore, the mean sinkage must be 3 in. Now, from our equation for mean sinkage:

$$\frac{W}{\text{T.P.I.}} = \text{Mean sinkage}$$

We have:

$W = \text{T.P.I.} \times \text{Mean shrinkage}$
$W = 50 \times 3$
$W = 150$ tons

Thus, we know the weight required; the solution for the remaining part of the required data is the same as in the first problem. The final draft in the second problem would be: fwd., 19 ft 06 in.; aft, 19 ft 06 in.; mean, 19 ft 06 in.

Calculating Trim with Large Changes of Displacement

When a ship is laid out during the planning stage of the cargo operation, the weights should be so distributed longitudinally that the final trim will be within an acceptable range. The methods of calculating trim and changes of trim discussed above have assumed that comparatively small parts of the ship's total cargo capacity were being used; therefore, small changes in displacement were envisioned. When loading a complete or even a half or quarter shipload, there will be a large change in displacement. A large change in displacement makes it very important that the correct $MT1$ and tipping center be used. T.P.I. will not be used for finding mean sinkage. Mean sinkage is determined accurately only by finding on a deadweight scale the correct drafts for the different deadweights involved.

First method: moments about the tipping center. Two methods may be used to calculate the trim when large changes of displacement are involved. The first method is based on the concepts already discussed. Care must be used to choose the $MT1$ for the final condition and the average position of the tipping center between the initial and final condition. In this first method, the trimming moments are taken about the tipping center.

Table 4-1 is a form that may be used to record the data when loading a large number of weights throughout the ship so that the net effect of all the trimming moments can be calculated. The first question that arises in the use of this form when determining the moments about the tipping center involves the position of the tipping center. Data from the hydrostatic curves of the Mariner type ship appear in Table 4-4. It can be seen that the tipping center (column G) is constantly changing its position. The best that can be done is to pick the tipping center for the average position. It must be pointed out that the geometrical centers of all compartments and tanks must be known relative to the midship section. The centers of such compartments and tanks must then be computed relative to the tipping center that has been calculated.

The T.P.I. is not used to determine the final mean draft. Using a mean T.P.I. would introduce a large error into the calculations. The mean sinkage is determined by referring to the deadweight scale, as mentioned above.

The final $MT1$ value must be used. This is obvious because it has already been pointed out that we assume all the weight to be loaded over

TABLE 4-1

FORM FOR CALCULATION OF CHANGE IN TRIM ABOUT TIPPING CENTER (T.C.)[a]

Compartment	Weight, tons	Distance from Amidships, ft	Distance Fwd. of T.C., ft	Distance Aft of T.C., ft	Trimming Moments	
					Fwd., ft-tons	Aft., ft-tons
		(This fact is needed because T.C. position is given from the midship section of the ship as a reference line.)				

Total weight _____   Total trimming moments: Fwd. ____ Aft ____

$$\frac{\text{Net trimming moments either fwd. or aft}}{\text{Final } MT1} = \text{Change of trim}$$

Final $MT1$ = _____
Mean position of T.C. = _____

[a] With large changes of displacement, use deadweight scale to obtain *final* mean draft. Apply this to original condition and then apply change of trim to obtain *final conditions*.

the position of the mean tipping center first. After calculating the effect on mean sinkage, we imagine the weight is shifted so as to produce the trim. The reader should study the elements of the solution of the problem represented by Fig. 4-11*a*.

Second method: moments about longitudinal B.   A more accurate and more commonly used method of calculating trim involving large changes of displacement will be explained next. Before attempting this explanation, however, the reader is reminded that all of the previous methods involving trim considered moments about the tipping center, or longitudinal center of flotation, of the ship. However, it was stated at the beginning

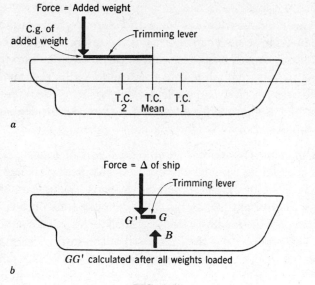

FIG. 4-11.

of the discussion on trim that trimming moments may be considered also to act about the center of buoyancy. In the first method, we considered the total weight involved as acting through a distance from the tipping center. In this second method, we consider the weight as causing a shift of G. This shift of G creates the trimming lever GG' as illustrated in Fig. 4-11b. The weight acting through the distance GG' is the *displacement* of the ship, and, thus, the *trimming moment* is formed. Figure 4-11 shows two diagrams. Diagram a points out the trimming moment of the first method. Diagram b points out the trimming moment of the second method. The reader should not proceed without a clear understanding of these points.

The second method requires that the final longitudinal position of G be calculated along the same lines as the vertical position of G was calculated in the study of transverse stability. From hydrostatic curves or suitable tables, the longitudinal position of B is obtained for the condition of even keel. The distance of G from B is the trimming lever, and the weight involved in the trimming moment is the displacement. Once we have the trimming moment, we simply divide by the $MT1$, as before, to obtain the change of trim.

Inasmuch as the longitudinal position of G must be calculated, it is obvious that the location of the geometrical center of every cargo compartment and tank must be known with reference to some longitudinal point. The midship section can be used as the reference point, but if this is done, it is necessary to bring plus and minus quantities into the calculations that unnecessarily complicate the solution. It is recommended that either the forward or after perpendicular be chosen as the reference point.

This method is thoroughly discussed in the section dealing with the trim and stability booklet for the Mariner type ships.

Change of trim with second method and partial loads. The second method of calculating trim may be used for calculating change in trim for partial loads as well as the first method. Suppose a ship capable of lifting 10,000 tons deadweight has all but 3000 of those tons on board, and the problem is to plan the loading of the last 3000 tons so that a suitable trim is obtained. The procedure is the same as that for starting with a light ship except that the longitudinal positions of B and G must be determined in a different manner. If the ship is on an even keel, the location of B can be determined by looking it up on suitable tables or from the hydrostatic curves for the ship. However, if the ship has a trim by the head or stern, the location of B (and therefore of G also) is obtained by using the equation for determining the change of trim by the second method. We have:

$$\frac{GG' \times \Delta}{MT1} = \text{Change of trim}$$

Solving for GG':

$$GG' = \frac{MT1 \times \text{Change of trim}}{\Delta}$$

All of the factors in the right-hand term are known. The value of GG' is applied to the position of B as if the ship had been on an even keel at the mean draft. B must be either forward of or aft of the latter position, depending on whether the ship has a trim by the head or stern, respectively. The distance that it is forward or aft will be equal to GG'.

A record of the distances forward and aft of G' at which the 3000 tons are disposed is kept, and the final position of G' (call it G'') is calculated. Thus $G'G''$ becomes the trimming lever, and the weight involved is the final displacement. The change of trim and final draft are determined as outlined previously.

Trim and Stability Booklet for Mariner Type Ships

Some of the data contained in a trim and stability booklet prepared by the Maritime Administration, Division of Preliminary Planning, are presented below in a series of tables and one curve, Fig. 4-12. Most of these tables are self-explanatory to the officer with a basic knowledge of trim and stability concepts. However, a brief explanation will be set forth here along with a rather detailed summary of the final solution for *GM* and the calculated draft as it appears in Table 4-9.

All ships should have information of this type available to the operating personnel afloat and ashore. The main lack in the past has been complete hydrostatic data such as appears in Table 4-4. The reader should note that this is simply an extension of the familiar deadweight scale to the point where it contains some really valuable information about the ship. That is, it is valuable if used consistently and correctly.

The tables and figure are as follows:

Table 4-2: A list of the ship's principal characteristics.

Table 4-3: A trimming table for quick solutions of effect on trim when loading, discharging, or shifting small weights.

Table 4-4: The hydrostatic properties of the ship. The usual dead-weight scale given on most capacity plans includes only the information appearing in columns *A, B, D,* and *E*. Without the information contained in columns *C, F,* and *G*, it is not possible to make many trim and stability calculations.

Table 4-5: Tank capacities and free surface effects of individual tanks. Instructions for obtaining the total free surface correction are included. Note that the correction is always subtractive from the *GM*.

Table 4-6: This very useful table gives the gain in *GM* obtained by ballasting any given tank at various displacements.

Figure 4-12: The required *GM* curve. The *required GM* may be defined as that *GM* that will prevent the ship from having a negative stability in case any one compartment has been flooded and will prevent heeling that might result in flooding of any other, undamaged compartment. Normal operating conditions are assumed. It should be obvious that actual conditions on a given ship can modify the value of the required *GM*. For example, according to the curve, the required *GM* for a fully loaded Mariner type ship (mean draft 29.9 ft) is 2.2 ft. However, if the ship were loaded completely with lumber below decks, this *GM* would be in excess of what might be considered a minimum safe *GM*.

TABLE 4-2

TABLE OF PRINCIPAL CHARACTERISTICS

Length, overall	563 ft 7¾ in.
Length, between perpendiculars	528 ft 0 in.
Length, 20 stations	520 ft 0 in.
Beam, molded	76 ft 0 in.
Depth to main deck, molded at side	44 ft 6 in.
Depth to 2nd deck, molded at side	35 ft 6 in.
Bulkhead deck	2nd deck
Machinery	Turbine
Designed sea speed	20 knots
Shaft horsepower, normal	17,500
Shaft horsepower, maximum	19,250
Full load draft, molded	29 ft 9 in.
Full load displacement	21,093 tons
Light ship displacement	7,675 tons
Light ship vertical position of center of gravity	31.5 ft
Light ship L.C.G. aft F.P.[a]	276.5 ft
Passengers	12
Crew	58
Grain cubic	837,305 ft^3
Bale cubic	736,723 ft^3
Reefer cubic	30,254 ft^3
Fuel oil (double bottoms + settlers)	2,652 tons
Fuel oil (deep tanks)	1,156 tons
Fuel oil, total	3,808 tons
Fresh water	257 tons
No. of holds	7
Gross tonnage	9,215
Net tonnage	5,367

[a] L.C.G., longitudinal center of gravity; F.P., forward perpendicular.

Table 4-7: This table provides a simplified means of determining the proper amount of double bottom tankage to meet the requirements of one-compartment damage for any indicated condition of loading. Interpolation should be made between the figures given to obtain the required double bottom tankage.

Table 4-8: These are four working forms to be filled out to obtain data necessary to calculate the *GM* and trim conditions. They have been filled out with an example. The amounts of dry cargo, reefer cargo, fuel oil or salt water, and fresh water are entered in the appropriate tables. The summary of each item is entered in the correct space on Table 4-9.

TABLE 4-3

TRIM TABLE FOR THE C4-S-1A:
TABLE OF CORRECTIONS IN INCHES TO DRAFT FORWARD AND AFT FOR EACH 100 TONS LOADED AT ANY DISTANCE FROM AMIDSHIPS

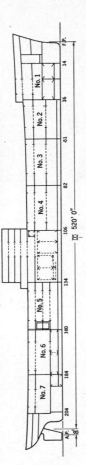

Example: Find the change in trim after loading 100 tons in No. 2 hold (160 ft forward amidships)

Initial draft	Forward 19 ft 6 in.	Aft 20 ft 6 in.
Correction	Forward +7.6 in.	Aft −4 in.
New draft	Forward 20 ft 2 in.	Aft 20 ft 2 in.

30 ft 0 in. draft — Amidships

Fwd: +9.0, +9.0, +8.5, +8.5, +8.2, +7.9, +7.6, +7.4, +7.1, +6.8, +6.5, +6.2, +6.0, +5.7, +5.4, +5.1, +4.8, +4.6, +4.3, +4.0, +3.7, +3.4, +3.2, +2.9, +2.6, +2.3, +2.0, +1.8, +1.5, +1.2, +0.9, +0.6, +0.4, +0.1, −0.2, −0.5, −0.8, −1.1, −1.3, −1.6, −1.9, −2.2, −2.5, −2.7, −3.0, −3.3, −3.6, −3.9, −4.1, −4.4, −4.7, −5.0, −5.3

Aft: −5.3, −5.0, −4.8, −4.6, −4.3, −4.1, −3.8, −3.6, −3.4, −3.1, −2.9, −2.7, −2.4, −2.2, −1.9, −1.7, −1.5, −1.2, −1.0, −0.8, −0.5, −0.3, 0, +0.2, +0.4, +0.7, +0.9, +1.1, +1.4, +1.6, +1.9, +2.1, +2.3, +2.6, +2.8, +3.0, +3.3, +3.5, +3.8, +4.0, +4.2, +4.5, +4.7, +4.9, +5.2, +5.4, +5.7, +5.9, +6.1, +6.4, +6.6, +6.8, +7.1

20 ft 0 in. draft — Amidships

Fwd: +11.1, +10.8, +10.4, +10.1, +9.7, +9.4, +9.0, +8.7, +8.3, +8.0, +7.6, +7.3, +6.9, +6.6, +6.2, +5.9, +5.5, +5.1, +4.8, +4.4, +4.1, +3.7, +3.4, +3.0, +2.7, +2.3, +2.0, +1.6, +1.3, +1.9, +0.6, +0.2, −0.1, −0.5, −0.8, −1.2, −1.6, −1.9, −2.3, −2.6, −3.0, −3.3, −3.7, −4.0, −4.4, −4.7, −5.1, −5.4, −5.8, −6.2, −6.5, −6.9, −7.2

Aft: −7.2, −6.9, −6.6, −6.2, −5.9, −5.6, −5.3, −4.9, −4.6, −4.3, −4.0, −3.6, −3.3, −3.0, −2.7, −2.3, −2.0, −1.7, −1.4, −1.0, −0.7, −0.4, −0.1, +0.3, +0.6, +0.9, +1.2, +1.6, +1.9, +2.2, +2.5, +2.8, +3.2, +3.5, +3.8, +4.1, +4.5, +4.8, +5.1, +5.4, +5.8, +6.1, +6.4, +6.7, +7.1, +7.4, +7.7, +8.0, +8.4, +8.7, +9.0, +9.3, +9.6

0 — 50 — 100 feet

Notes: (1) The corrections have been computed for the two drafts 10 ft apart to facilitate interpolation, but in practice it will be accurate enough to refer to the table nearest the ship's draft. (2) When discharging, use the table as with loading and change the plus and minus signs.

TABLE 4-4

DEADWEIGHT SCALE, HYDROSTATIC PROPERTIES, C4-S-1a

A	B	C	D	E	F	G	A
Mean Draft Bottom of Keel	Total Displacement, Salt Water tons	Trans-verse KM— Molded, ft	T.P.I.	MT1 ft-tons	L.C.B.[a] Aft F.P., ft	L.C.F.[b] Aft F.P., ft	Mean Draft Bottom of Keel

Deadweight scale (values read against adjacent vertical scales):

A (Draft)	B (Displacement)	C (KM)	D (T.P.I.)	E (MT1)	F (L.C.B.)	G (L.C.F.)	A (Draft)
30	21,000	31.4	70	1950	269	282	30
29		31.3		1900		281	29
28	20,000	31.2	69	1850		280	28
27	19,000	31.1	68	1800	268	279	27
				1750		278	
26	18,000		67	1700		277	26
25	17,000	31.05		1650	267	276	25
24	16,000	31.1	66			275	24
23	15,000	31.2	65	1600	266	274	23
22		31.3 / 31.4		1550		273	22
21	14,000	31.5 / 31.6	64	1500		272	21
20	13,000	31.8 / 32.0	63		265	271	20
19	12,000	32.5		1450		270	19
18		33.0	62		264	269	18
17	11,000	33.5		1400		268	17
16	10,000	34.0 / 34.5	61				16
15		35.0 / 35.5		1350		267	15
14	9,000	36.0	60		263	266	14
		37.0					
13	8,000	38.0	59	1300		265	13
12	7,800 / 7,600 / 7,400	Light ship					12

[a] L.C.B., longitudinal center of buoyancy.
[b] L.C.F., longitudinal center of flotation.

136

TABLE 4-5

FREE SURFACE CORRECTION AND TANK CAPACITIES, C4-S-1a[a,b]

Tank	Frames		97% Fuel Oil, tons	100% Salt Water, tons	Column A i slack	Column B i 97%	Vertical Center of Gravity	L.C.G.– F.P.
Double bottom 1	₵	14–24	48.2	52.8	106	67	4.5	39.9
Double bottom 1A	₵	24–36	81.9	89.8	464	204	4.8	64.9
Double bottom 2	P	36–57	71.2	78.1	428	158	2.7	106.6
	S	36–57	71.2	78.1	428	158	2.7	106.6
Double bottom 3	₵	57–82	227.6	249.5	3777	944	2.5	161.6
	P	57–82	55.6	61.0	300	120	3.0	169.2
	S	57–82	55.6	61.0	300	120	3.0	169.2
Double bottom 4	₵	82–106	224.1	245.7	3626	943	2.5	222.0
	P	82–106	128.1	140.5	1138	364	2.6	223.8
	S	82–106	128.1	140.5	1138	364	2.6	223.8
Double bottom 5	₵	106–127	196.2	215.1	3173	825	2.5	278.3
	P	106–134	178.0	195.2	2048	676	2.6	288.3
	S	106–134	180.0	197.4	2048	676	2.6	288.3
Double bottom 6	₵	134–160	242.3	265.7	3928	1021	2.5	354.4
	P	134–160	87.0	95.4	615	221	2.8	348.2
	S	134–160	87.0	95.4	615	221	2.8	348.2
Double bottom 7	P	160–184	94.6	103.7	768	269	2.7	412.4
	S	160–184	94.6	103.7	768	269	2.7	412.4
Deep tank 1	₵	14–24	125.3	137.4	134	130	16.5	40.3
Deep tank 1A	₵	24–36	257.6	282.5	945	680	16.8	65.1
Deep tank 2	P	106–113	100.7	. . .	20	20	19.1	260.8
	S	106–113	100.7	. . .	20	20	19.1	260.8
Deep tank 3	P	113–119	86.1	. . .	17	17	19.1	277.0
	S	113–119	86.1	. . .	17	17	19.1	277.0
Deep tank 6	P	160–172	201.2	220.7	1242	634	11.4	401.2
	S	160–172	201.2	220.7	1242	634	11.4	401.2
Deep tank 7	P	172–184	128.8	141.2	618	358	11.7	430.7
	S	172–184	128.8	141.2	618	358	11.7	430.7
Deep tank 8	P	184–190	50.5	55.4	68	58	9.6	454.0
	S	184–190	50.5	55.4	68	58	9.6	454.0

Table 4-5 (*Continued*)

Tank		Frames	100% Fresh Water, tons	100% Salt Water, tons	Column C i slack	Vertical Center of Gravity	L.C.G.–F.P.
Fore peak	₵	Stem–14		110.8		11.7	17.1
Aft peak	₵	204–218		93.0		24.9	506.8
Deep tank 4	P/S	120–127	123.7	. . .	5575	21.3	296.0
Deep tank 5	P/S	127–133	108.4	. . .	4789	20.9	312.0
Dist. water	₵	106–109	24.9		59	39.5	255.8

a Notes: Fuel oil at 37.23 ft³/ton, 97% full. Fresh water at 36.0 ft³/ton, 100% full. Salt water at 35.0 ft³/ton, 100% full.

b Free Surface Correction Procedure: (1) Add quantity in column A for tanks slack. (2) Add quantity in column B for tanks 97% full. (3) Add quantity in column C for fresh water tanks. (4) If any tank is empty, or pressed up with water, use zero for that tank. (5) Divide sum total by the ship displacement in tons to obtain free surface correction in feet.

L.C.G.–F.P. = distance between the longitudinal center of gravity and the forward perpendicular.

₵ = symbol for centerline of ship.

P = port.

S = starboard.

i = moment of inertia of the free surface water plane divided by 35.

Table 4-9: This is a form for summarizing all the data and solving for estimated *GM* available, item 6, and the drafts forward and aft, items 14 and 15 respectively.

Explanation of solution given on Table 4-9

1. The mean salt water draft, obtained from the ship's displacement using Table 4-4, is 25.7 ft.

2. The *KM* for the displacement also taken from Table 4-4, is 31.05 ft.

3. The *KG* is obtained by calculation. Divide total vertical moments (495,801 ft-tons) by the ship's displacement (17,732 tons); *KG* equals 28.0 ft.

4. *KM* less *KG* equals the calculated *GM* before correcting for any free surface. Thus: 31.05 − 28.0 = 3.05 ft.

5. The virtual rise of *G* due to free surface (F.S.) is obtained by taking the sum of the free surface moments, each moment being taken from Table 4-5 and entered on Table 4-9, and dividing this sum by the ship's displacement. Thus: 18,566/17,732 = 1.05 ft.

6.   The correction for free surface is subtracted from the *GM* to obtain *GM* available. Thus: 3.05 − 1.05 = 2 ft.

7.   Compare the *GM* available to the *GM* required. The latter is obtained from Fig. 4-12 and found to be 1.76 ft. The *GM* available is adequate. The *GM* available may be too great in some cases; if it is, the rolling period will be short; this is undesirable (see the discussion on magnitudes of *GM*s, p. 116).

8.   The longitudinal position of the center of gravity of the ship is calculated. It is obtained by taking the sum of the longitudinal moments

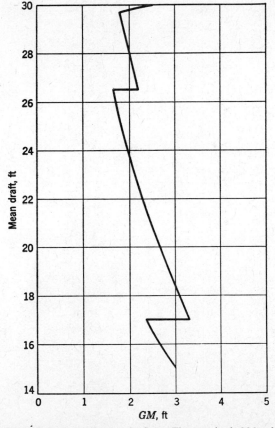

FIG. 4-12. Required *GM* curve for the C4-S-1a. The required *GM* values given in this diagram must be maintained if the ship, under *average* operating conditions, is to sustain damage in any one compartment without reaching a condition of negative stability after damage and without heeling that might result in flooding of any other, undamaged compartment.

Table 4-6

Gain in GM by Ballasting(ft), C4-S-1a

Displacement 100 tons	Tank 100 tons	D.B.[a] 1 52	D.B. 1A 89	D.B. 2 156	D.B. 3 371	D.B. 4 526	D.B. 5 607	D.B. 6 456	D.B. 7 207	D.T.[b] 1 137	D.T. 1A 282	D.T. 6 441	D.T. 7 282	D.T. 8 110
85		0.05	0.05	0.20	0.40	0.60	0.65	0.55	0.20	−0.10	−0.15	0.05	0.0	0.0
90		"	"	"	0.45	"	0.70	"	"	−0.05	−0.10	0.10	0.05	"
95		"	0.10	"	"	0.65	"	"	0.25	"	"	0.15	"	0.05
100		"	"	"	"	"	0.75	0.60	"	0.0	−0.05	"	0.10	"
105		"	"	"	0.50	0.70	0.80	"	"	"	0.0	0.20	"	"
110		"	"	"	"	"	"	"	"	"	"	0.25	0.15	"
115		"	"	"	"	"	"	"	"	0.05	0.05	0.30	"	0.10
120		"	"	"	"	"	0.85	"	"	"	"	"	0.20	"
125		"	"	"	"	"	"	0.65	0.30	"	0.10	0.35	"	"
130		"	"	"	"	"	"	"	"	"	"	"	"	"
135		"	"	"	"	"	"	"	"	"	"	"	"	"
140		"	"	"	"	"	"	"	"	"	0.15	"	"	"
145		"	"	0.25	"	"	"	"	"	"	"	"	0.25	"

140

150	..	..	0.40	..	0.10	..	..	..	..	..	..	..
155	..	..	..	..	..	..	..	..	..	..	..	..
160	..	..	..	..	..	..	..	..	..	..	..	..
165	..	..	..	..	..	..	..	..	..	..	..	..
170	..	..	..	..	..	..	..	..	..	..	..	..
175	..	..	..	0.20	..	..	..	..	..	..	..	..
180	..	..	..	..	..	..	..	..	..	..	..	0.10
185	..	..	..	..	..	..	..	..	..	..	..	..
190	..	..	..	..	..	..	..	..	..	..	..	..
195	..	..	..	..	..	..	..	..	..	..	..	..
200	..	..	..	..	..	..	..	..	..	..	..	..
205	0.15	..	..	..	..	..	..	..	..	..	..	..
210	..	0.30	..	..	..	..	..	..	..	..	..	..
213	..	..	..	..	..	..	..	..	..	..	..	..
215	..	..	..	..	..	..	..	..	..	..	..	..

[a] D.B., double bottom.

[b] D.T., deep tank.

TABLE 4-7

DOUBLE BOTTOM TANKAGE REQUIREMENTS IN TONS TO MEET ONE-COMPARTMENT
DAMAGE FOR NORMAL CONDITIONS OF LOADING[a]

Total Cargo Plus D.T. 1, 1A, 6, 7, & 8 (Column 1 + 2 + 3)	Excess of Hold Weight over Upper 'Tweendeck Weight in Tons						Additional D.B. Tankage per 100 tons of Deck Cargo
	(Column 3—Column 1)						
	+1500	+1000	+500	0	−500	−1000	
1,000		0	0	75	475	850	150
2,000	0	0	0	800	1225	1600	140
3,000	0	150	550	950	1350	1750	130
4,000	0	325	675	1050	1400	1775	120
5,000	50	400	750	1100	1425	1775	110
6,000	100	400	725	1050	1350	1650	100
7,000	50	350	650	950	1275	1600	90
8,000	0	200	500	800	1100	1400	80
9,000	0	0	325	650	1000	1600	70
10,000	0	250	500	800	1050	1325	60
11,000	0	50	325	575	825	1100	50
12,000	0	0	0	275	625		

The forms shown below may be used to determine the required double bottom tankage from the above table.

Upper 'Tween deck Layer	Col. 1, tons	Lower 'Tween-deck Layer	Col. 2, tons	Hold Layer	Col. 3, tons
No. 1 main deck	160	No. 1 3rd deck	130	No. 1 deep tank centerline	
No. 1 2nd deck	180	No. 2 3rd deck	369	No. 1A deep tank	
No. 2 2nd deck	291	No. 3 3rd deck	621	No. 2 tank top	271
No. 3 2nd deck	418	No. 4 3rd deck	641	No. 3 tank top	546
No. 4 2nd deck	401	No. 5 26' 6″ flat, dry and reefer	349	No. 4 tank top	650
No. 5 2nd deck	416	No. 5 3rd deck, dry and reefer	321	No. 5 tank top	406
No. 6 2nd deck	384	No. 6 3rd deck	703	No. 6 deep tank P/S	127
No. 7 2nd deck	250	No. 7 3rd deck	366	No. 7 deep tank	
				No. 8 deep tank	
Total	2500	Total	3500	Total	2000

TABLE 4-7   (*Continued*)

Summary

Item	Tons
Total column 1	2500
Total column 2	3500
Total column 3	2000
Total columns 1 + 2 + 3	8000
Total column 3 − column 1	−500
Required tankage (from table)	1100
Deck cargo in tons	200
Required D.B. tankage for deck cargo	160
Total required D.B. tankage	1260

a D.B., double bottom; P/S, port and starboard.

about the forward perpendicular (F.P.) and dividing by the ship's displacement. Thus: 4,799,348/17,732 = 270.6 ft aft of the F.P.

9.   The position of the longitudinal center of buoyancy (L.C.B.) is obtained from Table 4-4 for the ship's displacement. In this case it is 267.2 ft aft of the F.P.

10.   The trimming lever is the distance between the longitudinal center of gravity (L.C.G.) and the L.C.B.; thus, in this case it is 270.6 − 267.2 = 3.4 ft. The trim will be by the stern if the L.C.G. is *aft* of the L.C.B. or by the head if *forward*.

11.   The $MT1$ is obtained from Table 4-4. In this case, it is 1700 ft-tons.

12.   The actual trim or change in trim from an even keel is obtained by dividing the trimming moment by the $MT1$. The trimming moment is equal to the product of multiplying the trimming lever by the ship's displacement. In this case, trimming moment = 17,732 × 3.4 = 60,288.8 ft-tons. The change in trim then becomes 60,289/1700 = 35.45 in.

13.   It is assumed that the ship started with an even keel, with a mean draft of 25 ft 8.4 in. To obtain the draft forward and aft, the change in trim must be applied to this mean draft. Inasmuch as the trim in this case is by the stern, we add part of the 35.45 in. to obtain the draft aft and subtract to obtain the draft forward. If the longitudinal center of flotation (L.C.F.)

TABLE 4-8
LOADING TABLE [a]

Dry Cargo

Hold	Bale Cubic	Tons	KG	Moment	L.C.G.– F.P.	Moment
No. 1 main deck	16,085	160	55.6	8,896	59.2	9,472
No. 1 2nd deck	18,140	180	45.2	8,136	54.8	9,864
No. 1 3rd deck	12,210	130	31.9	4,147	56.6	7,358
No. 2 2nd deck	29,255	291	43.0	12,513	104.4	30,380
No. 2 3rd deck	34,592	369	29.1	10,738	105.3	38,856
No. 2 tank top	25,476	271	13.1	3,550	106.2	28,780
No. 3 2nd deck	42,000	418	41.3	17,263	161.3	67,423
No. 3 3rd deck	58,150	621	28.3	17,574	161.6	100,354
No. 3 tank top	51,375	546	12.7	6,934	162.7	88,834
No. 4 2nd deck	40,255	401	40.3	16,160	221.5	88,822
No. 4 3rd deck	60,020	641	27.7	17,756	221.9	142,238
No. 4 tank top	61,140	650	12.5	8,125	223.1	145,015
No. 5 2nd deck	41,775	416	40.5	16,848	356.5	148,304
No. 5 26 ft 6 in. flat centerline	16,388	175	30.8	5,390	350.2	61,285
No. 5 3rd deck centerline	16,022	171	21.4	3,659	351.0	60,021
No. 5 tank top	38,135	406	10.9	4,425	353.6	143,562
No. 6 2nd deck	38,610	384	41.0	15,744	416.5	159,936
No. 6 3rd deck	65,850	703	26.9	18,911	415.5	292,097
No. 6 deep tank P/S	11,930	127	11.2	1,422	402.6	51,130
No. 7 2nd deck	25,095	250	41.8	10,450	469.6	117,400
No. 7 3rd deck	34,220	366	28.4	10,394	469.4	171,800
Total	736,723	7676	28.5	219,035	255.7	1,962,931

Reefer Cargo

Hold	Reefer Cubic	Tons	KG	Moment	L.C.G.– F.P.	Moment
No. 5 26 ft 6 in. flat P/S	16,256	174	30.7	5342	354.4	61,666
No. 5 3rd deck P/S	13,998	150	21.8	3270	353.4	53,010
Total	30,254	324	26.6	8612	353.9	114,676

Table 4-8 (*Continued*)

Fuel Oil or Ballast

Tank	F.S.	Tons F.O., S.W.	*KG*	Moment	L.C.G.– F.P.	Moment
No. 1 D.B. centerline			4.5		39.9	
No. 1A D.B. centerline	204	82	4.8	394	64.9	5,322
No. 2 D.B. P/S			2.7		106.6	
No. 3 D.B. centerline			2.5		161.6	
No. 3 D.B. P/S			3.0		169.2	
No. 4 D.B. centerline	943	224	2.5	560	222.0	49,728
No. 4 D.B. P/S			2.6		223.8	
No. 5 D.B. centerline	825	196	2.5	490	278.3	54,547
No. 5 D.B. P/S $\frac{1}{2}$	4096	179	2.6	465	288.3	51,606
No. 6 D.B. centerline	1021	242	2.5	605	354.4	85,765
No. 6 D.B. P/S	442	174	2.8	487	348.2	60,587
No. 7 D.B. P/S	538	189	2.7	510	412.4	77,944
No. 1 D.T. centerline			16.5		40.3	
No. 1A D.T. centerline			16.8		65.1	
No. 2 D.T. P/S $\frac{2}{3}$	40	134	19.1	2559	260.8	34,947
No. 3 D.T. P/S $\frac{2}{3}$	34	117	19.1	2235	277.0	32,409
No. 6 D.T. P/S			11.4		401.2	
No. 7 D.T. P/S			11.7		430.7	
No. 8 D.T. P/S			9.6		454.0	
Fore peak			11.7		17.1	
After peak			24.9		506.8	
Total	8143	1537	5.4	8305	294.6	452,855

(*1286 tons F.O. in double bottom*)

Fresh Water

Tank	F.S.	Tons F.W.	*KG*	Moment	L.C.G.– F.P.	Moment
No. 4 D.T. P/S	5,575	124	21.3	2641	296.0	36,704
No. 5 D.T. P/S	4,789	108	20.9	2257	312.0	33,696
Distilled water	59	25	39.5	988	255.8	6,395
Total	10,423	257	22.9	5886	298.8	76,795

[a] D.B., double bottom; D.T., deep tank; S.W., salt water; F.O., fuel oil; P/S, port and starboard; F.S., free surface; F.W., fresh water.

TABLE 4-9

Item	Tons	KG	Moment	L.C.G.–F.P.	Moment	F.S.
Light ship	7,675	31.5	241,763	276.5	2,122,138	
Crew and stores	50	43.7	2,185	276.5	13,825	
Lube oil	13	25.8	335	317.5	4,128	
Fuel oil and salt water	1,537	5.4	8,305	294.6	452,855	8,143
Fresh water	257	22.9	5,886	298.8	76,795	10,423
Dry cargo	7,676	28.5	219,035	255.7	1,962,931	
Reefer cargo	324	26.6	8,612	353.9	114,676	
Deck cargo	200	48.4	9,680	260.0	52,000	
Total	17,732	28.0	495,801	270.6	4,799,348	18,566

1. Mean S.W. draft (see Table 4-4) — 25.7
2. KM (see Table 4-4) — 31.05
3. KG — 28.0
4. GM — 3.05
5. Correction for F.S. — 1.05
6. GM available — 2.00
7. GM required (see Fig. 4-12) — 1.76
8. L.C.G.–F.P. — 270.6
9. L.C.B. (see Table 4-4) — 267.2
10. Trim lever, aft — 3.4
11. $MT1$ — 1700
12. Trim in inches, aft — 35.45
13. L.C.F.–F.P. (see Table 4-4) — 276.9
14. Draft fwd. — 24 ft 1¾ in.
15. Draft aft — 27 ft 1¼ in.

Legend: Dry or reefer cargo; Fresh water; Fuel oil; Salt water

146

was located exactly amidships, we could find the *change of draft* forward and aft by dividing the change of trim by 2.[6] But if the L.C.F. is not amidship, the change of *draft* forward is equal to the change in trim times the ratio of the L.C.F.–F.P. to the length of the ship between perpendiculars (L.B.P.). In this case the L.C.F.–F.P. is found to be 276.9 ft, which is 12.9 ft aft of the midship section of the ship.

14.   The change of draft forward is equal to $(276.9/528) \times 35.45 = 18.6$ in. (subtracted). Thus, 25 ft 08.4 in. − 1 ft 06.6 in. = 24 ft 01.8 in. draft forward.

15.   The change of draft aft is equal to the difference between the change of trim and the change of draft forward, and it is additive. Thus, $35.45 - 18.60 = 16.85$ in., and 25 ft 08.4 in. + 1 ft 04.85 in. = 27 ft 01.25 in. draft aft.

The use of trimming tables.   Trimming tables constructed as illustrated in Table 4-3 are generally found below the profile view of the ship on the capacity plan. The tables are constructed to show the change in *draft* forward and aft when a weight of 100 tons is placed on the vessel. The effects on the drafts forward and aft can be determined by looking in the box directly below the point on the ship where the weight is placed. If the weight being considered is more than or less than 100 tons, the effect is increased or decreased proportionately. For example, if a 300-ton weight is loaded, the figures in the boxes are multiplied by 3.

In each box there is a plus or minus sign. These signs indicate how the values in the boxes are to be applied to the initial draft figures. These signs are for a loading process. Therefore, if the weight is discharged the signs must be changed. The changes in draft in the boxes take into consideration the change in mean draft due to mean sinkage.

Because the $MT1$ changes with draft, a set of numbers for each draft in increments of 1 ft would be better than for just the two drafts illustrated. It will be noted that the two drafts are 10 ft apart. This makes it possible to interpolate in calculating the change in drafts for a mean draft lying in between or slightly beyond the two drafts listed. Choosing the table for the nearest draft generally yields sufficient accuracy for most operating conditions.

---

[6] The longitudinal center of flotation is the tipping center of the ship, and corresponds to the center of the area of the water plane at which the ship is floating. As the ship trims it rotates about a transverse axis passing through this point. It is abbreviated L.C.F.

Longitudinal Stresses

The following discussion is not intended to be a thorough presentation of the involved subject of strength calculations as applied to the hull of a ship. A summation of the elements of the problem and their relationship to the responsibilities of the ship's officer is our prime objective. It is hoped that the discussion will also clarify the approach to the problem that the ship's officer should take.

The Strength Equation

Let us say that we have the problem of determining the stress in pounds per square inch on a ship's deck or bottom under a given set of conditions, wherein the type of sea that is running and the distribution of weight in the ship are specified.

Without deriving the equation here, we will accept the following mathematical statement as truth, for we need to use it in our solution of the above problem.

$$M = \frac{I}{y} \times p$$

where M = the bending moment of the hull girder.

I = the moment of inertia of the hull girder.

y = the distance from the neutral axis of the hull girder to the most remote member of the hull girder.

p = a measure of the resistance offered by the material of which the hull girder is made; hence, it is given in units of *stress, pounds per square inch.*

The section modulus. The ratio of I to y is also known as the section modulus of the hull girder. It depends on the form and distribution of material in the construction of the hull girder. I/y may be notated as S. The section modulus of the ship is calculated by the naval architect, and if we are to have the necessary data to calculate the answer to the problem stated above, we must know S for the hull of the ship.

Bending moment and stress of material. Two other factors remain in our equation: M, the bending moment of the hull under our specified conditions of sea and load, and p, the stress of the material. The latter is what we must find ultimately; hence, our discussion will not revolve around the problem of determining M in our equation.

Finding the bending moment. The first step is to determine the weight of the ship per linear foot starting from one end of the vessel. This is the weight in tons of each foot of length and includes the weight of the ship's structure and all the deadweight items aboard her, such as cargo, fuel, water, and stores. This is a difficult task.

The next step is to calculate the buoyancy of the ship per linear foot of length. This is a relatively easy task. All that must be done to obtain the buoyancy of a linear foot is to calculate the volume of the underwater portion of the hull for a 1-ft slice longitudinally and divide by 35. At this point, let us note that *total* weight must always equal the *total* buoyancy; however, the weight at any one point may exceed or be less than the force of buoyancy acting on that point of the hull.

Let us consider a single slice of the hull as illustrated in Fig. 4-13. If the force of buoyancy on this particular slice exceeds the weight or vice versa, the edges of the slice will experience a shearing stress. If the next and each succeeding slice experience this shearing stress, a force acting through a distance exerts itself on the hull of the ship. A force acting through a distance is termed a moment, and in this case it is called a *bending moment* because as the hull girder resists the force, it bends. If the shearing stress is large or acts over a considerable distance, the bending moment will be large.

The Strength Curves

The buoyancy and weight in tons per linear foot are plotted as curves as shown in Fig. 4-14a. Next, the difference between the weight and buoy-

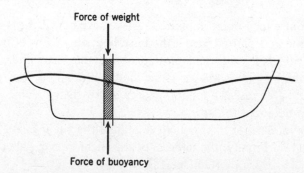

FIG. 4-13. If a particular slice of the ship's hull is considered as indicated in this diagram, it becomes obvious that unless the weight and buoyancy forces are equal, this slice is resisting a shearing stress tending to push it upward or downward. The ship profile is depicted as floating in a trochoidal curve with a height $\frac{1}{20}$th of the length of the ship; this is the so-called *standard sea* as defined by the naval architect.

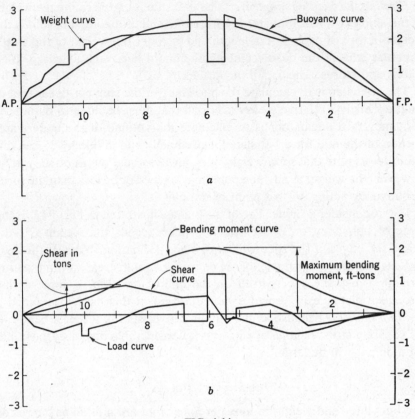

FIG. 4-14.

ancy values at every point is plotted. An excess of weight is assigned a negative value and plotted below the base line; an excess of buoyancy is assigned a positive value and plotted above the base line. The curve resulting from this plot is known as the load curve, and is shown in Fig. 4-14*b*. The total area under it (counting area below the base line as negative area) should be equal to zero.

Next, the integral of the load curve, called the shear curve, is plotted (see Fig. 4-14*b*). Finally, the integral of the shear curve, called the *bending moment curve,* is plotted (see Fig. 4-14*b*).

All of the curves shown in Fig. 4-14 are simply for purposes of illustration. The conditions depicted parallel those of a ship in still water. The strength curves for a ship in a standard wave with crest amidships would have more buoyancy amidships and less at the ends. Conversely, if the ship were in a wave with the trough amidships, the buoyancy would be reduced amidships and be increased at the ends. The weight curve would, of course, remain the same, unless the load conditions on the ship were changed.

The abscissas on all five curves are in units of length. The ordinates on the weight and buoyancy curves are in units of tons per linear foot. The ordinates on the shear and load curves are in tons, and the ordinate of the bending moment curve is in units of foot-tons.

Going through this process, the maximum bending moment as well as the bending moment at any given point may be obtained. With this value, the stress, tension, and compression on the hull bottom and deck plating can be calculated, providing we know the section modulus, because:

$$p = \frac{M}{S}$$

where M = the bending moment, which we have calculated.
S = the section modulus for the hull, which we must obtain from the design naval architect.

The naval architect specifies structural members for all decks and other parts of the hull to withstand the tension and compression stresses that are found to exist in them when the ship is placed in a standard sea under a standard load condition. This is done for both a hogging and a sagging condition.

Standard Sea and Standard Load

The standard sea is one in which the height is $\frac{1}{20}$th the length of the ship and in the form of a trochoidal curve. For a hogging condition, the crest of the sea is assumed to be amidships. The buoyancy amidships would then be at a maximum. For a sagging condition, the trough of the sea is assumed to be amidships. The buoyancy would then be at a minimum amidships.

The standard load for a ship with machinery spaces amidships assumes the midship tanks to be empty, the end tanks full, and all cargo compartments full. This is a poor loading pattern for such a ship, for it results in an initial hogging tendency.

The standard load for a ship with machinery spaces aft assumes the cargo spaces to be full and the end tanks empty. This will create an initial sagging tendency in such a vessel.

Obviously, the load and the sea can be exaggerated so as to produce excessive stresses; however, the naval architect does not design the ship to withstand the worst possible conditions, but rather what might be termed "reasonably bad" conditions. Then a suitable safety factor is worked into the calculations for determining the final scantlings used to construct the ship.

It should be apparent that if the ship is loaded with an extreme concentration of weight in the center hatches or in the end hatches and subsequently the ship meets with extremely heavy seas, the stresses on the hull girder will become dangerously high.

The Beam Theory

We have been referring to the hull of the ship as a beam or girder. The ship is really a complicated built-up shape, and much of the theory of stresses on the hull is based on the assumption that hull stresses are quite similar to beam stresses. The experience of naval architects has proved that this assumption is a reasonable one. Figure 4-15*a* illustrates a beam heavily loaded on its midlength. This parallels a ship in the sagging condition. Note that the steel of the upper decks is under a compression stress, and the bottom is under tension. Figure 4-15*b* illustrates a beam supported at its midlength and loaded on its ends. This parallels a ship's hull in the hogging condition. Note that the steel of the upper decks is under

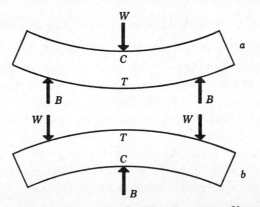

FIG. 4-15. The beam theory. W = excess of weight; B = excess of buoyancy; C = steel in this area under compression; T = steel in this area under tension.

tension stresses, and the bottom is under compression. If these tension or compression stresses become greater than the ship is designed to withstand, the plating will fail. Under tension, the plating will crack or tear apart. Under compression, the plating will buckle.

Practical Approach to the Problem

The ship's officer must make certain that his ship is never loaded in such a fashion that extreme hogging or sagging will occur. In actual practice, however, it is difficult for the ship's officer to obtain data to make calculations concerning the bending moments of the vessel as she sails each voyage. What, then, should be the practice to remove this "blank spot" in the average ship's officer in his knowledge of his ship? Three solutions are set forth below:

a. The shipowner should see that every ship is supplied with necessary data, such as strength curves, section modulus, geometrical locations of all spaces, and other information, to facilitate the calculations of bending moments and possibly the stresses. Then, instructions in how to use the data should be provided where necessary. All ship's officers should then prepare tabular methods of calculating the hogging and sagging moments under conditions of loading for each voyage. Such a tabular method has been worked out for the T-2 tanker by the American Bureau of Shipping and is presented later (Table 4-10). With the proper data and background, any ship's officer can make up a similar set of tables and diagrams to work out the hogging and sagging tendencies for his particular ship. If a tabular method has been devised for the ship upon which he is sailing, the ship's officer should work with it until he is thoroughly acquainted with it. Then he should use it consistently. The hogging and sagging tendencies of his ship in waves of standard proportions should become as familiar to the ship's officer as the sailing drafts of his ship.

b. If his ship is equipped with a mechanical device for the calculation of hogging and sagging bending moments, the ship's officer should be thoroughly familiar with its use and the significance of the data it provides him.

c. If his ship is equipped with an on-board computer that monitors enough stress points, the ship's officer should be very familiar with the instructions for operation and should correlate the findings with those of tabular and mechanical devices, if available. In addition, a correlation can be made in port before sailing by observing the drafts fore, aft, and mean. All vessels should be inscribed with midships draft marks to facilitate verification of these measurements.

Approximating the longitudinal distribution.   If the time, the data, or the materials to take advantage of any of the previously mentioned suggestions are lacking, an estimation of correct longitudinal distribution for ships with their engine room spaces amidships can be made. The method is based on the same premise as is the method for roughly estimating the vertical distribution. The amount of weight that should be placed in each longitudinal compartment is equal to the ratio of the cubic of that compartment to the cubic of all the cargo compartments. This may be expressed by the following equation:

$$\frac{v}{V} \times T = t$$

where v = the volume of the compartment in question.
V = the volume of all cargo compartments.
T = the total weight to be loaded.
t = the weight to be loaded in the compartment in question.

Caution must be exercised when using such a method. If used for ships with the engine room spaces aft, it would lead to excessive sagging moments.

Tabular method for checking on longitudinal load distribution.   The method presented here was developed by the American Bureau of Shipping to determine the adequacy of any load distribution, with respect to satisfactory design conditions, for the T-2 tanker on a hogging and sagging wave of standard proportions. Similar tabular methods are available for newer tankers.

The results obtained from this simplified method are approximate. It has been determined by comparison with numerous direct bending moment calculations that the degree of approximation is of an acceptable magnitude for displacements from 10,000 to 22,000 long tons and trims varying from 30 in. forward at the deep displacement to 170 in. aft at the light displacement.

The light ship weight with crew and stores and the longitudinal distribution of these weights for the purposes of this study are average for this class of vessel.

The method consists of determining *numerals* based on the longitudinal distribution of weight along the hull. These hogging and sagging numerals are obtained by multiplying the load in each space on the vessel by a factor depending on the longitudinal location of the weight, adding the product to the numeral for light ship with crew and stores, and deducting three times the deadweight divided by 100. To get the maximum benefit from this method, these calculations of hogging and sagging numerals should be made not only for the completed loading but also through the entire range of proceedings as the loading progresses or as changes in loading are made.

The numerals should be determined for both hogging and sagging. If either numeral exceeds 100, it indicates that the structure is subject to conditions that are more severe than the standard conditions. In many conditions of operation, especially in ships less than full and homogeneously loaded, it is possible to distribute the load in such a manner as to give a value much less than 100. For the purpose of operation, particularly in heavy weather, it is desirable to obtain the lowest possible numeral.

The diagram in Fig. 4-16 gives the hogging and sagging factors for an added load of 100 tons at any longitudinal location on the vessel.

Table 4-10 lists the hogging and sagging factors for each tank space on the vessel as read from the diagram in Fig. 4-16, and should be suitable for most conditions of load distribution. Should an item of appreciable weight, such as a deck load, be added to the vessel such that its longitudinal center is considerably removed from any of the tank centers, the hogging and sagging factors for this weight should be read from Fig. 4-16 and inserted in the table along with the weight of the item in long tons divided by 100.

Table 4-11 is an example worked out for a vessel with a cargo distribution that gives an *unfavorable* numeral.

Table 4-12 is an example worked out for the same vessel with the same cargo redistributed to give a favorable numeral.

A second tabular method for calculating stresses.   This variation of the previous method is a method for estimating the stresses on the T-2 tanker due to hogging and sagging. Table 4-13 illustrates the form to be used. Some comments regarding the form will clarify its use. At the top of the form in column 3, the number 217.70 is entered. This is the displacement of the ship in hundreds of tons assumed for design purposes. The numbers 10,500 and 14,600 in columns 5 and 7 respectively indicate the stress on

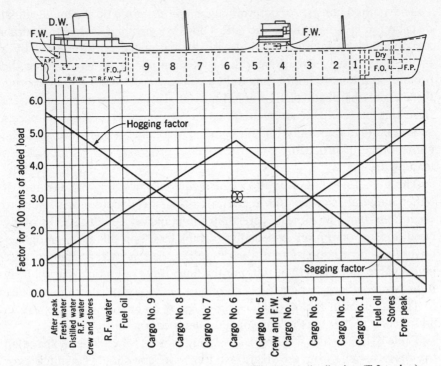

FIG. 4-16. Diagram for determining numerals from load distribution (T-2 tanker).

the deck when loaded under the design conditions. The distribution of the 21,770 tons is as shown in column 1.

The steps to be followed in using the table are as follows:

1. In column 2, list the actual distribution of the weights in the problem ship being considered.
2. Subtract column 1 from column 2. If column 1 is greater than column 2, place a minus sign before the result in column 3; if column 1 is less than column 2, place a plus sign before the result in column 3.
3. Observing rules for algebraic signs, multiply column 3 by column 4 and place the result in column 5.
4. Observing rules for algebraic signs, multiply column 3 by column 6 and place the result in column 7.
5. Add algebraically the numbers in columns 5 and 7. These sums will be the stresses on the weather deck in pounds per square inch (psi).

TABLE 4-10

TABLE FOR DETERMINING HOGGING AND SAGGING NUMERALS FROM LOAD DISTRIBUTION, BLANK FORM (T-2 TANKER)

Line No.	Item	Weight, long tons 100 (a)	Hogging Factor (b)	Hogging Numeral (c)	Sagging Factor (d)	Sagging Numeral (e)
1	Light ship and crew and stores	54.12	. . .	88.8	. . .	29.7
2	F.P. tank	. . .	4.82	. . .	0.65	. . .
3	F.O. forward	. . .	4.31	. . .	1.26	. . .
4	F.W. amidships	. . .	2.21	. . .	3.77	. . .
5	F.O. aft	. . .	3.90	. . .	2.52	. . .
6	R.F.W. forward D.B.	. . .	4.20	. . .	2.27	. . .
7	R.F.W. after D.B.	. . .	4.86	. . .	1.68	. . .
8	Distilled water	. . .	5.06	. . .	1.52	. . .
9	F.W. aft	. . .	5.19	. . .	1.40	. . .
10	A.P. tank	. . .	5.37	. . .	1.25	. . .
11	Cargo no. 1	. . .	3.92	. . .	1.73	. . .
12	Cargo no. 2	. . .	3.54	. . .	2.18	. . .
13	Cargo no. 3	. . .	3.00	. . .	2.82	. . .
14	Cargo no. 4	. . .	2.46	. . .	3.47	. . .
15	Cargo no. 5	. . .	1.91	. . .	4.11	. . .
16	Cargo no. 6	. . .	1.49	. . .	4.65	. . .
17	Cargo no. 7	. . .	2.08	. . .	4.11	. . .
18	Cargo no. 8	. . .	2.69	. . .	3.59	. . .
19	Cargo no. 9	. . .	3.29	. . .	3.06	. . .
20	Subtotals (lines 2 through 19)	. . .		. . .		. . .
21	Subtotals (line 1 + line 20)			. . .		. . .
22	Deadweight correction = 3 × (line 20 in column a); enter in columns c and e			. . .		
23	Displacement ÷ 100 (lines 1 + 20)	. . .				
24	Resultant numerals (lines 21 − 22)			. . .		. . .

157

TABLE 4-11
TABLE FOR DETERMINING HOGGING AND SAGGING NUMERALS FROM LOAD
DISTRIBUTION, EXAMPLE SHOWING UNFAVORABLE DISTRIBUTION (T-2 TANKER)

Line No.	Item	Weight, long tons 100 (a)	Hogging Factor (b)	Hogging Numeral (c)	Sagging Factor (d)	Sagging Numeral (e)
1	Light ship and crew and stores	54.12	. . .	88.8	. . .	29.7
2	F.P. tank	. . .	4.82	. . .	0.65	. . .
3	F.O. forward	. . .	4.31	. . .	1.26	. . .
4	F.W. amidships	0.70	2.21	1.5	3.77	2.6
5	F.O. aft	5.60	3.90	21.8	2.52	14.1
6	R.F.W. forward D.B.	1.56	4.20	6.5	2.27	3.5
7	R.F.W. after D.B.	0.70	4.86	3.4	1.68	1.2
8	Distilled water	0.30	5.06	1.5	1.52	0.5
9	F.W. aft	0.30	5.19	1.5	1.40	0.4
10	A.P. tank	. . .	5.37	. . .	1.25	. . .
11	Cargo no. 1	. . .	3.92	. . .	1.73	. . .
12	Cargo no. 2	13.72	3.54	48.6	2.18	29.9
13	Cargo no. 3	22.84	3.00	68.5	2.82	64.4
14	Cargo no. 4	23.21	2.46	57.1	3.47	80.5
15	Cargo no. 5	23.23	1.91	44.4	4.11	95.5
16	Cargo no. 6	23.23	1.49	34.6	4.65	108.0
17	Cargo no. 7	23.21	2.08	48.3	4.11	95.4
18	Cargo no. 8	23.08	2.69	62.1	3.59	82.9
19	Cargo no. 9	3.00	3.29	9.9	3.06	9.2
20	Subtotals (lines 2 through 19)	164.68		409.7		588.1
21	Subtotals (line 1 + line 20)			498.5		617.8
22	Deadweight correction = 3 × (line 20 in column a); enter in columns c and e			−494.0		−494.0
23	Displacement ÷ 100 (lines 1 + 20)	218.80				
24	Resultant numerals (lines 21 − 22)			4.5		123.8
						Even keel

TABLE 4-12

TABLE FOR DETERMINING HOGGING AND SAGGING NUMERALS FROM LOAD
DISTRIBUTION, EXAMPLE SHOWING DISTRIBUTION TO OBTAIN FAVORABLE NUMERAL
(T-2 TANKER)

Line No.	Item	Weight, long tons 100 (a)	Hogging Factor (b)	Hogging Numeral (c)	Sagging Factor (d)	Sagging Numeral (e)
1	Light ship and crew and stores	54.12	. . .	88.8	. . .	29.7
2	F.P. tank	. . .	4.82	. . .	0.65	. . .
3	F.O. forward	. . .	4.31	. . .	1.26	. . .
4	F.W. amidships	0.70	2.21	1.5	3.77	2.6
5	F.O. aft	5.60	3.90	21.8	2.52	14.1
6	R.F.W. forward D.B.	1.56	4.20	6.5	2.27	3.5
7	R.F.W. after D.B.	0.70	4.86	3.4	1.68	1.2
8	Distilled water	0.30	5.06	1.5	1.52	0.5
9	F.W. aft	0.30	5.19	1.6	1.40	0.4
10	A.P. tank	. . .	5.37	. . .	1.25	. . .
11	Cargo no. 1	6.77	3.92	26.5	1.73	11.7
12	Cargo no. 2	20.74	3.54	73.4	2.18	45.2
13	Cargo no. 3	22.84	3.00	68.5	2.82	64.4
14	Cargo no. 4	22.21	2.46	54.6	3.47	77.1
15	Cargo no. 5	17.03	1.91	32.5	4.11	70.0
16	Cargo no. 6	2.00	1.49	3.0	4.65	9.3
17	Cargo no. 7	19.91	2.08	41.4	4.11	81.8
18	Cargo no. 8	22.08	2.69	59.4	3.59	79.3
19	Cargo no. 9	21.94	3.29	72.2	3.06	67.1
20	Subtotals (lines 2 through 19)	164.68		467.8		528.2
21	Subtotals (line 1 + line 20)			556.6		557.9
22	Deadweight correction = 3 × (line 20 in column a); enter in columns c and e			−494.0		−494.0
23	Displacement ÷ 100 (lines 1 + 20)	218.80				
24	Resultant numerals (lines 21 − 22)			62.6		63.9
						Even keel

TABLE 4-13

TABLE FOR ESTIMATING HOGGING AND SAGGING STRESSES

	Full Load (1)	Any Condition (2)	Change, Hundreds of Tons (3)	Hogging		Sagging	
				Change per 100 Tons (4)	Change in Stress, psi (5)	Change per 100 Tons (6)	Change in Stress, psi (7)
Full load			217.70		10,500		14,600
A.P. tank	0			+470		−350	
F.W. aft	29			+432		−318	
Distilled tank	36			+410		−296	
R.F.W. aft	73			+370		−263	
Crew STS aft	40			+330		−232	
R.F.W. fwd.	166			+238		−148	
F.O. aft	692			+175		−97	
Cargo no. 9	1714			+50		+10	
Cargo no. 8	1803			−65		+112	
Cargo no. 7	1813			−182		+218	
Cargo no. 6	1814			−302		+322	
Cargo no. 5	1814			−218		+218	
Cargo no. 4	1812			−110		+98	
Cargo no. 3	1784			0		−25	
Cargo no. 2	1620			+102		−142	
Cargo no. 1	529			+178		−230	
Crew and F.W.	81			−160		+155	
F.O. fwd.	728			+236		−317	
Stores fwd.	20			+330		−402	
F.P. tank	0			+355		−430	
Totals							

The maximum allowable stress is 18,000 psi. If the sum in either column 5 or column 7 is above 18,000, some changes should be made in the longitudinal distribution of the weights. Another example of this second tabular method, for a jumboized T-3 tanker, is shown in Table 4-14, with 8 tons/in.^{2} being the limiting stress.

On-board computers.   On-board computers such as the LOADMAX-200 (see Fig. 4-17) compute all stability and trim calculations. In addition, stress points are monitored on the vessel and indicators show where limits are exceeded. There is also a graphic display of shear force/bending moment. Data is entered via a keyboard, and most on-board computers have a printer so that a permanent record may be obtained. Thus, a history of various loadings plus figures for the most optimal load may be maintained.

Checking on the Hull Deflection

Simply noting where the actual waterline amidships is as compared to the true mean draft affords a check on the amount that the hull is deflected. Draft marks amidships are found on many bulk carriers; therefore a check on the waterline's position can be made by inspection from the dock. A practical check on hull deflection for all ships may be made by subtracting the mean draft amidships from the hull's depth to obtain the freeboard. Next, a plumb line is prepared to equal the freeboard in length. With the upper end of the plumb bob at the statutory deck line, the position of the lower end is noted. If the lower end of the plumb line is in the water, the ship is sagged. The amount of the deflection is equal to the difference between the actual freeboard and the calculated freeboard. It can be determined by noting the number of inches the line must be shortened to make the plumb bob level with the waterline. This should be done on both sides of the ship, using the mean of the data received. The ship is hogged, in the above test, if the end of the plumb line is out of the water.

Figure 4-18 is a diagram of a barge in a sagging condition. Observe that the vessel is floating on an even keel with waterline *WL*. If we read the drafts forward and aft and divide by 2, we obtain a mean draft equal to *M*. The draft amidships is equal to *A* because of the sagging condition. In the method outlined above for checking on hull deflection, our plumb bob would be immersed in the water to the depth of *D*. If the plumb bob were this distance above the waterline, the hull deflection would be due to hogging.

TABLE 4-14
TABLE FOR ESTIMATING LONGITUDINAL STRESS IN T-AO (JUMBO) TANKERS

Applicable to: AO (Jumbo) 105, 106, 107, 108, 109

USNS *Pawcatuck* (TAO-108)   Calc. by R. J. Meurn   Date 1/20/80
First Officer

Description of Load	Load—Tons		Change from Full Load (B)−(A) 100 (C)	Change in Stress—Tons/in.^{2}			
				Per 100 Tons,		T-AO Cond	
	Full Load (A)	T-AO Cond (B)		Hogging (D)	Sagging (E)	Hogging (C) × (D) (F)	Sagging (C) × (E) (G)
Fwd peak tank	0.0	300.00	+3.00	0.116	−0.115	+0.35	−0.35
Ship's fuel oil	374.0	0.00	−3.74	0.084	−0.122	−0.31	+0.46
Ship's lube oil	706.0	47.00	−6.59	0.064	−0.102	−0.42	+0.67
Cargo tank 1 (DFM & Avgas)[b]	1084.9	690.00	−3.94	0.031	−0.070	−0.12	+0.28
" 2 (DFM)	1850.1	2368.00	+5.18	0.003	−0.041	+0.02	−0.21
" 3 (DFM)	2270.0	2559.00	+2.89	−0.029	−0.010	−0.08	−0.03
" 4 (JP-5)[b]	2282.6	2516.00	+2.33	−0.059	0.020	−0.14	+0.05

`` 5 (JP-5)	2338.1	2508.00	+1.70	−0.089	0.051	−0.15	+0.09
`` 6 (JP-5)	1922.4	1522.00	−4.00	−0.124	0.086	+0.50	−0.34
`` 7 (DFM)	2161.4	2540.00	+3.79	−0.124	0.101	−0.47	+0.38
`` 8 (DFM)	2465.9	2578.00	+1.12	−0.093	0.071	−0.10	+0.08
`` 9 (DFM)	2466.0	1230.00	−12.36	−0.064	0.041	+0.79	−0.51
`` 10 (DFM)	1870.8	1776.00	−0.94	−0.037	0.015	+0.03	−0.03
Ship's (DFM)	275.0	250.00	−0.25	−0.010	−0.013	+0.003	+0.003
Cargo tank 11 (F.O.)	479.6	590.00	+1.10	−0.006	−0.017	−0.007	−0.02
Fuel oil service tanks	1139.9	914.00	−2.25	0.018	−0.041	−0.04	+0.09
Reserve feed water	48.0	48.00	0	0.036	−0.059	0	0
Reserve feed water	71.0	71.00	0	0.081	−0.104	0	0
Fresh water tanks	53.0	53.00	0	0.103	−0.126	0	0
Aft peak tank	0.0	300.00	+3.00	0.114	−0.137	+0.342	−0.41
Total critical loads	23858.7	22840.00	Total change			+0.198	+0.202
Light ship + non critical loads	10061.4	10061.4	Correction for Δ			−0.600	−0.200
Total displacement	33920.1	32901.40	Stress full load			4.58	7.37
			Total stress[a]			4.18	7.37
			Stress by computer			4.20	7.40

[a] Total stress should not exceed 8.0 tons/in.^{2}.

[b] DFM, diesel fuel marine; JP-5, jet fuel (Grade 5).

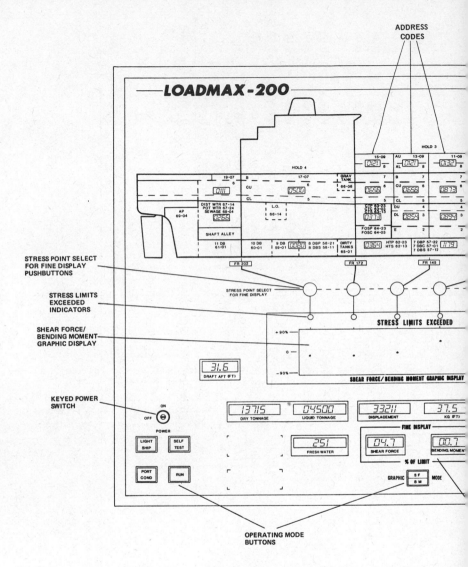

FIG. 4-17. On-board computer. Courtesy of Raytheon, Inc.

164

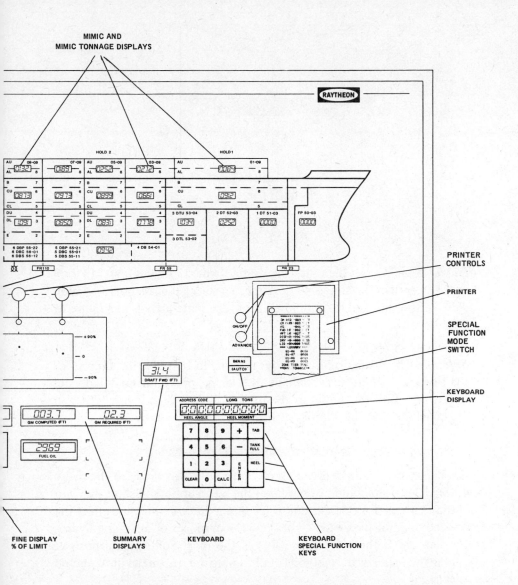

MIMIC AND
MIMIC TONNAGE DISPLAYS

RAYTHEON

HOLD 2   HOLD1

PRINTER
CONTROLS

PRINTER

SPECIAL
FUNCTION
MODE
SWITCH

ON/OFF

ADVANCE

(MAN)
(AUTO)

31.4
DRAFT FWD (FT)

KEYBOARD
DISPLAY

ADDRESS CODE   LONG TONS

HEEL ANGLE   HEEL MOMENT

003.7   02.3
GM COMPUTED (FT)   GM REQUIRED (FT)

2969
FUEL OIL

7	8	9	+	TAB
4	5	6	−	TANK FULL
1	2	3	ENTER	HEEL
CLEAR	0	CALC		

FINE DISPLAY   SUMMARY   KEYBOARD   KEYBOARD
% OF LIMIT   DISPLAYS   SPECIAL FUNCTION
                                                          KEYS

165

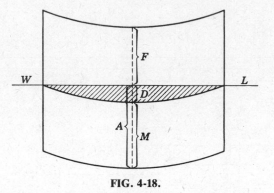

FIG. 4-18.

NUMERICAL EXAMPLE

Given: A tanker has the following drafts: fwd., 13 ft 00 in.; aft, 18 ft 00 in. The depth of the ship is 40 ft. After measuring the freeboard on both sides of the ship and taking the average, the mate arrives at a figure of 24 ft 00.0 in.

Required: Is the ship hogged or sagged? If so, how much?

Solution: The mean draft *amidships* is equal to 15 ft 06.0 in.; therefore, the freeboard amidships should be 24 ft 06 in., providing no hull deflection exists. But, the actual freeboard is 6 in. less than it should be, so we conclude that the ship is sagging 6 in.

True Mean Draft and True Displacement

In the numerical example given above, reference was made to the mean draft *amidships*. It should be noted that the mean draft amidships may *not* be the true mean draft of the ship if we define the *true mean draft* as the draft used to enter the deadweight scale for purposes of obtaining hydrostatic data about the ship. Since this is the most common purpose of obtaining the mean draft, methods of obtaining the true mean draft should be discussed.

The corrections that we will show as necessary to consider will be:

1. Correction for trim.
2. Correction to perpendiculars.
3. Correction for hull deflection.
4. Correction for density.

Correction for trim. This correction can best be considered by refer-ring the reader to a diagram such as Fig. 4-19. First we consider the ship on an even keel with waterline *WL* and mean draft *KO*. We will consider the center of flotation at some point other than on the midship section, because if the center of flotation is on the midship section, there is no correction to the mean draft for out of trim conditions. Therefore, we assume the tipping center of the ship to be at *A*.

Now, if weights on the ship are moved so that the ship trims by the stern, she will assume a new waterline W_1L_1 that passes through point *A*. Her displacement will not have changed, but it is obvious that the mean draft amidships will increase to *KP*. However, the true mean draft for purposes of entering the deadweight scale or hydrostatic curves is still equal to *KO*, which is also equal to *RA*. Hence, it can be seen that the distance *OP* is the correction that must be applied to the mean draft obtained in the usual manner to obtain the true mean draft. In this case the correction would be subtracted.

From the above discussion and the diagram of Fig. 4-19, it becomes apparent that it is important to know where the tipping center lies with respect to the midship section of the ship. It will be noted that the correc-tion may be added or subtracted, depending on where the tipping center is (forward or aft of the midship section) and on the trim of the ship (by the head or by the stern).

Calculating the correction for trim. In Fig. 4-19 the following is evi-dent from trigonometry:

$$\tan \theta = \frac{\text{Trim of the ship}}{\text{Length of the ship}}$$

Also:

$$\tan \theta = \frac{OP}{OA}$$

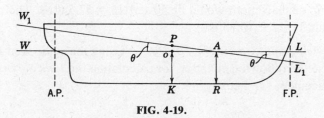

FIG. 4-19.

where OP = difference between mean draft amidships and true mean
        draft or, in other words, the correction.

OA = distance of tipping center from midship section.

Solving for OP:

$$OP = \frac{OA \times \text{Trim of ship}}{\text{Length of ship}}$$

The length of ship used in this equation is the length between perpendiculars, notated L.B.P. OP will be subtracted from the mean draft amidships
to get the true mean draft if the tipping center is forward of the midship
section and ship is trimmed by the stern or if the tipping center is aft of the
midship section and the ship is trimmed by the head. The correction (PO)
is added when the inverses of the two above situations exist. OP, as
calculated above, is in feet and should be changed to inches by multiplying by 12 to obtain the effect on displacement, as shown in the following
example.

NUMERICAL EXAMPLE

Given: A ship is 500 ft long between perpendiculars. The tipping center
is 12 ft forward of the midship section. Drafts: fwd., 16 ft 00 in.; aft, 20
ft 00 in.

Required: True mean draft.

Solution: Let x equal the correction to the mean draft to obtain the true
mean draft. This correction is subtractive.

$$x = \frac{12 \times 4 \times 12}{500} = 1.15 \text{ in.}$$

As can be seen from the above example, the correction will not amount
to much unless the ship is badly out of trim. Assuming a T.P.I. of 50 tons,
the error in displacement would be $50 \times 1.15 = 57.5$ tons. The true mean
draft would be 17 ft 10.85 in.

Change of displacement for 1-ft trim.   Using the same idea as before,
we can use the following equation for calculating the *tons* of correction to
the displacement for 1 ft of trim:

$$c = \frac{\text{T.P.I.} \times 12 \times d}{L}$$

where c = correction to displacement in tons.

L = L.B.P. in feet.

d = distance of tipping center from amidships in feet.

The sign given the correction depends on the trim and position of the tipping center.

	Tipping Center Aft.	Tipping Center Fwd.
Trimmed by the stern	Add	Subtract
Trimmed by the head	Subtract	Add

Accuracy of correction for trim equation.   The above equation is only an approximation of the true correction. The error lies in assuming the hull form below the waterline to be the same *with trim* as it is on an *even keel.*

There is no simple solution to obtaining a true displacement correction for badly out of trim conditions. If the correction for 1 ft of trim is multiplied by the number of feet of trim, an error will result. For up to 5 ft of trim this error will not be significant.

Shift of the tipping center.   When a ship trims by the stern, the tipping center actually shifts aft of the location where it appears if on an even keel at the given true mean draft; if trimmed by the head, it shifts forward. This shift becomes significant only when a ship is trimmed heavily one way or another. For tankers, which sometimes operate with large trims, the naval architect frequently calculates the exact corrections for 5 and 10 ft of trim. Table 4-15 shows such data for the T-2 tanker. As can be seen in this table, the error that would result by multiplying the 1-ft-trim correction by 10 to obtain the correction for a 10-ft trim is so that it is best to ignore the correction for such abnormal trim conditions, unless, of course, such a table is available. For intermediate trims and drafts, it is necessary to interpolate to obtain the exact correction.

Correction to perpendiculars.   Another source of error when seeking the true mean draft is caused by the draft marks being placed along lines

TABLE 4-15
HYDROSTATIC DATA (T-2 TANKER)

Draft, ft	Tipping Center from Amidships, ft	T.P.I.	Change in Displacement for 1-ft Trim by the Stern, tons	Change in Displacement for 5-ft Trim by the Stern, tons	Change in Displacement for 10-ft Trim by the Stern, tons
8	5.8 fwd.	59	8.2 (minus)	31 (minus)	40 (minus)
10	5.3 fwd.	60½	7.6 (minus)	32 (minus)	39 (minus)
12	4.8 fwd.	61	7.0 (minus)	30 (minus)	37 (minus)
14	4.1 fwd.	61½	6.0 (minus)	26 (minus)	31 (minus)
16	3.2 fwd.	62	4.7 (minus)	19 (minus)	24 (minus)
18	2.3 fwd.	62½	3.4 (minus)	10 (minus)	13 (minus)
20	1.2 fwd.	63	1.8 (minus)	1 (plus)	8 (plus)
22	0.2 aft	64	0.3 (plus)	12 (plus)	34 (plus)
24	1.8 aft	64½	2.8 (plus)	22 (plus)	57 (plus)
26	3.6 aft	65½	5.6 (plus)	33 (plus)	74 (plus)
28	5.5 aft	66	8.7 (plus)	43 (plus)	88 (plus)
30	7.6 aft	67	12.1 (plus)	54 (plus)	100 (plus)

other than the true perpendiculars of the ship. It is assumed by many operating personnel, when taking the drafts forward and aft, that these marks are on the true perpendiculars. This is manifested by the fact that they add the draft values as read, divide by 2, and call the result the mean draft. On a ship with a raked bow, it is impossible to put the draft marks along the perpendicular. Hence, when the ship is *not on an even keel,* the true draft at the position of the perpendicular is something other than what is read.

The correction that must be applied to the draft as read to obtain the draft at the perpendicular is a function of the ratio of trim of the ship to the length of the ship and the distance of the draft mark from the perpendicular in question. Thus, the correction may be determined by the following equation:

$$c = \frac{T \times d \times 12}{L}$$

where c = the correction to the draft as read in inches.
   T = trim of the ship in feet.
   L = L.B.P.
   d = distance from perpendicular to the draft mark in feet.

Figure 4-20 illustrates the elements of this problem. The sign of the correction is as follows:

Trimmed by the Stern	Trimmed by the Head
Add correction aft	Subtract correction aft
Subtract correction fwd.	Add correction fwd.

Because the signs change at either end, if the draft marks at both ends are equidistant from their respective perpendicular no correction need be made.

All ships should be equipped with tables showing these two corrections for various conditions of trim and draft so that officers reading the drafts and recording the mean drafts in the log book can record accurate data.

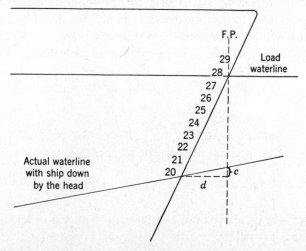

FIG. 4-20.   Diagram illustrating correction to perpendicular. The actual waterline has been given an exaggerated trim by the head to accent the correction to the perpendicular. The correction c should be added to the draft to obtain the draft at the forward perpendicular d equals the distance from the draft mark to the forward perpendicular; note that it is the length of a perpendicular dropped from the intersection of the actual waterline and the ship's stem to the F.P.

The correction for trim and the correction to perpendiculars are not required if the ship is on an even keel. In the majority of cases they are not significant, but their values should be known.

Effect of hull deflection on displacement reading. As a result of checking on the hogging and sagging as outlined above, deflections of the hull of 2 or 3 in. will be found to be quite common. Occasionally, deflections of 6 to 8 in. will be experienced. Deflections of 1 ft or more have been reported.

In the face of these facts, the question of what the ship's true displacement is when hogged or sagged arises. Entering the deadweight scale with the mean draft corrected for the errors already mentioned will not give the true displacement when the hull is deflected. Figure 4-20 clearly illustrates that if we take the drafts fore and aft and correct as outlined for the mean, the draft will be above the true mean if hogged and below if sagged. When the hull is deflected, the real mean draft lies somewhere along the distance *D*, still referring to Fig. 4-18.

Displacement correction for hogged or sagged condition. A practical means of obtaining a very close approximation of the correction for displacement due to hog or sag is first to correct for the mean draft as outlined above, then use this mean draft to enter the deadweight scale and pick out a displacement correct for everything except hull deflection. The correction for hull deflection is then calculated with the equation:

$$c = 9.6 \times \text{T.P.I.} \times D$$

where *c* = correction in tons.
    D = hull deflection in feet.
T.P.I. = tons per inch immersion for mean draft.

The correction is added if the ship is sagging and subtracted if it is hogging.

The theory of this approximate correction is based on calculating the volume of the cross-hatched portion of Fig. 4-18, assuming that the hull is deflected in a parabolic curve and that the curvature of the hull bounding the outer surface is also a parabolic curve. With the volume obtained divided by 35, the result is the displacement correction.

Correction for density. When a ship is floating in water that is brackish, another correction to the displacement as read from the deadweight

scale must be made. This is called the *correction for density* and is necessary because the deadweight scale is prepared assuming the ship is floating in salt water. The correction amounts to the displacement as read from
the deadweight scale multiplied by the difference between the specific
gravity of the water in which the ship is floating at the time of reading the
drafts and the specific gravity of sea water, which is taken as 1.026. This
can be expressed in the following equation:

$$c = \Delta \times (\text{S.G.}_1 - \text{S.G.}_2)$$

where c = the correction in tons.
  S.G.$_1$ = specific gravity of sea water (1.026).
  S.G.$_2$ = specific gravity of the water in which the ship is floating.
  Δ = displacement as read from the deadweight scale using the true
    mean draft.

The density correction is always subtractive.
Obviously, the ship must be equipped with a suitable hydrometer for
measuring the density of the water in which the ship is floating.

Draft correction for density. Because of the difference in density
between pure sea water and fresh or brackish water, the ship's mean draft
will change when going from one to the other. This change in draft or
correction may be found by a number of equations.

To obtain the fresh water draft if the salt water draft is known, multiply
the latter by the ratio 36/35. If the fresh water draft is known, the salt
water draft can be found by multiplying the fresh water draft by 35/36. The
results will be approximate, but accurate enough for most practical purposes.[7]

A general equation based on the specific gravity differences and the
T.P.I. concept is as follows:

$$c = \frac{\Delta \cdot \text{Difference in specific gravity}}{\text{T.P.I.}}$$

From this equation, assuming a specific gravity of 1.026 for salt water and
1.00 for fresh water, an equation for the difference in draft between fresh

[7] 35 ft³ of salt water weighs 1 long ton and 36 ft³ of fresh water weighs 1 long ton (approximately); hence the source of this useful ratio.

and salt water is:

$$c = \frac{\Delta}{40\text{T.P.I.}}$$

where Δ = displacement in tons.
    c = difference in draft in inches.
T.P.I. = tons per inch immersion.

Concentration of Weights

The weights may be distributed transversely, vertically, and longitudinally so that no damage can be caused to the ship, yet there remains one aspect of the problem of handling weight distribution on the ship that has not been discussed. This is the problem of loading weights so that the total weight on any deck, or part of any deck, is not so great that damage will be done to the ship structure.

Deck Load Capacities

The deck load capacities of all decks of the ship must be known if a decision is to be made concerning this problem. The deck load capacities are given on most capacity plans. If they are not recorded there, the ship's officer should make every effort to find these values and place them on the plan. The data are given in pounds per square foot and may be defined as the average load that the deck can support. This does not mean that if the deck load capacity is exceeded on any single square foot, the deck will fail to support it. It is a figure that can be used to determine the design load of the deck of a given compartment or part of a compartment. For example, the deck load capacity of the weather deck of the average merchant ship is on the order of 350 lb/ft². This means that if a heavy lift is being loaded on deck that has a base 100 ft² in area, the left could weigh as much as 35,000 lb without endangering the structure. In other words, divide the area over which the load will be spread into the number of pounds in the load; if the result is equal to or less than the deck load capacity, there is no reason for shoring up. If the result is greater than the deck load capacity, shoring up under the deck may be necessary.

Although the foregoing indicates that only the area over which the load is actually resting should be used in determining the deck load, this allows for a considerable margin of safety. The load is distributed over an unknown area extending out beyond the precise limits of the container. Some engineering data suggest that it is safe to assume that the supporting area extends out from the container one-half its length and breadth all around. Thus the area that contributes to the support of the land is four times the area of the base of the container. Obviously, this assumption will lead to trouble when carrying loads that extend over very large areas of the deck, such as over the entire beam of the ship. The assumption may be used with caution, however, when considering single small containers.

NUMERICAL EXAMPLE

Given: An American Racer class vessel has a no. 4 lower hold deck capacity of 528 lb/ft².

Required: Could you load 100 tons of boiler plate that covers an area of 8 ft × 40 ft?

Solution:

$$\text{lb/ft}^2 = \frac{\text{tons} \times 2240 \text{ lb/ton}}{W \times L \text{ (area)}} = \frac{100 \times 2240}{8 \times 40} = 700 \text{ lb/ft}^2$$

Therefore, the answer is no, unless dunnage is used to distribute the weight over a larger area.

Height Limitation

When loading small dense units of cargo, such as steel billets, lead ingots, or tin plate, the height of the stowed cargo block must be limited. If such cargo is loaded too high, the deck load capacity will be exceeded and the ship structure will be damaged.

Two things must be known to calculate the height in feet up to which it is safe to load any given cargo. First, the deck load capacity must be known. Second, the stowage factor of the cargo must be known. The stowage factor is defined as the number of cubic feet required to stow one long ton of the cargo without any broken stowage. Knowing these values, the following equation may be used:

$$h = \frac{c \pm f}{2240}$$

where h = the maximum height to load the cargo.

c = deck load capacity.

f = the cargo's stowage factor.

2240 = the number of pounds in 1 long ton.

NUMERICAL EXAMPLE

Given: A stowage factor of 14 and a no. 4 lower hold deck capacity of 528 lb/ft^2.

Required: How high can railroad steel be loaded?

Solution:

$$h = \frac{c \times f}{2240} = \frac{528 \times 14}{2240} = 3.3 \text{ ft}$$

Utilizing Free Space over Dense Cargo

Obviously, if the cargo is very dense it cannot be stowed very high. Cargo with a stowage factor in the order of 10 can be stowed only 2 or 3 ft high in 'tweendeck areas, whereas in the lower hold the height will be limited to about 7 ft. This leaves several feet of free space over such cargoes that cannot be used if the dense cargo is stowed to the limit of the deck load capacity. We therefore violate our third principle of stowage in that we do not obtain maximum utilization of all available cubic.

When two cargoes are available, one with a very low stowage factor and the other with a high stowage factor, it may be desirable to limit the height of the heavy cargo below the full limit of the deck so that the free space on top can be filled with the lighter cargo. If general cargo is to be stowed over such heavy concentrated cargo, an average stowage factor may be estimated and the height of the dense cargo limited accordingly.

Calculating the Height Limit of Dense Cargo when Overstowing with Light Cargo

We consider a column 1 ft square and equal in height to the cargo compartment. This column must be filled with a mixture of light and heavy cargo such that its total weight equals the deck load capacity of the compartment. Note that the cubic capacity of this column is equal to the height of the compartment. Hence, we have a problem somewhat similar to the full and down loading problem.

In this problem, we may work with densities or stowage factors. In the general solution that follows and the numerical example, densities are used because the values are considered easier to work with. Our problem is to determine the number of cubic feet that each cargo type must occupy out of the total cubic in the 1-ft^2 column. This number of cubic feet is also the height of each cargo type in order to fill the space and use up all the deck load capacity.

Deriving a general solution

Let a = density of light cargo.

b = density of heavy cargo.

h = number of cubic feet in a 1-ft^2 column in the compartment.

c = deck load capacity in lbs/ft^2.

x = cubic feet of light cargo in the column h.

y = cubic feet of heavy cargo in the column h.

Then:

$$x + y = h \qquad\qquad (1)$$

and

$$ax + by = c \qquad\qquad (2)$$
$$(-)ax + ay = ah \qquad\qquad (3) = a \times (1)$$
$$\overline{by - ay = c - ah}$$

$$y = \frac{c - ah}{b - a}$$

$$x = h - y$$

But since column h is 1 ft square, x and y may also be read as units of height alone.

NUMERICAL EXAMPLE

Given: Steel billets with a stowage factor of 12 are to be stowed in a compartment 12 ft high with a deck load capacity of 400 lb/ft². General cargo with an average stowage factor estimated at 160 is to be stowed over the steel.

Required: How high should the steel be tiered to allow the free space over the steel to be filled with the general cargo and not exceed the deck load capacity?

Solution: First solve for the density of the steel billets and the general cargo.

$$2240/12 = 186 \text{ lb/ft}^3 = \text{density of the steel}$$

$$2240/160 = 14 \text{ lb/ft}^3 = \text{density of the general cargo}$$

Let x = cubic feet of general cargo and y = cubic feet of steel. Then:

$$x + y = 12$$

$$
\begin{aligned}
14x + 186y &= 400 \\
(-)\underline{14x + 14y = 168} \\
172y &= 232
\end{aligned}
$$

$$y = 1.35 \text{ ft}^3 \text{ steel}$$

$$x = 10.65 \text{ ft}^3 \text{ general cargo}$$

Therefore, the steel should be tiered 1.35 ft high and the general cargo 10.65 ft high.

(2) PROTECTING THE CARGO

The first objective and principle of good stowage have been thoroughly discussed and it should be apparent that to meet these a ship's officer must **know his ship**. The second objective and principle objective of good stowage can be met fully only by officers who **know their cargo**. Knowing your ship is an easier task than knowing your cargo. In the first body of knowledge there are certain fundamental, unchanging, and consistent principles of mathematics and physics to be mastered. This takes a little time and effort, but once understood there is an end to the required effort for practical operating purposes. In the second body of knowledge, many of the factors involved change from shipping route to shipping route and even from ship to ship. The problems may vary from time to time on the same ship. The ship's officer, to know his cargo, must always be alert to the changing commodities received on board. Changing commodities bring new problems, and it is an absolute necessity continually to seek empirical information to know what is permissible.

To know what is the *best* thing to do under a given set of conditions is not always possible; in some cases the true answer to a question may not be known by anyone. There is no organized program of research to determine the exact answers to many of the problems met with in practical cargo operations. Each ship is a laboratory of a sorts, but the facts that emerge from company investigations of poor out-turns of cargoes are not publicized through any medium; indeed, in some cases the facts are guarded with great care. The most skillful cargo officers are those with years of worthy experience and gifted with good judgment. But the best experience and the best judgment will pay higher dividends if backed with a good foundation of facts gained from the experience and the judgment of those who have passed the same way before.

Segregation of Cargo

One of the fundamental requirements in the protection of the cargo is the proper segregation of the various types. This also is one of the most difficult requisites to meet when carrying full loads of general cargoes. (It offers almost no problems on ships carrying only one or two items.)

Segregation refers to the stowage of cargoes in separate parts of the ship so that one cannot damage the other because of its inherent characteristics. Wet cargoes must be kept away from dry cargoes. Generally, certain areas of the ship will be specified for the stowage of wet items when the ship is laid out during the planning stage of the cargo operation. In the same way, other areas will be specified for dry, dirty or clean cargoes.

Although segregation is called for in the case of odorous and delicate cargoes, special sections are not specified for their stowage. Each time the ship is laid out, care must be taken not to make a gross error in this respect. Segregation of light and heavy cargoes is necessary with respect to their vertical positions. Heavy items must always be given bottom stowage in any compartment. Refrigerated cargoes must be given stowage in spaces especially equipped to handle them, and segregation among items under this single category must also be given attention. Finally, the stowage of any dangerous cargoes must be in strict accordance with the segregation required by the provisions of the regulations issued by the U.S. Coast Guard, the Department of Transportation, and the International Maritime Organization (IMO) (see Chapter 6).

Wet cargo, as used in this discussion, refers to items that are liquid but in containers. With such cargo comes the possibility of leakage, and the stowage should be such that in case of any leakage the liquid will find its way to the drainage system without damaging any other cargo. Wet cargo does not refer to bulk liquid commodities; the latter requires obvious special stowage in deep tanks. Examples of wet cargoes are canned milk, beer, fruit juices, paints, lubricating oil, and so on.

Dry cargo refers to the general class of items that cannot possibly leak and furthermore can be damaged by leakage from the wet cargo. This category includes flour, feed, rice, paper products, and many more items.

Dirty cargoes are those commodities that are exceptionally dusty and always tend to leave a residue behind them. The residue, of course, will contaminate other cargoes. Examples are cement, antimony ore, charcoal, and lamp black.

Clean cargoes are those that leave no residue, are not likely to leak, and generally will not cause any damage to any other cargo, but are themselves highly vulnerable to contamination.

Odorous cargoes are those commodities that give off fumes that are likely to *taint* certain susceptible cargoes if they are stowed in the same or even in adjacent compartments. Examples are kerosene, turpentine, ammonia, greasy wool, crude rubber, lumber, and casein. Some odorous cargoes may taint more delicate cargoes and yet are susceptible to tainting themselves.

Delicate cargoes are those that are highly susceptible to damage by tainting from the odorous types. Examples are rice, flour, tea, and cereals.

Fragile cargoes are those commodities that are susceptible to breakage. Examples are glass, porcelain, and marble.

Hygroscopic cargoes are cargoes that have the ability to absorb or release water vapor. Examples are all grains, wood products, cotton, wool, sisal, jute, paper, and other products of an animal or vegetable origin.

To discuss segregation thoroughly, a discussion of commodities is necessary. A worthwhile coverage of the subject of commodities cannot be undertaken in this book. The intention here has been to point out the necessity of giving segregation the serious consideration it warrants when planning the stowage. The ability to adjudge correctly whether cargoes need segregation comes only with experience. It is also necessary to consider the fact that cargoes may possess several inherent characteristics and may, for example, be classified as not only dry but also delicate and clean.

Dunnaging

A second requirement in the stowage of cargo so as to protect it is the correct use of dunnage. The word dunnage as it is used in relation to modern cargo stowage refers to the wood that is used to protect the cargo. The common dunnage board is a 1- by 6-in. piece with a length of 10 to 12 ft. The lumber used for dunnage is classed technically as number 4 or 5 stock in the board or rough merchandise grades. For special uses such as when stowing heavy lifts or steel products, heavier lumber is used, such as 2- by 10-in. deals, 6- by 8-in. timbers, or split pieces of cord wood. Refrigerated cargoes require strips of common building lath as dunnage, which measure $\frac{3}{8}$ by $1\frac{1}{2}$ by 48 in. In all cases, the dunnage should be dry and clean. Dunnage that has been contaminated by previous cargo or is wet from any cause should not be used. If it is badly contaminated, it should be discarded. Once wetted by salt water, its use may cause more damage than can be justified by any savings made by attempting to salvage it. Green wood is a very poor risk for use as dunnage.

Type of Wood

For some uses, it makes little difference what type of wood is used. For example, any wood can be used if the dunnage is to be used in stowing steel rails or providing drainage under bags of antimony ore. However, for dunnaging sugar, hides, marble, and certain other commodities, only well-seasoned dry clean pine, spruce, or fir should be used. If oak, redwood, or mahogany is used with the latter group of commodities, staining will result. Generally it is poor practice to accept anything but dry clean pine, spruce, or fir, because the dunnage is used for various cargoes and the operation cannot depend on what type of dunnage is in the hold. As the cost of lumber increases, more use is being made of plywood. Regardless of the type or quality, the wood, especially cheaper grades, must be inspected to insure it is dry and unstained.

Amount of Dunnage on the Ship

The average general cargo carrier requires approximately 100,000 board-ft of dunnage, which will weigh about 150 tons. Depending entirely on the type of cargo being stowed, the amount specified above may be increased or decreased by 25%. In some cases, the amount of dunnage may be much less than 75,000 board-ft, but it will rarely be over 125,000. The average general carrier requires replenishment of her dunnage to the amount of about 25,000 board-ft after each voyage because of the need for condemnation of part of the supply during the voyage. Dunnage control should be regulated by the operations department and not left entirely up to each ship. Ship's officers should give dunnage (see Fig. 4-21) the same care as the cargo, for two reasons: First, the dunnage is stowed with the cargo, so that unless the dunnage is clean and dry the cargo will be damaged by it; second, the cost of dunnage wood has increased.

The Use of Dunnage

Dunnage is used to protect cargo by preventing:

1. Contact with free moisture.
2. Condensation.
3. Crushing.
4. Chafage.
5. Spontaneous heating.
6. Pilferage.

Dunnage is also used to facilitate rapid and systematic discharge. The following discussion will include an explanation of how each of the above six results are obtained through the use of dunnage.

Preventing contact with free moisture. The term *free moisture* has been used in this category to eliminate confusion with damage caused in some instances by the transfer of hygroscopic moisture. No amount of dunnaging will prevent the latter type of trouble; the only effective means

(a)

(b)

FIG. 4-21. (a) Bag stowage on a floor of dunnage three tiers high. One dunnage board has broken at a knot, seen in the lower right-hand corner. Before bags are stowed over this floor, this break in the floor should be made solid. Another interesting point in this view is the bag-on-bag stowage, which clearly shows the spaces left between the bags for ventilation purposes. Finally, the bags have "dog ears," which enable easy handling without the use of hooks. (b) A floor of dunnage over two different levels of cargo making a level platform on which to start the stowage of a different type of cargo. This dunnage is being used to strike a level that will reduce crushing from uneven pressures both on the cargo being stowed and the cargo below.

183

of preventing such damage in general cargo is by proper segregation. Free moisture refers to water in the liquid form that might be present in a cargo hold as a result of a leak in the hull plating, an adjacent tank, deck, or openings into the hold or of heavy condensation. Also included in this category, insofar as dunnaging is concerned, is liquid from any wet cargo.

The first dunnage used in the stowing of most cargoes is used for this purpose. This is the dunnage laid on the deck upon which the first tier of cargo is stowed. To provide for drainage under the cargo, the first dunnage should be laid with the length toward the drainage system and spaced about 6 in. apart. There should be at least two tiers of dunnage, the bottom tier laid to provide the drainage and the top tier to support the cargo. Thus, on a modern ship with drain wells aft and running transversely, the first tier of dunnage would be laid fore and aft. The second tier should always be perpendicular to the first and may be spaced 2 or 3 in. apart. On ships with side bilge systems, the first dunnage laid in the lower holds should run diagonally, slanting aft from the centerline.

If bags are being stowed on top of such a floor, the dunnage should be spaced not more than 1 in. apart. Wide spacing of dunnage floors upon which bags are to be stowed results in the bottom bags being split and a heavy loss of contents through leakage. If the commodity in the bag requires ventilation, spacing the dunnage a little is advantageous. If the commodity does not require ventilation, the floor will be better if made solid.

Necessity for multiple layers of dunnage.   When laying the bottom tier of dunnage where there is a high probability of drainage from some source, the number of tiers of dunnage running toward the drainage system should be increased. Five tiers may be laid with a sixth tier on top as a floor with some cargoes that need maximum circulation of air as well as drainage. The ship's officer should not be timid about requiring that dunnage be laid correctly. *It is a basic requisite of proper care and custody,* and can prevent costly damage. Longshoremen the world over are very apt to lay dunnage floors in a slipshod and highly inadequate manner. Common deficiencies are laying it so that the best drainage is not provided, laying an insufficient number of tiers, and improper spacing. The ship's officer must be in the hold of the ship to prevent gross negligence in this initial step in stowing cargo.

Vertical dunnage: To prevent contact with steel members of the ship and thereby preclude the possibility of wetting with condensation, vertical dunnage is installed between the cargo and all steel members. The members referred to are the stanchions, frames, ladders, transverse bulkheads, partial longitudinal bulkheads, and ventilator shafts running through the space. This dunnage may be put into position by standing it up against the vertical member, and temporarily held in place by one container of cargo being stowed against the bottom. Another way is to stand the dunnage up in place and tie with rope yarn, then wrap heavy paper around the dunnage and member together and tie with more rope yarn. The rope yarn serves the purpose of holding the dunnage in place until the cargo is safely stowed.

The dunnage drainage floor and the vertical dunnage to separate the cargo and the steel members complete the dunnage used to prevent contact with free moisture.

Dunnage to prevent condensation. The dunnage that prevents condensation is the dunnage that is placed in the cargo to facilitate the circulation of air currents. It is by thorough ventilation that high-dew-point air is removed from hold interiors, and this is necessary to prevent condensation. When laying the floor upon which to stow certain cargoes, such as fish meal in bags, deep tiering is utilized to help circulation under the cargo and thus remove moisture-laden air. Such floors may be five tiers deep with a sixth layer on top upon which to stow the cargo. Blocks of cargo may be separated vertically by laying crisscross dunnage separation at convenient intervals, normally about 5 ft. Venetian ventilators, which are made from dunnage wood, may be worked into some cargoes for ventilation purposes. The venetian vent is made with two pieces of dunnage measuring 1 by 6 in. and about 10 ft. long. These two boards are separated by slanting pieces of lumber measuring about 1 by 2 in. and nailed to the side to form a support and separation for the two solid pieces and to form openings along the vent's entire length. Figure 4-22 is an illustration of a venetian ventilator. This type of fitting is used so frequently with rice cargoes that occasionally it is referred to as a *rice ventilator.*

Use of the venetian vent or rice ventilator. The venetian vent is inserted within bagged cargoes. At 5-ft intervals, the vents are laid longitudinally and transversely to form continuous air channels in both directions. Generally two longitudinal lines are laid running in the same vertical plane as the hatch carlings (fore and aft hatch coamings), and two transverse lines are laid in the same vertical plane as the hatch end beams.

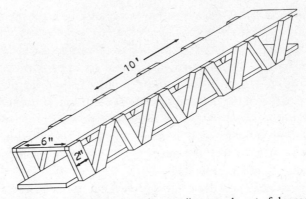

FIG. 4-22.   A venetian vent or rice ventilator made out of dunnage.

These four lines intersect at the four corners of the hatch, and at these points a vertical vent is fitted. This arrangement continues for the full depth of the cargo. When the cargo is stowed, such a system provides continuous internal movement of air and is an effective aid in the prevention of heavy condensation. If the ship is operating on a route where the outside temperatures may fluctuate between two extremes or if the hatch is extremely long, the vents may be inserted at more frequent longitudinal and transverse intervals.

Sweat battens.   The permanent dunnage attached to the frames of the ship is also an aid in ventilation. This dunnage is known as cargo battens or sweat battens and may be secured either vertically or horizontally. It is generally made of 2- by 6-in. lumber held in place by clips that are bolted to the inboard flange of the frame. The sweat battens prevent the cargo from filling up the frame spaces and hence these spaces are reserved to act as vertical air ducts.

Deep dunnage floors, crisscross dunnage floors, venetian ventilators, and the sweat battens complete the dunnage used to prevent condensation.

Dunnage to prevent crushing.   Dunnage that is used to prevent crushing of cargo is that which is placed in the hold to prevent the cargo from shifting, to spread weights evenly so that pressure is equalized on lower tiers in deep holds, and to maintain levels.

Eliminating voids.   Cargo seldom fits exactly into a hold. When stowing athwartships, a void nearly always appears at either side or along the centerline of the ship into which no container will fit. It is a standard rule never to leave a void in a stowed cargo block; such voids must be filled with something. The best thing to place in such voids is cargo that is small and durable, known as *filler cargo*. However, if filler cargo is not available, dunnage must be used. The dunnage may be used to build up bracing between the two sides to prevent movement of the cargo, or dunnage may be piled directly into the void. The need for dunnage to fill in such voids often occurs in the wings of compartments where there is much curvature, such as the end lower holds. This dunnage is used for two reasons: first, for filling in the void; and second, for maintaining a level tier. Figure 4-23 illustrates such a use of dunnage. Filling voids is a precaution against shifting, which results in crushed cargo.

Toms, shores, and braces. Large pieces of cargo that stand alone and blocks of cargo that are not supported on one of their sides require shoring or bracing to prevent shifting and possible crushing.

Shoring is the process of using 6- by 8-in. timbers or similar pieces known as shores to secure cargo. The shore runs from a low supporting level *up* to the cargo at an angle.

FIG. 4-23. Dunnage to fill wing voids. This view shows dunnage being used to fill up a triangular void in the wings to form a level surface upon which the next tier of drums will be stowed. Also shown here is a well-laid dunnage floor. Athwartship stripping has been laid down first, and then longitudinal pieces over the first layer form a solid bed. These longitudinal pieces tie the stowed block together and reduce the probability of the bulkhead being constructed falling over. This is well-stowed cargo.

Bracing is the process of using timbers to secure cargo by running the brace horizontally from a support to the cargo. The timbers used for this purpose are known as braces or, more correctly, as *distance pieces*.

Tomming is the process of using timbers to secure cargo by running the the timber from an upper support *down* to the cargo either vertically or at an angle. The timbers used for this purpose are known as *toms*.Tomming is used to secure top tiers of cargo that are not secured by other cargo being stowed over them. Cargoes that need tomming, especially in the 'tweendeck areas, are steel products, such as rails or billets; pipes; bombs; and similar items. In the lower holds, tomming may be employed, but the overhead clearance may be so great that it is better to use lashings.

Shoring and bracing are usually all that can be done when securing deck cargo. Tomming can be used on deck only when the cargo is below the level of a mast house or other deck structure and a tom can be fitted into position. Tomming should be used whenever possible, even if shoring is also used.

Below decks, bracing or tomming is preferable to shoring. Shoring transmits a lifting effect to the cargo that is undesirable. Shoring may loosen in heavy weather unless carefully wedged and supported. Tomming presses downward against the cargo, and when the ship is heaving in a heavy seaway the rising of the cargo only presses the tom more tightly against the overhead support. The shore, however, works with the sea and the cargo may be lifted upward. When it settles, the shore may fit loosely and eventually its supporting effect may be completely lost.

Examples of shoring, tomming, and bracing are illustrated in Figs. 4-24 to 4-27.

Peck & Hale shoring net.   The shoring or tomming of cargo shown in Figs. 4-24 to 4-27 clearly shows interference with the free space, which it might be more convenient to leave clear. It is also quite obvious that a good deal of lumber is consumed along with the labor of constructing the shoring. It may also happen that cargo may be stowed next to a block of cargo in the wings for part of a voyage but that during another part the adjacent cargo will be removed. In such a case, shoring must be erected at an intermediate port or the cargo must be broken down by shifting some of it into the hatch square. This requires labor and materials, and in some parts of the world the materials for erecting the shoring may not be available.

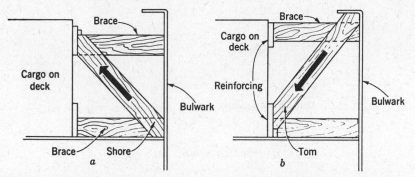

FIG. 4-24. A sketch of deck cargo being secured with shores (*a*) and toms (*b*). Bracing (or distance pieces) are used in both cases. Note that when a force, as from a sea striking the opposite side, pushes the cargo against the shore, the shore tends to lift the cargo up. If the cargo is lifted upwards, the shoring itself, braces, and lashings may be loosened and eventually cause the cargo to shift. The tom, on the other hand, forces the cargo downward toward the deck. Tomming is better than shoring, but a combination of both should be used. Lashings should always be used also. Wherever a brace, shore, or tom exerts pressure against the cargo, reinforcing should always be used.

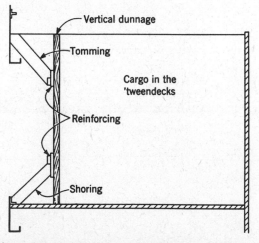

FIG. 4-25. Tomming and shoring of cargo stowed in the wings of the 'tweendeck space. It is important to place reinforcing wherever pressure will be exerted by the toms or shores, and it is also important to cut the vertical dunnage so that it fits snugly between the deck and overhead in the space or to use wedges to take up any slack. If the cargo being shored up is not in lengths, such as cartons or cases, the reinforcing shown here should run longitudinally to exert pressure all along the face of the cargo block; also, the vertical dunnage pieces should be more frequent, although each vertical piece may not need to be secured with a shore and tom. Initiative and resourcefulness must be displayed by the carpenters and the officers to erect good shoring with a minimum of time and materials.

189

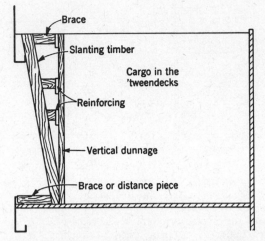

FIG. 4-26. An easy way of shoring a bulkhead of cargo in the 'tweendecks, providing timbers of about 4 by 6 in. are available. The slanting timber should be cut to just fit with the greatest dimension running athwartships. The braces are not likely to be displaced, because they will work downward with the working of the cargo and tend to tighten themselves.

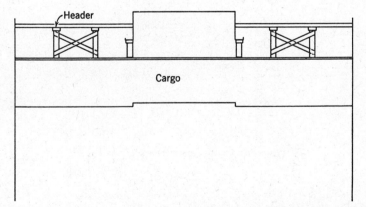

FIG. 4-27. Tomming over cargo of such a nature that it might be displaced by heavy working of the ship in a seaway, such as cylinders, bombs, pipes, or steel billets. It is important to tie the toms together with dunnage wood as shown and to use cleats on either side of the upper end of the tom nailed to the header. Wedges should be driven under the lower ends of the toms to make them fit tightly. This view is looking forward or aft. The headers run longitudinally, with the toms spaced at approximately 6-ft intervals.

190

The Peck & Hale shoring net, as illustrated in Figs. 4-28 and 4-29, is an answer to these problems. It provides a means of support for the cargo bulkheads without the use of shoring or tomming.

Determining the length of a shore.   The length along the center of the stock can be calculated by trigonometry, if, of course, measurements *A*

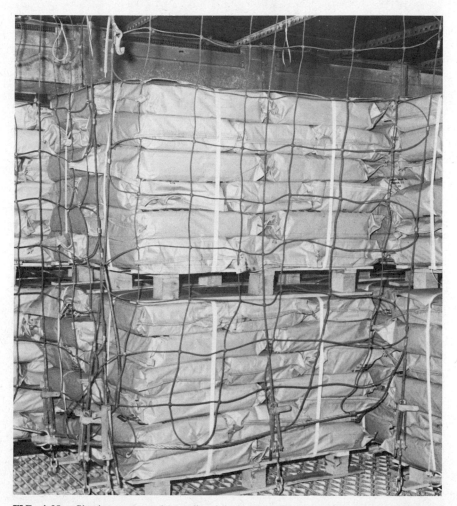

FIG. 4-28.   Shoring nets used to split triple hatches for general cargo and/or vehicles. Courtesy of Peck & Hale, Inc.

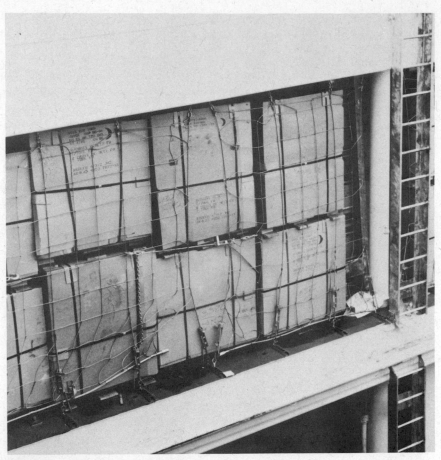

FIG. 4-29. Peck & Hale net shoring system used to secure palletized stowage in wing of hatch. Courtesy of Peck & Hale, Inc.

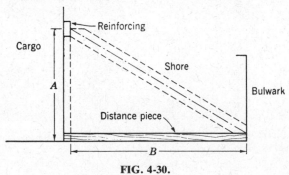

FIG. 4-30.

192

and *B* (see Fig. 4-30) are known. A more rapid and very practical method using a carpenter's square is as follows:

a.   Measure distance *A* from the center of the reinforcing to the deck, or to whatever height the lower end of the shore may extend. In this example we are assuming that the lower end is supported on the brace or distance piece. Measure distance *B* from the edge of the anchorage to the face of the reinforcing. In this example we will assume *A* = 4 ft and *B* = 6 ft 09 in.

b.   Lay off on a carpenter's square, using the ratio of 1 in. to 1 ft, measurements *A* and *B* as shown in Fig. 4-31.

c.   Measure the distance between points *A* and *B* in inches and convert to feet. This is the length of the shore along the center of the stock, in this case, $7\frac{7}{8}$ ft (see Fig. 4-32).

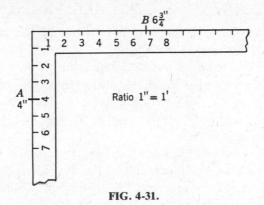

FIG. 4-31.

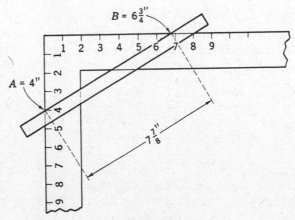

FIG. 4-32.

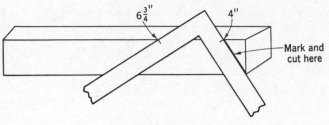

FIG. 4-33.

Cutting the shore.   The shore should be cut so that flat bearing surfaces are provided and so that sharp points are avoided. The shore should not be cut to the desired length until after one end has been cut to the proper angle as described below.

    a.   Lay the square along the shore as indicated in Fig. 4-33, making sure that the measurements 4 in. and 6¾ in. lie along the same line. Mark and cut the shore as shown in the figure.

    b.   Measure the center of the stock and mark a line at right angles to the cut just made, starting at the center of the stock (Fig. 4-34). When this last cut has been made, one end of the shore has been cut. The length of the shore is now reckoned from the point at this end and along the center of the stock (see Fig. 4-35).

    c.   Along the center of the stock measure the length of the shore, 7⅞ ft (7 ft 10½ in.), and mark off a perpendicular at the other end of the stock.

    d.   Place a carpenter's square at the center point with the same measurements on the same line as before. This time make the cutting line on the *other side of the square* (see Fig. 4-36).

    e.   Mark a right angle from the center point of the last cut (Fig. 4-37) and cut along the line indicated in the figure. This is the last cut. The shore is now properly cut to a length of 7 ft 10½ in.

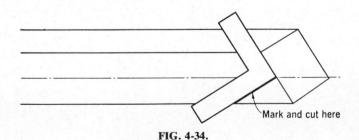

FIG. 4-34.

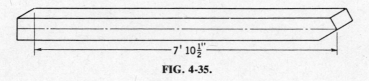

FIG. 4-35.

Steel strapping as shoring. Blocks of cargo may be secured in any desired position without the use of shores or toms by using steel strapping that virtually lashes the block to the ship's side or to a bulkhead, as shown in Fig. 4-38.

Figure 4-39*a* shows three steel straps secured behind dunnage to three separate portable beam clamps. These clamps can be set into position on frames, beams, or stiffeners within a few minutes and removed easily. This is the preparation necessary prior to loading cargo that will be secured with the steel strapping. Another preparatory step would be to do the same thing at another and opposite bulkhead, as shown in Fig. 4-39*b*.

With the preliminary steps completed, the cargo is stowed. Vertical dunnage strips are placed along the face of the stowed cargo block at 3- or 4-ft intervals. Other lengths of steel strapping are spliced to the ends previously prepared. Finally, the ends are brought together and set tightly against the block by using the stretcher, and secured by use of the crimper (see Fig. 4-46).

This procedure will secure a block of cargo of any size or shape in any position so long as a bearing surface is available and the preliminary steps are taken. No shores, toms, or braces are needed.

Dunnage floors. Dunnage floors, solid or spaced, are laid between tiers of cargo in cardboard containers in deep lower holds. This is very important if the inside containers are known to be fragile or prone to leakage. A floor is not needed between every tier. It is recommended that the first floor be laid after three tiers, the second floor after another three tiers, the third floor after six additional tiers, and a fourth floor after still

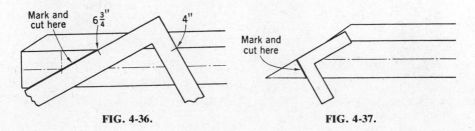

FIG. 4-36.   **FIG. 4-37.**

(a)   (b)

FIG. 4-38.   (a and b) Cargo secured with steel strapping.

another six tiers. No more than four floors will be needed. The first two floors are the most necessary. The purpose of all these floors is to prevent the spread of damage in case one of the inside containers fails. For example, if a bottle or can of beer began to leak in a carton stowed in the third tier up from the bottom of a hold in which cartons were stowed from the deck all the way up, what would happen if no dunnage floor were used? The beer would wet the carton and reduce its strength, and in all probability the adjacent inside containers would be called upon to support the entire weight of all the cases between them and the surface of the stowage. Without the support of the outside container and with one inside

(a)   (b)

FIG. 4-39.   (a and b) Steel straps secured to frames, with the opposite ends left slack until after the cargo has been stowed.

container gone, the adjacent bottle would break and the one adjacent to that would break in turn. This procedure would be accelerated in heavy weather. Eventually, a large amount of damage would be caused in the area of the first broken bottle or collapsed can. The stowed block would sag in toward the damaged area, and damaged cargo would be found throughout the bottom layers. The inside containers would not stop failing until the damage reached the upper layers, where the superimposed weight would be lessened.

Now if dunnage had been laid between the third and fourth tiers, the weight on the bottles adjacent to the broken bottle would have been partially taken up by the bridge of dunnage over the third tier. In fact, an entire carton could be left out of the third tier without causing any damage. The floor of dunnage over the third tier would simply bridge the gap, and there would be a negligible increase in pressure on other containers in the tier. Thus, it is seen that floors of this type prevent crushing.

As the stowage progresses vertically when stowing odd-sized and -shaped cases and crates, steps will be formed in the cargo block. If dunnage is not placed in the stowage so as to spread the pressure of the upper tiers equally on the lower tiers, there is danger of some containers being crushed. What this amounts to is the use of dunnage to maintain a level surface upon which to stow the next crate or case. When stowing small crates on top of a larger crate, dunnage is needed to transmit the weight to the edges of the large crate instead of to the area inside the perimeter of the crate. This is of more importance in lower holds that are deep than in 'tweendeck spaces (see Fig. 4-21b).

The dunnage used to fill voids, maintain levels, floor off, shore, brace, and tom the cargo completes the dunnage used to prevent crushing.

Preventing chafage.   Chafage of cargo is a source of damage when containers rest against the edges of structural members or other dunnage. Vertical dunnage placed against the permanent sweat battens is used to prevent small containers from becoming hung up on the upper edges of the battens. If this should happen as the ship works in a seaway, the sides of the cardboard containers may be sheared off or bagged cargo may be torn. In general, the dunnage placed in the cargo to prevent chafage is not extensive. The use of dunnage for the other purposes set forth herein precludes chafage. There are exceptions, however, as mentioned above, and the ship's officer should be on the alert for situations that call for additional dunnage to preven chafage.

Preventing spontaneous heating.   Some cargoes require dunnage to provide air channels through the stowed cargo block in order to carry away heat generated by the cargo. If such heat is not carried away, the temperature will gradually rise and serious damage may result. Refrigerated cargoes of ripening fruit are the best example. Ripening fruit creates a large amount of heat, and cool air currents must pass through such cargoes or areas will become comparatively warm and the ripening process will be accelerated, with resulting damage in the form of overripe product at the discharge port. Dunnage placed between every tier of the cargo in the form of strips of building lath provides the necessary air channels. It is important that the lengths of the strips lie in the direction of the air currents.

Other examples of cargoes requiring such dunnaging are onions, fish meal, and charcoal. These latter cargoes do not use building lath. They may use 1- by 6 in. dunnage, crisscross floors, or even the venetian ventilator. As can be seen, the dunnaging used to prevent spontaneous heating is the same as that used for the prevention of condensation. In most cases, dunnage used to aid circulation of air fulfills both objectives.

Pilferage.   Dunnage for the protection of cargo from pilferage is mentioned for the sake of completeness. This dunnage takes the form of a rough fence or barrier around cargo that is subject to pilferage but for some reason cannot be stowed in a special locker.

Additional dunnage uses.   Dunnage is also used to separate cargoes. In this form it aids the discharge process rather than protecting the cargo as in all of the above examples. When used as separation, strips of dunnage or solid floors are laid between layers of the cargo. Separation is necessary when the same type of cargo is destined for separate ports or for different consignees at the same port. Clear and definite separation makes the cargo easier to discharge in consignee blocks. In the same way, the separation lessens the probability of overcarrying the cargo.

Lashings

Cargo becomes damaged when it is not secured in its stowed position. Dunnage may be used for this purpose, as outlined in the preceding section. Another means of securing cargo, and one that is especially useful for securing on-deck cargo, is the use of lashings.

Lashings should be of steel chain, wire rope, or steel strapping. Lashings of fiber rope should never be used when attempting to secure cargo on the deck of a ship for a long voyage. Fiber rope stretches under stress and chafes easily. In emergencies or other circumstances it may have to be used. If used, it should be checked often, and when slackness appears additional line should be used to frap the original lashings tighter. Never remove the original lashings with the idea of replacing them.

Regardless of the material used, the first requisite for the use of lashings is the installation of pad eyes at points along the deck that will afford the best leads for securing cargo. This is a requirement that the ship's officer should take care of as soon as he knows where the cargo is to be stowed. This applies whether the cargo is being stowed below deck or on deck, so long as it is going to be necessary to lash it down.

Chain lashing is secured by using a shackle at the lower or deck end of the chain, passing the chain over or around the cargo in whatever manner seems to be most appropriate for the job on hand, and securing the other end by the use of a pear link, pelican hook, and turnbuckle. The pear link provides a movable link to which the turnbuckle can be attached without having to use the opposite end of the chain. The chain will not always be the exact length needed and it is too expensive to cut it to fit each time it is used; hence, it is necessary to have this adjustable method of securing one end. The use of a pear link is illustrated in Fig. 4-40. After securing the pelican hook and the pear link, the turnbuckle may be set up by inserting a piece of wood in the barrel and twisting. Once the lashing is tight, the wood should be left in the barrel and lashed to something to prevent the turnbuckle from backing off and becoming slack. For the same purpose, a short stick should be inserted through the pear link on one side of the turnbuckle.

Chain is used when securing extremely heavy objects on deck, and for lumber deck loads. The most common size is $\frac{3}{4}$ in. in diameter, with a 36-in. turnbuckle and shackles of 1 or $1\frac{1}{2}$ in.

Wire rope is secured by using clips to form eyes where needed; the eyes are secured to the pad eyes with shackles. The wire rope is passed between the deck and the cargo or over the cargo and finally attached to a turnbuckle with a pelican hook. Wire rope is more workable than chain and can be pulled tight by hand very easily, after which the clips can be applied. When the clips are on, the rope is seized, then cut. The turnbuckle is set up to tighten the lashing; the same precaution against backing off that was mentioned in the case of chain lashings must be taken.

The most common type of wire rope used for this purpose is 6 by 19,

FIG. 4-40. (*a*) The turnbuckle in this photo has a pelican hook on one end and a pear link directly attached to the other end. The pear link is at the upper end with the chain set in the narrow end of the link. To eliminate the possibility of the chain links pulling past the pear link, the slack should be led back and tied with rope yarn. After the lashing has been set up by twisting the turnbuckle barrel, the wood should be left in the barrel. (*b*) Peck & Hale adjustable lashings used to secure a truck. The lashings are anchored by D rings. Courtesy of Peck & Hale, Inc.

⅝-in., plow steel wire. Old cargo falls should be saved for use as lashings of this type. Three-quarter-inch wire rope is used whenever the object being secured is heavy enough to warrant it.

Peck & Hale Device

A patented device for lashing automobiles or trucks in position is manufactured by Peck & Hale, Inc. (Fig. 4-41). This device is an adjustable cable secured to the bumper Y brackets at the front and to spring shackles or Y brackets at the rear of the car. A deck ring is required to which the lower hood of the cable is secured. All attachments are made from underneath to the car, and no tools are required. Four cables are used to secure each car, as shown in Fig. 4-42. The cable is ¼ in. in diameter. Cars secured with this device require no 4 in. × 4 in. lumber for chocking. This is a very important fact, not only because of the expense of the lumber but because of the labor required to set the chocking in place during the stowing process and the additional labor of breaking up the chocking when discharging. Also eliminated is the confusion and congestion caused by the workmen during installation or removal. There is also the danger of damage to the automobiles as the carpenters work in the closely stowed areas. Figure 4-43 shows a method of providing anchorage for the deck ends of lashings when stowing automobiles.

Steel Strapping

Steel strapping has, during the past 40 years, become a widely used type of lashing on ships. Such strapping was used by railroads for many years prior to its use on board ships. The Signode Company manufactures a number of special fittings that enable quick, safe, and easy application of steel straps as lashings on ships. The straps for use below decks are 1¼ in. wide and 0.05 in. thick, with a breaking strength of 7500 lb. A heavy-duty strap for use on deck is 2 in. wide and 0.05 in. thick, with a breaking strength of 11,000 lb. This compares with breaking strengths for ¾-in. carbon steel chain of 28,000 lb, ⅝-in. wire of 13,000 lb, and for ¾-in. wire of 47,000 lb on the average for improved plow steel.

The strapping comes in rolls that are kept in specially built frames that make it easy to transport the strap supply about the ship and easy to remove the required amount during installation. The strapping may be rove through a *triangular link,* an *offset hook,* or a portable *beam* or *frame clamp.* All of these fittings present a flat bearing surface to the strap. The

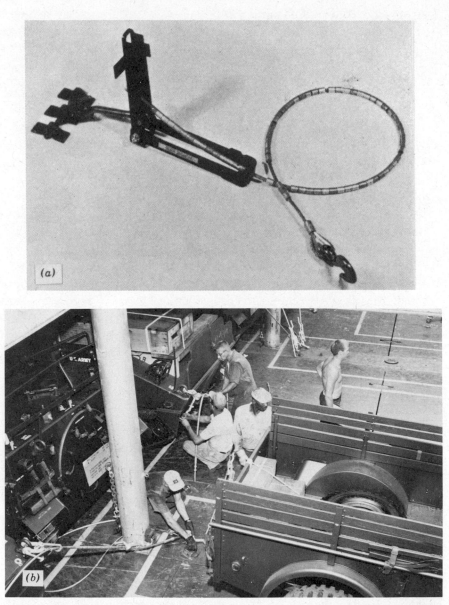

FIG. 4-41. (a) Peck & Hale adjustable cable. The adjustable cable consists of only two parts, a toggle and a beaded wire rope. The toggle has a jaw for receiving the beads. When the handle is closed, tension is applied and the beads are locked in position. The beaded assembly with movable $\frac{1}{2}$-in. beads provides a means for immediate adjustment to any length desired. Any number of beads may be furnished, depending on the amount of adjustability needed. (b) The Peck & Hale cable system in use. Courtesy of Peck & Hale, Inc.

FIG. 4-42.   Vehicles secured with Peck & Hale adjustable lashing. Note the absence of any lumber cribbing. Courtesy of Peck & Hale, Inc. (U.S. Army Photograph.)

strap is rove through the eye of whatever fitting is being used, then over the cargo to another fitting and back up to the middle of the cargo block, where the ends are secured. Alternatively, two ends may be secured to fittings and then brought together over the cargo. It depends entirely on the judgment of the officer in charge of the operation.

Before securing the ends, tension is applied with a hand tool known as a *bulkhead stretcher* or with socket wrenches (see Fig. 4-44). When suitable tension has been achieved on the strapping, *seals* are placed over the joint and then crimped together with *sealer*. A tool known as the *cutter* is used to cut the strapping at convenient lengths. A fitting known as a *winch tightener* or *anchor spindle* used in conjunction with a removable socket wrench is used to set up the tension in case of slack during the voyage. To prevent the winch from backing off under stress, the socket wrench is placed on the rotating spindle so that the handle bears against the strap. Figure 4-45 shows the use of an offset hook as an anchoring

FIG. 4-43. Peck & Hale adjustable cable being used for stowing automobiles. Because of the lack of pad eyes or ring bolts in the deck, a ¾-in. chain has been stretched from one frame across the ship to the opposite member. Courtesy of Peck & Hale, Inc.

(a) *(b)*

FIG. 4-44. *(a)* Coil box and cutter. *(b)* Tension being applied to strapping with rachet socket wrenches on the spindle of the winch tightener. When these wrenches are removed, the socket wrenches seen in Fig. 4-47 are put on. Courtesy of Signode Co.

device. Figure 4-46 shows the application of the seals with the crimper and the type of turnbuckle used; the winch tightener with socket wrench can be seen by the tin can. Figure 4-47 shows steel strap lashings applied to secure a deck load of a diesel locomotive; note also the distance pieces and cribbing used to assist in securing the carriage.

The use of the winch tightener is not necessary if the turnbuckle is inserted in the lashing.

FIG. 4-45. Two wire clips and a piece of ⅝-in. wire from a short strap, which is used to anchor the offset hook. Two-inch steel strapping is passed around the hook.

FIG. 4-46. Securing steel strapping. The man who is standing is using the crimper to secure seals that have been placed over the two parts of the lashing. The cutter is seen in the lower left hand corner. The man kneeling has his right hand on the stretcher. The winch tightener and socket wrenches are near the tin can. The lashing in the background has a spindle in place; the lashing being worked on has a turnbuckle in place. Both serve the same purpose. Courtesy of Signode Co.

FIG. 4-47. Locomotive carriage lashed with 2-in. steel strapping. The winch tightener with socket wrench has been used instead of turnbuckles. Courtesy of Signode Co.

(3)   MAXIMUM USE OF AVAILABLE CUBIC

Controlling Broken Stowage

The stowage of cargo so that the greatest possible amount of the cubic capacity of the ship contains cargo is primarily a matter of controlling the *broken stowage*. Broken stowage has been defined as that space within a loaded ship that is not occupied by cargo. Therefore, the space that contributes to the broken stowage is space between containers of irregular shape containers with curvature, space filled with dunnage, and the space over the last tier into which no cargo can be fitted for one reason or another. Broken stowage is expressed as a percentage of the total available cubic. Broken stowage on uniformly packaged commodities will average about 10%; that on general cargo will average about 25%. Both of these values may be affected by several variables. Some of these variables, as shall be seen, are within the control of the shipowner's organization; some are not. The control of broken stowage starts with the laying out of the ship and continues throughout the actual stowage process.

Use of Filler Cargo

Filler cargo is defined as small durable packages or pieces of cargo that may be stowed in the interstices or voids between larger pieces. As has already been pointed out, it is a violation of basic principles to leave a void anywhere within a stowed cargo block. Hence, if a void occurs, it either must be filled with dunnage, braced or shored with dunnage, or filled with filler cargo. Unless filler cargo is used, the broken stowage is increased. The *smallness* of these pieces of durable cargo is in comparison with the cargo in which the voids appear. For example, when stowing large frames of structural steel, wooden boxes of foodstuffs may fit in and around the pieces of steel in such a way as to qualify as filler cargo. On the other hand, when one is stowing cases of foodstuffs in the wings of a lower hold and a row has to be shut out because of the curvature, smaller pieces of boxed merchandise or bundles of pipe may be used as filler cargo. Thus, whatever an officer judges to be suitable for stowage in void spaces becomes filler cargo.

The efficient use of filler cargo is dependent on the terminal manager's administrative policies with respect to it. The first step is to have all personnel alerted to the need for spotting shipments that are suitable for such use. Next a system must be set up whereby such cargo is assembled to be ready for use when needed during the loading process. This must be organized or the use of filler cargo will be haphazard.

Choosing Cargo to Fit the Hold

When laying out the ship, decisions are made as to what cargoes go where. During this stage of the cargo operations, a ship may be "blown up" (given an excessively high broken stowage percentage). This can be caused by the decision to stow large crates and cases in the end lower holds, where small curved items should go. Planning to stow drums for the first port of call in a 'tweendeck where only nine-tenths of a drum can be stowed in the space remaining over the last tier will result in much broken stowage. The latter mistake could be rectified during the actual loading process (as many such mistakes are), providing a readjustment is possible or additional cargo can be obtained for stowage over the drums. Generally this cause of broken stowage is pretty well controlled; in practice, however, the principle of choosing cargo to fit the hold should always be kept in mind.

Skill of the Longshoremen

The final factor in the control of broken stowage is probably the most variable. The skill, industry, and interest of the longshoremen are definitely important factors in whether a ship is stowed compactly. The use of excessive amounts of dunnage in lieu of filler cargo and the failure to stow some items in neat and uniform rows and tiers are two ways that longshoremen can do the greatest amount of harm insofar as the control of broken stowage is concerned.

When stowing drums or crates and cases of odd sizes, the extra effort needed to two-block adjacent containers may prevent many longshoremen from doing the job right. A large number of loosely stowed containers not only will increase the broken stowage but may set the stage for disastrous shifting during heavy weather. When stowing bagged goods that are flowing into the ship through chutes, longshoremen may step aside and allow a large number of the bags to fall at random; after a time, and before the chute becomes jammed, a few of the uppermost bags can be set straight. There may be no indication of the improper stowage except as reflected in the rise in broken stowage percentage for the compartment. The poor stowage will come to light, of course, when the cargo is discharged.

During the actual stowage of the cargo, the walking boss and gang boss can do much to affect the value of the broken stowage. The judicious use of dunnage and filler cargo and the manner in which individual containers are placed in stowage are all ways of controlling broken stowage. Still another is the measuring of the holds before stowage of bales or any items that may be stowed in such a way that more than one vertical dimension can be obtained from the container. If the hold measurements are taken and compared with the container's measurements, a decision can be made quickly regarding how the containers should be placed in the hold so that the last tier will come as close as possible to the top of the hold. Inasmuch as the space over the last tier contributes to broken stowage, it can be seen that this results in better control of the value.

In summary, the best use of available cubic is obtained by paying attention to three factors: (1) the use of filler cargo; (2) the choice of cargo to fit the hold; and (3) the skill of the longshoremen.

(4) RAPID AND SYSTEMATIC LOADING AND DISCHARGING

Prevent the Long Hatch

The rules for providing for rapid and systematic loading and discharging are few. One of the first rules is to prevent the stowage of a disproportionate amount of cargo for any one port in any one hatch. A hatch that is so stowed is known as a *long hatch*. The ship's time in port is controlled by the maximum number of gang hours in any one hatch; therefore the work should be divided evenly among all hatches. For example, if a five-hatch ship has 3000 tons for a given port and hatches 1, 3, 4, and 5 have 400 tons apiece whereas hatch number 2 has 1400 tons, the ship will be in port just as long as if all hatches had 1400 tons. In other words, it would take as long to discharge the 3000 tons as it would 7000 tons, provided the 3000 was stowed as stated above and the 7000 was divided evenly in all five hatches. If a given hatch is double rigged, as most large hatches are, that hatch can handle more than a single-rigged hatch.

Prevent Overstowage

Another obvious rule is to prevent cargo from being overstowed. *Overstowage* does not mean that the cargo necessarily has cargo stowed directly over it, but simply that cargo is blocking the discharge in one way or another. Overstowage is prevented by thorough planning before the loading operation commences. If cargo is overstowed, it necessitates the shifting of the blocking cargo before any discharge can take place. It may require that some cargo be completely discharged and then reloaded. All of this activity is costly in time and money and should be avoided.

Prevent Overcarriage

The third rule for obtaining rapid and systematic discharge encompasses three factors, all of which may be described as means of preventing *over-carried* cargo. Overcarried cargo is cargo that is inadvertently left in the ship and taken beyond the port of discharge on the ship's itinerary. Thus, it is obvious that overcarried cargo represents the antithesis of rapid and systematic discharge operations. It will be seen that the three devices that reduce the possibility of overcarried cargo are also natural aids to rapid discharge.

Port Marks

The first device is the use of *port marks*. Port marks are geometrical designs placed on cargo in various colors so that the destination can be noted at a glance without reading carefully all the printing upon the case. These port marks are in the form of green circles, blue diamonds, red crosses, or other such markings. When blocks of cargo are so marked in the hatch, the ship's officer at the port of discharge can more easily and with more certainty check on the completeness of the discharge. This check is necessary every time a port is discharged. The longshoremen should never be allowed to cover up a hatch until a ship's officer has personally inspected the hold to make certain all cargo for the port is out. This check must be made sometimes under adverse conditions, such as during foul weather in the middle of the night. The ship's officer in charge of the deck should be aware, of course, of the approximate time that the discharging is to be completed, and should be on hand to check the hatch at the time of completion. If the cargo is well marked with port marks, the check will be more definite.

In most cases, the port marks are placed on the cargo by a member of the dock force. During the stowage, these marks may be stowed so that they are not visible in the hold; this makes them useless for the above purpose. The remedy is to place additional marks on the cargo after stowage or to put the marks on the cargo in such a fashion that they will always be visible.

Block Stowage

The second device is to provide for block stowage insofar as possible, that is, to eliminate small segments of cargo in several locations in any one hatch and to make some attempt to assemble large blocks at one level in one compartment. This may not be possible because of the need for segregation, but an effort should be made to attain such stowage. Block stowage is accomplished through planning.

Related to this problem is the need for keeping consignee marks in blocks within any one port block. This is mainly useful to the dock force in segregating the cargo as it is being discharged. When discharging, the necessity of spotting the cargo on the dock by consignee for organized delivery may prove to be a difficult and time-consuming problem, especially if the cargoes with several consignee marks are discharged simultaneously. The segregation is a simple task if one mark comes out at a time until all cargoes bearing that mark are discharged. The planning and stowing of the cargo should provide for consignee marks being kept together and marked off in the hatch by some separation system.

Proper Separation

The last device to be mentioned with regard to the prevention of overcarriage is the proper separation of cargoes. *Separation* refers to the material used to separate blocks of cargo by port and/or consignee. If blocks are separated, the longshoremen are directed, more or less, through the cargo. If the separation is omitted or incomplete, confusion is the result, and some cargo may be inadvertently left in the ship.

Some cargoes need no separation because they are different in nature and mistaking one container for another is impossible. For example, when a ship is discharging bags of rice for one port and the adjacent cargo for the next port is barrels of flour, there is no possibility of mixing the rice and the flour. This may be termed "natural" separation.

Separation between cargoes of the same type with different consignee marks and in bags is obtained by the use of strips of burlap. Heavy paper may also be used. Dunnage boards widely separated are used, but dunnage worked into the stowage of bags may lead to chafing and tearing of the bags, with a consequent loss of contents.

When separating general cargoes, heavy paper with appropriate marks on the exterior is used. This refers to the separation of miscellaneous containers, such as small cartons and cases containing general merchandise bound for different ports but stowed adjacent to each other. Such stowage is most likely to result in the overcarriage of a few containers if port marks and separation are not given proper attention.

Lumber needs careful separation of consignee marks. This cargo may be separated by laying strips of rope yarn athwartship spaced about 6 ft longitudinally over the stowed mark. Painting stripes across the stowed block is also used, but should not be used when working with milled or surfaced lumber in the finishing grades. Wire and staples are used, but this method of marking off is not desirable because the staples left in the lumber obviously are liable to cause trouble when the lumber is being worked. Wire and staples are not as easy to handle as rope yarn or paint, and the efficiency of marking off is no greater.

Wire may be used when separating structural steel, steel pipe, or steel rails. Paint and dunnage strips may also be used for separating steel products.

In summary, provision for rapid and systematic loading and discharging depends on three factors: (1) preventing long hatches; (2) preventing overstowed cargo; and (3) preventing overcarried cargo.

(5) SAFETY OF CREW AND LONGSHOREMEN

Cargo should be stowed so that during the discharging process it is unlikely that unsafe areas will develop for the men working the cargo or the men working about the ship. An example of a situation in which discharging cargo may result in the formation of unsafe areas is when workers discharge a block of cargo before moving to another block even though high and perhaps unstable bulkheads are formed all around the square where the longshoremen are laboring. Instances have occurred also in which the manner of discharging either aggravated the situation or actually created it needlessly.

The objective of providing for the safety of crew and longshoremen is so obvious that it is often ignored. There should be no unstable bulkheads of cargo, any unsafe areas should be roped off, warning signs should be used and enforced, and proper lighting must be provided. The ship's officer must always keep in mind that he is responsible for the safety of crew and longshoremen aboard.

CHAPTER 5

PLANNING THE STOWAGE

PLANNING COMPLICATIONS

When faced with the problem of planning for the stowage, the officer drawing up the plan must deal with a number of factors that directly or indirectly complicate this task. It may help to clarify the entire problem to list these factors so that the reader can consider each in its proper relation to the others and be more aware of where the troubles really lie.

Diversity of the Cargo

The diversity of the cargo, in its inherent characteristics as well as its shape and weight, is first and foremost on the list. If all cargoes could be stowed together in any compartment or in any order vertically and longi- tudinally, the problem of planning would be reduced in difficulty by over

213

50%. The fact that cargoes must be separated in definite ways complicates their segregation. When there is only one cargo, or even when there are several that can be stowed together, planning for good stowage is greatly simplified.

The Number and Sequence of Ports

If all the cargo is going to one port, it is obvious that the planning problems are greatly reduced. Under such conditions, it is impossible to overstow or to overcarry cargo or to make long hatches.

It is interesting to note, however, that a situation making it difficult, if not impossible, to satisfy all the objectives of good stowage can be created by an itinerary of just two ports with just two cargoes. The reader might consider, for example, a ship booked to carry a full and down load of steel rails for port A and cartons of packaged cereals for port B, with the ship to call at port A first. How would such a cargo be stowed? Remember the necessity of correct weight distribution, segregation of light and heavy cargoes vertically, control of broken stowage, and rapid and systematic discharge. The booking of such a cargo would be a gross error, of course, on the part of the freight traffic department unless the sequence of ports could be reversed.

The Shape of the Hold

This factor must be considered in regard to the question of control of broken stowage. All factors of stowage may be well satisfied, but if the plan calls for large cases to be stowed in the lower holds of the end hatches, the broken stowage percentage will be high. If possible, bagged cargoes or smaller curved items should be stowed in such areas. This is an example of the give and take that must go on when planning the stowage of any cargo (see Fig. 5-1).

Obstructions in the Hold

Obstructions in a hold should be kept to a minimum by the designers. If obstructions exist in the hold, otherwise-perfect stowage may have to be changed because the cargo will not go in around the obstructions or, if it will, the broken stowage may rise an unwarranted amount. In this respect, some of the areas in the Mariner type ship are badly designed. For

FIG. 5-1. View of the curvature of the 17-ft flat at hold no. 7 on the Mariner type ship. This is above the level of the shaft alley, but it can be seen how much more complicated the stowage of cargo in this space would be with a shaft alley running down the center longitudinally.

an example of a 'tweendeck area with a number of complicating features (see Fig. 5-2). This figure shows the 4½-in. coaming, a requirement that has been eliminated by the design of a watertight flush hatch covering. This 4½-in. rise, with the inclined platform reaching out into the wings and fore and aft bays, makes the stowage of the cargo very difficult. Also, it reduces by 4½ in. the height of the largest packages that can be stowed in the wings. These facts reduce the commercial value of any ship. The fore and aft ventilation ducts running below the beams are an obstruction common only to the Mariner type. Athwartship ducts running up in the beam spaces and discharging the air in a fore and aft direction would make the latter obstruction unnecessary. Other obstructions protruding into the cubic of the compartment can be seen.

Heated Bulkheads

Bulkheads that are heated may make it necessary to stow cargo such as lard, wattle bark extracts, and similar items in other spaces when all other

FIG. 5-2.   Mariner type shiphold no. 3 second deck space, illustrating types of obstructions.

factors are best satisfied by placing them in the heated compartment. This fact, of course, requires a change in the plan and results in a complication.

Optional Cargo Shipments

Optional cargo is cargo that is loaded for discharge at any one of several ports at the option of the shipper. Such shipments are sometimes difficult to keep from being overstowed. They must be ready for discharge at all of several ports instead of only at one. Only a limited amount of optional cargo can be stowed without the necessity of shifting cargo for its discharge.

Unavailability of Cargo

Cargo may be booked and planned for stowage, but the shipper may fail to deliver it to the dock when it is time to load. Such a situation makes it necessary to change the entire plan. The degree of the change depends on the amount of cargo involved and other conditions.

THE TENTATIVE CARGO PLAN

Importance of the Tentative Cargo Plan

With only two cargoes and two ports creating difficult problems, as illustrated in the above hypothetical case, the reader can appreciate more readily the problems that may arise when stowing a hundred commodities for several ports. General cargo carriers operate under the latter conditions consistently. The operation may be further complicated by the need to load as well as discharge cargoes at intermediate ports on the itinerary.

The correct way to prevent costly mistakes from occurring during the loading process is to lay out the ship as far in advance as possible. The preliminary laying out of the ship is the plan for stowing the vessel made up from the engagement sheet or recapitulated data from such a source. This first plan is called the *tentative plan* to eliminate confusion with another plan known as the *final stowage plan*. There are terminals in operation where large cargo liners are loaded without the benefit of a formal tentative plan. However, to do so is contrary to the best operating principles. It is true, of course, that when the sample type of ship is loaded in regular and frequent intervals over runs that are almost identical year in and year out, the importance of the tentative plan will be less than when a ship is being loaded for the first time on a new run. With years of experience with the same ship, the same cargo in general, and the same run, the loading methods will be very similar; however, they will never be identical. The differences will always warrant a rough tentative plan at the least. With respect to weight distribution, the same mistake might be made for years and quite unnecessarily cause gradually increasing costs for repair and maintenance. The latter mistake would be rectified only if the final stowage plans were analyzed. Lacking this check and continuing to load the ships by intuition, the shipowner may operate his fleet with less-than-optimal efficiency.

Preparation of the Tentative Cargo Plan

The assistant superintendent is generally the individual responsible for laying out the ship. As mentioned above, he obtains the data for preparing the plan from copies of the engagement sheet kept by the booking clerk. The booking may not be complete when he starts to make up the plan.

This will complicate the task to some extent and necessitate a number of changes during the actual loading process. The object is to have some basis for making initial decisions when starting the loading operation and to have an instrument with which to brief all supervisory personnel about the operation.

As mentioned above, when the same ship receives similar shipments voyage after voyage to be loaded for the same consignees in approximately the same amounts, the tentative and final stowage plans, if analyzed constantly, should reflect the best answer for stowing the cargo on board in accordance with the five objectives of good stowage. The development of new products and changes in demand for the old will cause a gradual change in the types and amounts of cargoes found on old and established runs. The change will be so gradual that there will be little difference in the tentative plan between one voyage and the next, but there will be change. The need for analyzing the stowage from time to time will always be present on the general cargo carrier.

Provisions of the Tentative Plan

The tentative plan provides for loading cargo bound for the last port of call first, that is, on the bottom tiers. (The layers of cargo are referred to as *tiers*. The athwartship and fore and aft lines of stowage are referred to as *rows*.) The plan will show limitations on the stowage as to height and bulkheads fore, aft, and athwartship. The approximate amounts in tons and numbers of containers expected to be placed in the indicated spaces will also be shown. All of these data may be changed during the actual stowage in order to expedite the total operation.

Heavy cargoes must be loaded on the bottom tiers. A complication arises immediately when a heavy item must be loaded so that it can be discharged at one of the early ports on the itinerary. There are several ways to meet this problem. One of the most common is to split the hold transversely, leaving at least one section of the hatch available to the hook, and, if necessary, to subdivide the two parts further longitudinally. A second solution is to save a part of or the entire hatch square for the first port. A third device is to use the upper decks for the heavier pieces bound for the early ports. A fourth is to resort to on-deck stowage if it is allowable by the terms of the bill of lading. A fifth solution, and the least desirable, is to stow the heavy item under cargo bound for a later port.

This may be the only possible answer under some conditions, but it requires the shifting of cargo to get the load discharged and means added time and expense for the cargo operation. It should never be used without having exhausted every other possibility.

The tentative plan will provide for the correct distribution of weight vertically and longitudinally. This problem was discussed in detail in the preceding chapter.

The tentative plan will reflect the correct segregation of the cargo. The problem of segregation is that of stowing the cargoes so that one cannot cause damage to another, and this may be the most difficult problem to solve. Compromises will have to be made in this regard at times.

Finally, the tentative plan will provide for an amount of cargo in each hold such that no one hold will require a disproportionate number of gang hours of work to discharge the cargo. This calculation will take into consideration the number of sets of cargo gear at each hatch.

Using the Tentative Plan

The use of the tentative plan insofar as the assistant superintendent is concerned should be quite obvious. Once it is prepared, he can use it to judge whether the finished ship will be in accordance with the best principles of stowage. He can analyze it and discuss it with the members of the dock force, insuring that all supervisory personnel will be briefed about the operation. If for any reason a change is required, the old plan is a beginning point. A very important use of the plan is so that the assistant superintendent can tell how many gangs are needed from day to day and what gear is required. Many daily decisions will be made more quickly and accurately if based on a good tentative plan.

The ship superintendent or walking boss needs the plan to see how the ship is to be stowed. Instead of having continually to seek the assistant superintendent, he can make more intelligent decisions on his own. If changes must be made, he can always consult with the pier superintendent. Having the tentative plan, however, makes the entire operation run much more smoothly and guarantees better results.

Whereas the use of the tentative plan as outlined for the above officers within the shipowner's organization is fairly uniform from company to company, the use of the plan by the master of the ship and his officers varies from full use to a complete lack of use.

The Ship's Officers and the Tentative Plan

Cargo operations represent a high operational cost for a general cargo carrier. Every effort should be expended to make the procedures as efficient as possible under practical operating conditions. One of the things that can be done in this respect but is neglected in many cases is the briefing of the ship's officers regarding all the plans for loading the ship.

To begin with, a copy of the tentative plan should be presented to the master of the ship, so that this officer can make a preliminary analysis of it. The approximate draft, metacentric height, bending stresses, and segregation should be clearly notated on the plan. These are the bases for judging whether a given layout is good or bad. The master of a ship should know whether the ship he is to take to sea is going to be loaded as he would load her. It is the responsibility of the operations department to see that every ship master has the data needed to ascertain this fact. In some instances, the master of the ship has little if any influence on the way his ship is to be loaded; regardless of the reason, this is not right. Such cases may stem from the method of operation of the company or from the laissez faire attitude adopted by the master himself.

In every case, the master of the ship should make some analysis of the tentative plan for the stowing of his ship. He should either approve or disapprove of the stowage. If there are things of which he disapproves, there should be a clarification of the issue with the pier superintendent. In the final analysis, the master of the ship is the officer responsible for the way the ship is stowed. This fact must not be forgotten by anybody.

The master should brief all of the ship's officers in regard to the loading operation. This briefing should be in the form of a conference on the ship with the pier superintendent and walking boss present if possible. The chief officer should have a copy of the tentative plan, and after the loading conference the other deck officers should carefully note those items that will require checking during the actual operation.

Under this system, the ship's officers will be able to stop serious mistakes in loading before they occur. The shipowner will be able to utilize the years of experience of the officers with the cargoes they carry. Too frequently, the officers are completely uninformed about the operation taking place on their ship. All information about the proceedings is often held by the pier superintendent and walking boss, and not systematically passed on to the officers. Direct questions may be answered readily, but not enough information is given to enable the officers of the ship to assist fully in the cargo operation. Lacking full and complete data, the officers feel unfamiliar with problems that in many cases could be handled better if they were properly briefed.

The practice of briefing subordinates concerning any operation in which they are going to take part is a fundamental principle of leadership. Conducting a briefing makes all hands feel more as if they have a personal stake in the outcome and gives them a feeling of participation that is very important for the attainment of efficiency. This policy of thorough briefing, of course, should be set by the operations manager. If he fails to do so, the master of each ship should do what he can to enforce it. Under a cooperative relationship between the master and his officers, any ship will function in a more efficient manner than under an impersonal and uninformed relationship. Compliance with the three C's of Communication, Cooperation, and Coordination cannot be overstressed.

THE CAPACITY PLAN

When laying out the ship, certain information concerning the ship must be readily available. The data needed are incorporated on a plan of the ship known as the *capacity plan*. A full and complete capacity plan answers the questions of whether a given cargo shipment can be stowed in any given compartment or part of a compartment and what the effect on the ship will be.

The information that a capacity plan should have upon it is as follows:

Compartmentation. An inboard profile view of the ship showing all of the compartments of the ship; the holds, tanks, store rooms, quarters, and working spaces are shown.

Volumes. The volumes of all the above spaces are not specified. The volumes of the tanks and the cargo holds are given. The *bale* capacity and the *grain* capacity of each cargo compartment capable of receiving dry cargo are given. The bale capacity is the cubic capacity measured vertically from below the beam flanges down to the upper side of the tank top plating or the spar ceiling if any exists. Transversely the measurements are from inside the sweat battens, and longitudinally from inside any stiffeners or insulation on the bulkheads. The grain capacity is the cubic capacity measured from inside the plating of all six sides less the volume taken up by the structural members extending into the hold. In other words, the bale capacity is the volume of the hold that can be utilized when stowing bales and the grain capacity is the volume that can be utilized when stowing grain. The bales cannot go in between all the frame and beam spaces, but the grain will flow around such members.

The volumes of the complete hold may be recapitulated in a table on the plan, but the profile plan of the ship should be broken up into such units of space as are often used in actual stowage of cargo. Figure 5-3 is an illustration of how the profile plan of the ship may be subdivided to make this part of the capacity plan more useful.

Below the profile plan of the ship, plan views of the various decks appear. On these plan views, the cubics of the compartments should be broken up into smaller units. On a hatch that has three sections, the plan view of the decks should have the space divided up into fifteen separate parts. Such a division is shown in Fig. 5-4. When speaking of or making written notations about these spaces, the terms applied to areas 1 and 3 are *forward winged out,* or simply *forward wings.* If the position refers to only one wing, it should be designated by name, such as *forward port wing.* Area 2 is *forward amidships.* Areas 13, 14, and 15 are referred to in the same manner except that they are aft. The entire area on each side of the hatch in the wings is referred to as *wings abreast.* More explicit designation may be given by reference to the section of the hatch. For example, wings abreast would mean the cargo was stowed in areas 6, 7,

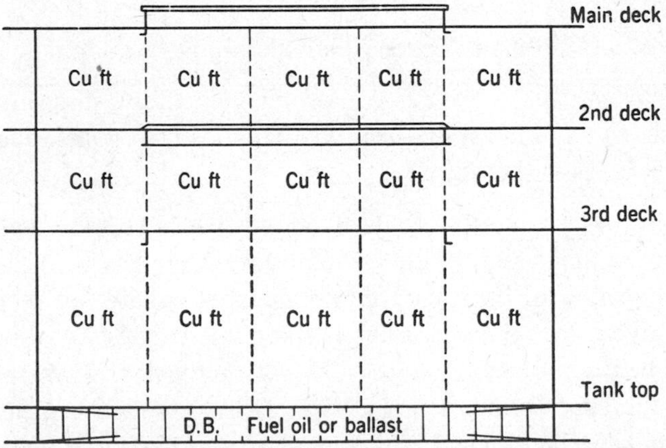

FIG. 5-3. Subdivision of the profile view. Subdividing the profile plan as shown here makes it easier to obtain data needed when making calculations relative to the stowage of cargo. This subdivision is not made by the naval architect, and so has to be done by the ship's officer.

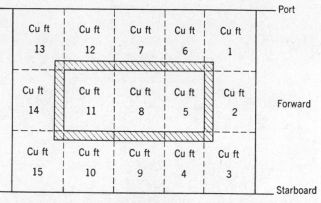

FIG. 5-4. Subdivision of the plan view. The plan provided by the naval architect may have too much detail on it to allow a clear subdivision as shown here. If so, the ship's officer should draw a plain outline and subdivide it for planning of the cargo stowage. The shaded area is a 3-ft safety area in which cargo should not be stowed if the compartment below will be open.

12, 4, 9, and 10. If cargo was stowed only in the approximate limits of area 10, this may be designated *starboard wing abreast aft*. Areas 5, 8, and 11 are designated the *hatch square*. It should be emphasized that there is no universally accepted terminology for designating all of these areas. These areas may simply be given numbers or referred to as "bays," a term used on many piers to designate certain stowage areas. The terms used need only be descriptive of the position to be acceptable.

Safety area. When cargo for one port is stowed in a 'tweendeck space and will remain there while it is necessary to work the hatches below, this cargo should not be stowed too close to the edge of the hatch opening. Cargo should be kept 3 ft back from the edge of the hatchway. It is a good practice to paint a white or red line this distance from the edge as a guideline for longshoremen (see Figs. 5-4 and 5-5). The space between this line and the edge of the coaming may be called the *safety area* or, in some cases, it is appropriately named *no-man's land*. The cubic of this space may sometimes be included in the hatch square cubic figure and sometimes in that of the appropriate areas surrounding the hatch. Just how it is handled is not too important, but the area in which it is included should be known.

This no-man's land serves to prevent longshoremen from falling into the open hatch when working around the space when opening or closing

FIG. 5-5. Number 4 hold of *S.S. Mallory Lykes* with 3-ft safety line shown. Courtesy of M/N W. R. Solis, Class of 1985, USMMA.

hatches, and also provides room for the ship's personnel to pass safely. Also, when cargo is being loaded or discharged, it lessens the probability that the cargo stowed in the 'tweendeck area will be struck in such a fashion that a container is damaged or knocked below so that a longshoreman is injured.

Clearances. The distance between the bottom of the hatch coaming and the deck, as well as the lengths, widths, and heights of all the compartments, should be notated on the capacity plan. The distance from the forward edge or after edge of a hatch coaming to the opposite lower end of the compartment below should be included, because this measurement will tell the planning officer at a glance whether a given length of pipe or timber will go into the compartment in question. It should be kept in mind that if a given length will just squeeze by the obstruction when being stowed in the beginning, then only one tier will be possible. As the cargo builds up, the drift, as this measurement is called, will be reduced. The length of clear deck run on the weather deck is also an important clearance and should be indicated on the plan. This measurement controls the size of single items that can be stowed on deck.

Deck load capacities. To enable the planning officer to know whether a given cargo can be stowed on a given deck insofar as the structural strength of the deck is concerned, all decks of the ship should have their capacities in pounds per square foot notated upon them. If the deck capacities are not given one may have to resort to the National Cargo Bureau rule of thumb of 45 times the height of the compartment in feet. Of course, this rule of thumb can be utilized only for the 'tweendecks and lower holds.

Geometrical centers. To make the capacity plan useful for calculation of trim and stability problems, the distances of the geometrical centers of all cargo compartments and tanks above the keel and from one of the perpendiculars should be given. The center of gravity of the contents of a compartment above the keel will not coincide with the figure given on the plan unless the compartment is filled with a homogeneous mass. When the compartment is not filled with a homogeneous mass or is only partially filled, an estimate must be made of the location of the center of gravity. For example, if the compartment is filled with a general cargo with the heavier items on the lower layers, the center can be assumed to be about one-third of the height of the compartment above the deck of the compartment. If half of the compartment's capacity is filled with one cargo, the center of the mass will be one-fourth of the height of the compartment above the deck of the compartment, and so on. Refer to Fig. 5-6 for a numerical example. For an explanation of the need for such data see Chapter 4.

Trimming table. Below the profile view of the ship, a trimming table will be found. This table enables the quick determination of the effect of a small change in longitudinal weight distribution on the trim of the ship. When calculating the final trim after loading a large amount of cargo, it is just as simple and more accurate to use the methods outlined in Chapter 4. The use of the trimming table is explained on p. 135. It is important to

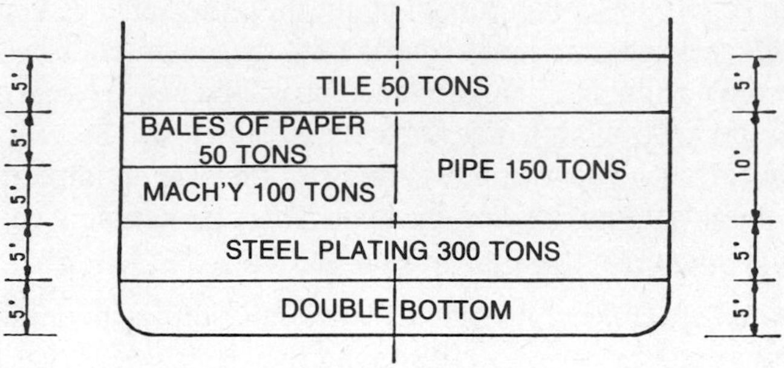

Item	Wgt	VCG	Moment
Stl. Pl.	300	7.5	2250
Mach'y	100	12.5	1250
Pipe	150	15.0	2250
Paper	50	17.5	875
Tile	50	22.5	1125
	650	11.92	7750

$$VCG = \frac{7750}{650} = 11.92 \text{ ft.}$$

FIG. 5-6. Vertical center of gravity (VCG) for no. 4 hold.

keep in mind that most of these tables are based on a unit of 100 tons (see Table 4-3).

Deadweight scale. (See Table 4-4.) The deadweight scale is found on all capacity plans; however, the data given with the scale are often incomplete for all the needs of the planning officer. This scale is simply a diagram of the draft of the ship marked off in increments of 1 in. The displacement of the ship and deadweight carrying capacity corresponding to the draft are always given. To make the deadweight scale more complete, the T.P.I., *MT*1, and *KM* of the ship should also be given. All of these data vary with the draft. The *deadweight carrying capacity* for any given draft is the *displacement* of the ship for the given draft less the *weight of the light ship*. The cargo deadweight carrying capacity cannot be recorded upon this scale, because this value varies with the amount of fuel, water, and stores on board. Under any given set of conditions, the cargo deadweight carrying capacity may be calculated if the number of tons of fuel, water, and stores on board is known. The latter value is subtracted from the deadweight carrying capacity taken from the deadweight scale to obtain the *cargo deadweight capacity*.

Also reflected on the deadweight scale is the freeboard of the ship. As explained in Chapter 4, the freeboard of the ship, as derived by taking the mean of the forward and after drafts, will coincide with that reflected on the deadweight scale only if the ship is neither hogged nor sagged. If the ship is hogged, the actual freeboard will be greater than that taken from the scale. If the ship is sagged, the actual freeboard will be less. This fact can be used as a check on whether a ship is hogged or sagged.

Capacities and arrangement of cargo gear. The capacity of each boom should be notated on the capacity plan so that the planning officer will have these important data at hand. Often a question regarding whether a given piece of gear can take a given load rigged in one way or another must be answered. For similar reasons, the length of each boom and rigging itself should be depicted upon the plan.

Miscellaneous data. Besides providing the data mentioned above, which are needed to plan quickly and thoroughly for the stowage of cargo, the capacity plan is also a source of such information as the principal dimensions of the ship, pitch of the propeller, boiler and engine data, and some of the information included on the profile and plan views of the ship.

With all these data, the original plan can be made up and the resulting *GM*, trim, and layout of the ship can be obtained. Furthermore, in case of a need for changing the stowage, the effects on the *GM,* trim, and bending stresses can be readily obtained.

It should be pointed out that the *KM* is omitted on almost all deadweight scales by naval architects. However, the *KM*s for each draft can be obtained from other sources and placed on the available plan of every ship by the ship's officers. If a capacity plan with these data already recorded on it is available from the builders, it should be requisitioned. If the capacity plan found on a ship when an officer joins her is discovered to lack many of the details mentioned above, one of his first jobs should be to seek the missing data and place them upon the plan. The officer doing this will know his ship much better and be able to function in a far more efficient and competent manner.

THE FINAL STOWAGE PLAN

The tentative stowage plan is a guide to loading the ship. When the ship is completely loaded, the stowage of the cargo will resemble the tentative plan but will never be precisely in accordance with it. The numerous changes, large and small, cannot be kept on the tentative plan so that it will serve as a final plan. It is necessary to construct a final stowage plan as the ship is loaded compartment by compartment. This plan must show in great detail the actual stowage of the ship. It must depict the vertical, longitudinal, and transverse limits of all ports and of as many of the individual marks as is practicable.

Use of the Final Plan

The final plan is used by the stevedore of the discharging port to plan the discharging operation. It is used by the ship's officers to know the stowage of their ship so that in case of an emergency they can estimate the seriousness of the situation and more intelligently plan for action to cope with it. Also, it is analyzed by the master to determine the metacentric height, the appropriate bending moments, and what might have been done to better the trim if it is extreme in any way, and to judge the quality of the stowage generally. Later, if he deems it warranted, the master may recommend to the operations department a change in past methods or offer constructive suggestions. He can do none of these things unless he has a good final stowage plan and analyzes it thoroughly.

Without an accurate, complete, and clear final stowage plan, there can be no efficiency in the discharging operation and the safety of the ship and crew is impaired greatly.

Construction of the Final Plan

The constructions of the tentative and final stowage plans are almost identical except that the tentative plan will not be so detailed. This section will describe the construction of a final plan and explain some of the notations. It must be pointed out that the notations are not standardized throughout the merchant marine. The variations are minor, and once the reader has a general idea of the way these plans are laid out, he can read all plans after a few minutes of study.

The stowage plan is generally constructed on a profile plan of the ship with the bow depicted to the reader's right. The stowage of all 'tweendeck areas is presented as a plan view; that of the lower holds, as an elevation. Now inasmuch as a profile view of the ship is used to draw the plan, a plan of the 'tweendeck hatchways must be drawn on the plan to be used as a reference in locating the cargo stowed therein. Figure 5-7 shows the blank form for one hold, assuming the ship has two decks and three sections to the hatchway. Obviously, nothing is proportional or to scale on the blank form. As the limitations of the stowage are drawn on the plan, some approximation of proportion should be attempted. For example, if a mark runs one half the distance from the forward bulkhead aft to the hatch edge, the plan should show that mark as running half the distance. This may not be possible to do when a number of marks or ports are stowed in the same area. It becomes impossible because the space between marks, when the division becomes small, is not large enough for legible printing on the actual plan.

Two types of plans are found in use. One of these may be described as the *commodity plan* and the other as the *block plan*.

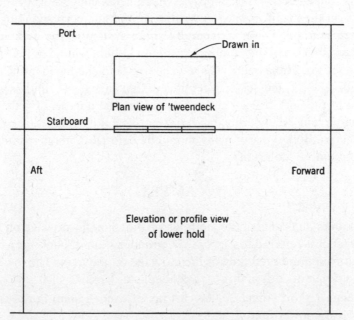

FIG. 5-7.

Commodity Type Plan

The commodity plan shows as much of the detail of the stowage as possible on the profile of the ship; that is, the limitation of all cargo is outlined on the plan. In the spaces prepared, the names of the commodities, the marks, the tons, and the number of packages are recorded. This is usually all the data that will be needed to plan for the discharge of the cargo or for action when fighting fire or in case of an emergency. This type of plan presents a *picture* of the stowage that can be readily interpreted. The officer sees the segregation immediately on viewing the plan.

Block Type Plan

The block plan shows the limitation of the cargo, but in the spaces on the profile of the ship numbers are placed. Each number refers the reader of the plan to a line on a table or list placed on the plan below the profile of the ship or on a separate sheet of paper. On this line, the cargo represented by the number is fully described. The block plan makes it possible to give more detail concerning the stowed cargo if it is desired. The only room needed on the face of the plan is that necessary to print a number, which is small indeed. Then almost as much detail concerning the cargo as appears on the manifest may be written on the line corresponding to the reference number. There is generally some system to the numbering of the cargo blocks: All the numbers in the no. 1 hold will be in the 100s, all those in the no. 2 hold in the 200s, and so on. To make the identification of items more easy and less susceptible to mistakes, the initials or first syllable of the port of destination may follow the number. Furthermore, the number may be underlined in a distinctive color representing the port of discharge. Both types of plans should be colored to make reading them easier and more foolproof.

EXAMPLE

To understand some of the methods of showing the stowage on a plan, the reader is referred to Fig. 5-8. The reader should follow step by step the stowage and recording of the stowage on the plan. The lower hold has a depth of 25 ft and the 'tweendecks a depth of 11 ft.

Stowage 1: Port Allen, 20,000 cartons of canned soup in the forward part of the lower hold.

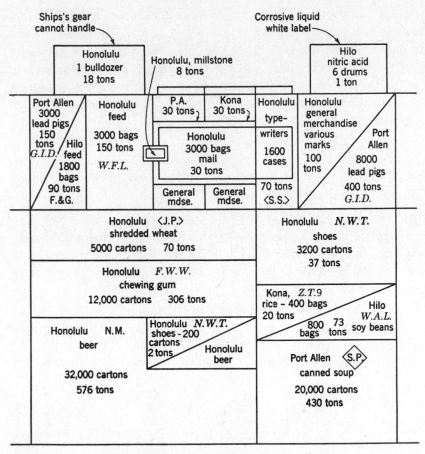

FIG. 5-8. The commodity type stowage plan.

Comments: Stowed $11\frac{1}{2}$ ft deep from the forward bulkhead aft to take in one section of the hatch and all the way across the ship. The mark is *S.P.* in a diamond.

Stowage 2: Honolulu, 32,000 cartons of canned beer in the after part of the lower hold.

Comments: Stowed $12\frac{1}{2}$ ft deep up to the edge of the hatch; then drops down about 4 ft at the square of the hatch and continues forward taking up two sections of the hatch. In the wings, it continues at a level of $12\frac{1}{2}$ ft. The mark is *N.M.*

Stowage 3: Honolulu, 200 cartons of shoes stowed in the square of the hatch over the beer 4 ft deep.

Comments: This view is an elevation; hence the wing stowage is separated from the hatch square stowage on the plan by the use of a slanting line through the common athwartship area. The cargo in the square of the hatch should come out first and it is shown above the slanting line to indicate this fact. The mark is *N.W.T.*

Stowage 4: Hilo, 800 bags of soy beans.

Comments: These soy beans are in the two wing bays and must come out after the Kona rice; hence the stowage is shown below the slanting line. The mark is *W.A.L.*

Stowage 5: Kona, 400 bags of rice.

Comments: Stowed in the way of the hatch square for a depth of $4\frac{1}{2}$ ft. Same depth as the soy beans except that it is amidships. It comes out first; hence it is on top of the slanting line. One section of the hatch is used. The mark is *Z.T.9.*

Stowage 6: Honolulu, 12,000 cartons of chewing gum.

Comments: Stowed 6 ft deep all the way across and including two sections of the hatch. The mark is *F.M.W.*

Stowage 7: Honolulu, 5000 cartons of shredded wheat.

Comments: Stowed $6\frac{1}{2}$ ft deep all the way across including two sections of the hatch. The mark is <J.P.>.

Stowage 8: Honolulu, 3200 cartons of shoes.

Comments: Stowed 9 ft deep all the way across to include one section of the hatch. The mark is *N.W.T.*

The stowage of block number 8 completes the lower hold.

The following stowage is in the 'tweendeck.

Stowage 1: Port Allen, 8000 pigs of lead.

Comments: These lead pigs would have to be stowed not more than 20 in. high; this would leave considerable space above them for lightweight cargo. But the 'tweendeck stowage is shown on the plan view. This corresponds to the problem of showing the stowage of a lower hold on an elevation when the cargo is not the same all the way athwartships. The slanting line is used; forward of the line, we place the cargo that is to come out last. Thus the Port Allen cargo is overstowed by the Honolulu. In 'tweendeck storage this is not done very often. Generally the cargo blocks run from the deck to the overhead. The mark is *G.I.D.*

Stowage 2: Honolulu, 100 tons of general merchandise.

Comments: Stowed forward of the hatch over the Port Allen lead. No marks have been listed because they are so numerous that the space doesn't permit it. The hatch list or manifest would have to be consulted to know exactly what the cargo comprises. The term *general merchandise* denotes less than carload lots of drugs, hardware, furniture, appliances, canned goods, dry goods, shoes, clothing, and tools. Note that with the general merchandise, the cargo stowed forward of the hatch in the 'tweendecks weighs 500 tons. Unless the area of the deck multiplied by the deck load capacity equaled 500 tons or more, this stowage would quite likely cause damage to the ship.

Stowage 3: Port Allen, 3000 lead pigs.

Comments: Stowed over a large area about 20 in. high like those forward. Total weight amounts to 150 tons. Since the Port Allen lead is aft of the slanting line and the stowage is aft of the hatch, it is overstowed by the Hilo feed. The mark is *G.I.D.*

Stowage 4: Hilo, 1800 bags of feed.

Comments: These bags of feed come out before the Port Allen lead. They are stowed over the lead in the same way that the Honolulu general merchandise is forward. The mark is *F. & G.*

Stowage 5: Honolulu, 3000 bags of feed.

Comments: Stowed from the edge of hatch aft to the Port Allen lead and Hilo feed and from the deck to the overhead and all the way across the ship. The mark is *W.F.L.*

Stowage 6: Port Allen, 30 tons of general merchandise.

Comments: Stowed in the port and starboard wing bays of the third section of the hatch. Space limits even an abbreviated description on both sides, so the two sides are used for a short notation of the cargo. The arrow on the port side refers the reader to the other side for more information.

Stowage 7: Kona, 30 tons of general merchandise.

Comments: Same stowage as given the Port Allen general.

Stowage 8: Honolulu, 1600 cases of typewriters.

Comments: These are stowed in the wing bays of the forward section and in the square of the hatch of the forward section from the deck to the overhead. The mark is <S.S.>.

Stowage 9: Honolulu, 3000 bags of U.S. mail.

Comments: One or two sections of the 'tweendeck square may be saved for mail received during the last hours of loading. Mail may be placed over other cargo in the square. It makes good beam filler cargo when space is at a premium.

Stowage 10: Honolulu, one millstone.

Comments: This is an 8-ton lift, and as such it requires special rigging for handling and must be conspicuously entered on the plan. It is only one item and the space on the plan is far out of proportion to what it occupies, but it is necessary to draw such items to the attention of all who use the plan so that rapid discharge is assured.

Stowage 11: Honolulu, one bulldozer.

Comments: On-deck cargo is shown in whatever way is most convenient with full notation. Heavy lifts must always be well marked. If ship's gear cannot handle the lift, some statement should inform the reader of this fact.

Stowage 12: Hilo, 6 drums of nitric acid.

Comments: The stowage of explosives and other dangerous articles must be illustrated clearly and accurately on the final cargo plan. Besides the ordinary data for other cargo, such cargo must have its classification and the label required clearly printed on the plan and underlined. Thus, nitric acid, which is classified as a corrosive liquid and carries a white label, is recorded as shown on Fig. 5-8. All hazardous cargoes must also be listed on the Dangerous Cargo Manifest.

The plan in Fig. 5-8 is of the commodity type. A hatch list is made up that gives more complete data about every shipment stowed in each compartment. Thus, with the commodity type plan and the hatch list, the stevedore discharging the ship knows all that is necessary.

The corresponding block plan would be as illustrated in Fig. 5-9. Below the profile drawing of the ship and stowage, a table such as Table 5-1 would be placed. By cross reference between the plan and the table, the same information would be obtained as from the other type of plan. The numbers' all being in the 200s means that the hold we have been discussing is the no. 2 hold. A combination of the two types of plans is also frequently seen, depending on the preference of the plans clerk.

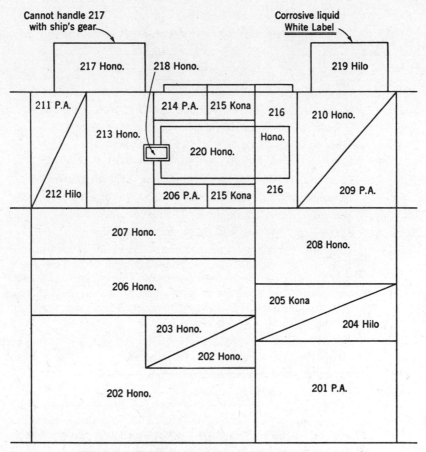

FIG. 5-9. The block type stowage plan. (See Table 5-1)

Preparation of the Final Plan

The final stowage plan is prepared by a member of the receiving clerk's staff known as the *plans clerk* or by the ship's officers. The procedure varies from company to company: In some companies, the ship's officers keep no records and accept the plan made up by the dock force. In others, one plan is made up by the ship's officers and another by the plans clerk. In certain cases the officers may keep only a rough copy of the stowage and use it to check the plan from the dock. In a third method of operating, only one plan is maintained, and this by the officers. Whenever a plan is

TABLE 5-1
CARGO DESCRIPTION FOR FIG. 5-9

Number	Description
201	Port Allen, 20,000 cases of canned soup, 430 tons, for *S.P.*
202	Honolulu, 32,000 cartons canned beer, 576 tons, for *N.M.*
203	Honolulu, 200 cartons shoes, 2 tons, for *N.W.T.*
204	Hilo, 800 bags of soy beans, 73 tons, for *W.A.L.*
205	Kona, 400 bags rice, 20 tons, for *Z.T.*9
206	Honolulu, 12,000 cartons chewing gum, 306 tons, for *F.W.W.*
207	Honolulu, 5000 cartons shredded wheat, 70 tons, for <J.P.>
208	Honolulu, 3200 cartons shoes, 37 tons, for *N.W.T.*
209	Port Allen, 8000 pigs lead, 400 tons, for *G.I.D.*
210	Honolulu, general merchandise consisting of: 20 cases drugs, for F.G.; 12 bags pipe fittings and 5 bundles pipe, for J.S.; 400 cartons shoes, for Y.P.; 8 cases household effects, for C.L.S.; 1000 cartons dry groceries, for K.K.; five refrigerators, for H.A.; 17 cartons radios, for H.P.; and miscellaneous Sears and Roebuck packages. Total: 100 tons.
211	Port Allen, 3000 pigs lead, 150 tons, for *G.I.D.*
212	Hilo, 1800 bags feed, 90 tons, for *F. & G.*
213	Honolulu, 3000 bags feed, 150 tons, for *W.F.L.*
214	Port Allen, general merchandise: 32 cases dry goods, for Z.T.; 64 cases hardware, for P.A.H.; 160 cases typewriters; for G.I.S.; 400 cartons dry groceries, for D.S.L.; 16 cases drugs, for R.X.; and miscellaneous Sears and Roebuck packages. Total: 30 tons.
215	Kona, 2000 bags feed, for G.G.; 200 cases typewriters, for K.S.; 1500 cases drugs, for K.G.H. Total: 30 tons.
216	Honolulu, 1600 cases typewriters, 70 tons, for <S.S.>
217	Honolulu, 1 bulldozer, 18 tons. **Ship's gear cannot handle.**
218	Honolulu, 1 large millstone with sling attached, 8 tons. Readily accessible in square of hatch aft.
219	Hilo, 6 drums nitric acid, 1 ton. **Corrosive liquid, white label.**
220	Honolulu, 3000 bags mail, 30 tons.

kept by the officers, the second or third officer will have the direct responsibility of entering the data on a master plan. The chief officer will confer with the officer keeping the plan from time to time. In this way he can keep abreast of the operation while taking care of other duties. In case of a change in the tentative plan or a development of any kind that is not in accordance with the best principles of stowage, the chief officer would be notified; then he would attempt corrective measures. If the chief officer is not satisfied with the corrective measures, he may make an appropriate log book entry, notify the master, and/or write a formal report concerning the incident.

Of the three means of keeping the final stowage plan, the second system is superior. The least desirable, of course, is the first, which places complete reliance on a plan kept by a member of the dock force. In spite of their experience and sincerity, plans clerks will from time to time make mistakes in the construction of the plan. Some of these mistakes may be serious.

One reason for the plans clerk making mistakes is that to a large extent, he takes the data for making the plan from notations on tally slips, and does not witness the actual stowage. As a result, the precise limitations of the stowage are not accurately depicted on the final plan. At times the notation on the tally slip may be in error; if so, the stowage shown on the plan will also be in error. Errors of this sort lead to embarrassing situations for the ship's officers. The officers are expected to know all about the stowage of the cargo on their ship. When they don't, they are automatically and justly judged incompetent.

The best way for a final stowage plan to be made for the use of the ship is to have the officers prepare it from actual observation of the stowage in the holds and data received from the receiving clerk's office. Thus the officer noting the stowage of cargo block number 201 in Fig. 5-9 would make a rough notation on a blank plan showing the limitations of the stowage. If the plan was of the commodity type, he would enter the type of cargo and probably the marks. Obviously, he would not know the exact number of cartons or the total weight; these data he would have to obtain from the receiving clerk's office.

Keeping a plan as outlined above guarantees that the officer will know the stowage well. It is not a difficult task, although it may seem difficult to anyone who has not actually performed it. But difficult or easy, the advantages are great. The ship's officers will have a clearer knowledge of the stowage, and in gaining it they will have circulated from hold to hold keeping a watchful eye on the entire operation. The actual procedure in keeping the plan requires that the officer start in the no. 1 hold and, after observing the stowage activities there, proceed to the no. 2 hold and finally all the way aft to the last hold. He should carry with him a small book with a cargo compartment drawn on each page. The book should be small enough to fit into a pocket. In this book, he should make notations of what he observes. When the ship has been completely observed, the officer transfers these data to a master rough plan. The blank that forms this plan may be kept in one of the officer's cabins or in the chart room. All officers should observe the progress of the plan's construction and check back in the holds from time to time to ascertain its accuracy.

One officer should be responsible for making hourly trips through the holds as described above and entering the information on the rough master plan. On some of the rounds through the ship, there may be no change from the last observation. But when a change does occur, the officer should be on hand to see where the old mark left off and the new mark started. If a tentative plan is available, all officers will know approximately when changes are to be made, and they can plan their work more intelligently. They can be on hand to check on the security of the cargo that is going to be stowed under and on the laying of dunnage or separation, and in general to see how the stowage of the last cargo is finished and the next cargo commenced.

If a plans clerk is also keeping a plan, it may be used to check the plan kept by the ship's officers. This, the second system mentioned above, is the best because it provides for a check on the accuracy of each.

Disposition of Final Plan

When the ship has been completely loaded, a smooth plan is made up from the rough original, duplicated, and the copies distributed. Copies should be sent ahead of the ship to all the discharging ports, and copies should also be distributed to the operations department and to all deck officers of the ship. During the final stages of the loading, the ship's officer keeping the plan may be unable to complete a smooth copy and duplicate it before the ship sails. If this is the case, there will be no copy to send ahead of the ship. In the first port of call, a plan may be dispensed with for the first half-day's work if a brief recapitulation of the top stowage is sent from the ship a few days before the ship arrives. This message can give the agents at the first port of call enough information to organize the first day's operation. On arrival, copies of the complete stowage can be distributed to port officials and other copies sent ahead to the ensuing ports on the itinerary.

If a plan is kept by the plans clerk, it will be finished a short time after the ship sails and can be sent ahead.

CHAPTER **6**

STOWAGE OF THE CARGO

PREPARATION OF THE CARGO SPACES

Cargo is stowed in the holds and tanks of the ship. In the hold areas, special compartments may be built, such as cargo lockers for valuable or easily pilfered items, magazines for the stowage of explosives, and insulated compartments for refrigerated cargoes. Prior to the loading of any cargo in any compartment, be it a container, tank, magazine, reefer space, special locker, or plain hold, the compartment must meet minimum standards of cleanliness. In some instances, the cleanliness of the hold is not of major importance, for example, when loading bagged cement. In most cases, however, an improperly prepared hold may result in heavy damage claims due to tainting or contamination. If the shipper can prove that the attention paid to the preparation of a hold area was less than that warranted by good seamanship and thereby create grounds for claiming

240

improper care and custody, he has a good chance of collecting damages. It is the duty of the chief officer to have all cargo compartments prepared for the receipt of cargo and to make log book entries concerning the efforts expended along these lines.

Standard Procedure for Cleaning Holds

Holds may be cleaned by the ship's crew or by ship service company gangs hired on a contract basis. The choice of the type of labor to be used is most commonly made by a member of the operations department on an economic basis. When ships are in places where cheap labor is available, then such labor should be used. However, when the choice is between the ship's crew members and a shoreside contractor in a port of the United States, the differential in cost may favor using the ship's crew. Another factor in favor of using the ship's crew may be good labor relations on board the ship; the decision should be given careful consideration.

The actual process of cleaning a general carrier's holds is simple and routine, providing that no special problem of cleaning is created by the carriage of extremely odorous or dirty cargoes. The equipment needed for each hatch is as follows: one dirt sling, two endless rope slings, enough brooms for all hands in the cleaning gang, rope yarns, and some cluster lights. The cleaning starts in the upper 'tweendeck. The first task is to clear a small space of all debris and start piling all the clean dunnage in a neat stack over two 2- by 4-in. battens spaced about 6 ft apart. If 2- by 4-in. battens are not available, a double height of 1- by 6-in. boards will suffice. After the usable dunnage has been salvaged, the condemned dunnage should be slung up in the rope slings. If two rope slings are not enough, use more. All the paper and other debris should be rolled up and placed in the dirt sling, which consists of an ordinary rope net sling with a lining of canvas to prevent the dirt from falling through. Every ship should have at least one dirt sling for every hatch, and perhaps one or two extra.

If the ship is at sea, the dunnage in the rope sling and debris in the dirt sling can be discharged over the side. This is accomplished by slinging the load up in the conventional manner except that one part leading to the cargo hook is removed and a rope yarn is used as a connecting device. The load is then swung out over the side and the rope yarn is cut with a knife. Thus, the load is safely dropped into the sea, but the sling is retained on the hook. In this manner, each compartment can be completely cleaned up before shifting to the next lower one.

If the ship is in port or in an area where such material cannot be dropped over the side, the debris, including the condemned dunnage, must be dumped on the weather deck of the ship. This should be done in such a way that the safety of crew members walking fore and aft is not impaired. If the pile of debris on deck gets too large, the loads should merely be slung up and left in the hold ready to be hooked up and dumped over the side when the opportunity presents itself. If the working circumstances are such that the decks and the holds must be cleared, the mate must make arrangements for trucks or railroad cars to come alongside and receive the debris to be discharged from the hold being cleaned (see Fig. 6-1).

When the good dunnage has been stacked up in one wing where it will be available as needed for the next cargo, the condemned dunnage slung up ready for discharge, and the paper, debris, and dirt swept up and placed in a dirt sling ready for discharge, the cleaning process is complete.

Washing the Interiors of Dry Cargo Holds

Occasionally it may be necessary to wash the interior of a dry cargo hold. Washing is always done *after* the hold has been cleaned according to the system described above as the standard procedure for cleaning holds. After carrying cargoes that leave behind heavy dust or glutinous residue, the simple sweeping outlined above would be insufficient to clean the hold if the next cargo were a delicate or clean type of commodity.

The washing procedure is simply that of using a wash deck hose below decks to direct a solid stream of water over all the surfaces of the hold and toward the drainage well. If possible, the washing should be done alongside a dock where fresh water can be used directly from a dock outlet. If the washing is done at sea, the preliminary wash may be with salt water, followed by a thorough rinse with fresh water. If the washing is done with salt water without a rinse of fresh water, the salt residue left after drying will pick up moisture from the air and may damage cargo coming in contact with it. Probably the worst feature of allowing the salt water to dry on the hold interior is the increase in the corrosion rate that will result.

A period of 2 or 3 days should be allowed before attempting to load another cargo in a freshly washed hold. This time is needed for drying. If the outside atmosphere is relatively dry and windy and a good stream of air is directed into the hold with the hatches removed for a maximum

FIG. 6-1. Debris on deck from hold cleaning operation.

exhaust circulation, the drying process should not take more than 36 hours. However, in the case of high outside humidity or heavy rain, the drying period may extend to several days. If the ship is equipped with a dehumidifying system, the hatches can be closed and the system placed on recirculation with dry air, which will result in one hatch being well dried out in 3 hours. If an exceptional condition arose in which every hatch of the ship had to be washed, the drying of all hatches could be accomplished in 1 day.

Deodorization Methods

If drain wells or parts of a given hold become contaminated by odorous cargo residue, they may have to be treated with a simple deodorizing wash. A solution consisting of $\frac{1}{2}$ lb of chloride of lime mixed with $2\frac{1}{2}$ gal of fresh water is quite effective. If the area is only slightly affected, it may simply be swabbed down with this solution. If it is very dirty, it should be washed with a solid stream of fresh water from a wash deck hose and then swabbed down with the chlorated lime solution. If the odor persists, sprinkle some of the powder over the area. High test hypochlorate (HTH) is a more powerful deodorizing agent that also comes in powder form. Compared with chlorated lime, smaller amounts of HTH are used when making up solutions for deodorizing holds. A tablespoonful of HTH added to a bucket of fresh water makes a fairly strong solution; no more than this should be used. This powder should not be left in the area, where it might contact cargo directly. If the solution is too strong, it may eradicate the odors of the past cargo but the odor from the HTH may cause damage itself.

Preparation for Bulk Commodities

The preparation of a dry cargo hold for the loading of bulk commodities generally involves more than the simple cleaning described above. One thing that always must be done is to prepare the drainage system to pass water but to retain the bulk commodity. In the case of the drain well plates in the lower hold, this involves either covering with burlap and inserting the edges underneath the plates or, more preferably, covering the opening with two layers of burlap and cementing the edge down. The cement that is applied to the edge should be flattened and not built up. If the cement is rounded off or built up in any way, there is more danger of its getting knocked loose and allowing the bulk commodity to fill up the drain wells. If this occurs, there is no way of pumping water out of the compartment in case of a leak from any source. Ignoring such a detail could mean the loss of the ship.

Additional protection for the burlap covering is afforded by placing a small wooden box about 6 in. high with the bottom removed over the drain well opening. This covering is referred to as a *high hat* in some trades. It is not absolutely necessary, but the added protection may be well worth the trouble.

When stowing bulk copra, the interior of the hold is entirely covered with fiber mats. Bulk coal requires no interior covering, but pipes must be inserted in the hold *before* the coal is loaded so that temperatures can be taken in the interior of the stowed cargo. In the case of bulk grain, a longitudinal bulkhead may be built along the centerline of the ship in the lower holds if more than a specified amount is loaded. If the grain is loaded so that it completely fills the hold, special feeders may be built in the 'tweendeck spaces.

The above examples are presented to show the type of special preparations that may have to be taken before loading certain cargoes.

Standard Procedure for Cleaning Tanks

On tankers, machines are used almost exclusively for the washing of the tanks. Hand washing methods are used for washing tanks on most dry cargo ships, but machines are used in some instances. A brief description of washing by machine will be given below.

When washing by hand, the temperature of the water cannot be as high as when using a machine. Consequently, the tanks are heated by steaming. Washing by machine eliminates the need for this prolonged steaming. The fact that steaming is not necessary is one of the principal advantages of washing by machine. Steaming a tank causes expansion and later contraction of the hull, which may result in hull damage, especially after being repeated many times. Steaming also accelerates corrosion of the tank interior. If a machine is used, the ship must be capable of delivering at least 180 gal of salt water/min to the machine at a temperature of 185°F under a pressure of about 180 lb/in.^{2}. This will supply one machine. If two machines are to be run simultaneously, the ship must have a heat exchanger and pump capable of doubling the above supply. Two types of machines are used. One is the Butterworth machine, pictured in Fig. 6-2, and the other is the Pyrate machine. They operate on the same principle. The water is discharged under high pressure and high temperature, and as it strikes the surface of the tank it cleans the surface of all oil, wax, and dirt. The machine rotates in two planes and the stream from each nozzle covers every inch of the tank that can be reached by a straight line from the machine's position in the tank during its operation. From this last statement, it can be seen that surfaces behind beam knees, bilge brackets, and web frames will not be struck by the solid stream of hot water. Therefore, these surfaces must be reached by repositioning the machine so that they are struck by the solid stream or by hand hosing and cleaning after the machine has been removed.

The machine may be lowered into a tank through a special hole cut through the deck for this purpose or through a manhole from one side. If lowered through a manhole, a block with a bull line or single whip to hold the machine in the center of the tank must be rigged. If the special hole is used, it should be cut through the center of the tank. Regardless of the method used to suspend the machine in the tank, the cleaning routine is

FIG. 6-2. Cutaway model of the Butterworth tank washing machine. The hose connected to the threaded upper end hangs free in the tank. The water turbine is connected to a shaft, which in turn rotates an eccentric crank. The eccentric crank operates a pawl that rotates a worm. This rotating worm meshes with a worm gear and by this means the nozzles are rotated in a vertical plane; at the same time a stationary gear rotates the entire assembly in a horizontal plane.

the same. A recommended method is to lower the machine to within 5 ft of the bottom of the tank and run it for about 30 min, draw the machine up to within 5 ft of the top of the tank and run it for about 15 min, and then lower it 5 ft and run it for another 15 min. Continue lowering and washing at 5-ft increments until within 5 ft of the bottom again. Run the machine at this lower level for 30 min to complete the process.

All the time that the machine is running, the pumps should be slowly pumping out the wash water. This keeps a steady stream of water running toward the suction foot and carries away some of the sludge and dirt that would otherwise be floating on the surface of the water in the tank. This sludge would adhere to the sides of the tank at the lower levels and make the cleaning job much more difficult. The reason for operating the machine at the lower position first is to warm up the tank and knock down the worst of the scale and dirt at the beginning of the operation.

The length of time and number of drops may be increased or decreased depending on the condition of the tank and the degree of cleanliness desired. Figure 6-3 shows the method of rigging up a machine to wash from different positions by the use of blocks and single whips. The machine is assumed to be entering the tank through a manhole, but the same thing can be done if it is lowered through a tank washing hole.

After washing as outlined above, the tank is opened and ventilated with a mechanical blower. The rigging of wind sails is not always practicable.

After ventilation, men must be sent into the tank to muck it out. *Mucking out* may be defined as the process of scraping up scale and sediment

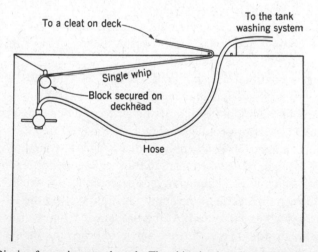

FIG. 6-3. Rigging for an improved wash. The objective here is to place the machine in a more advantageous position for washing than that afforded by being lowered straight through the deck opening. Instead of rigging a block as shown here, ¾-in. holes may be bored through the deck at suitable points and a ¾-in. wire on a reel attached to the washing machine. This is a system recommended by the U.S. Navy for positioning the machine to obtain a better wash. After the wash, the holes are tapped, plugged, and welded over.

left in the tank and placing it in a bucket that is then hauled out of the tank and dumped either over the side or on deck to be shoveled over the side when the ship is at sea. This process is done entirely by hand and is a slow job. It should be done thoroughly. The mucking out process is followed by wiping down the interior, spot cleaning the dirtier areas with a hand hose or by wiping with a solvent and rags.

The entire process of washing by machine can be divided into these three stages: (1) washing; (2) ventilating; and (3) mucking out. The first and last stages vary in length of time and care taken in carrying them out depending on the condition of the tank at the beginning and the cleanliness desired in the end.

Washing by Hand Hose

The majority of dry cargo vessels still are not equipped with the machines, pumps, and heat exchangers for cleaning tanks mechanically; therefore, the tanks must be washed by hand. Washing by hand starts with the steaming of the tank by turning on the steam smothering line. The steaming period may run for 4 to 8 hours depending on the condition of the tank before the process starts. If the tank has been carrying black fuel oil and is to be prepared for a product that is easily contaminated, the tank will require at least 6 hours of steaming.

After the steaming period, the tank must be allowed to cool for 1 to 2 hours. The cooling time will depend on the outside temperature, and no definite time limit can be specified.

As soon as possible, a man should be sent into the tank with a wash deck hose and washing by hand should commence. This man should have a line around his shoulders and under his arms and be tended by one man from the deck. A vapor-proof light should be lowered into the tank so that the man in the tank can see where he is directing the stream of water from the hose. This man must work from the top of the ladder leading into the tank, which is an awkward and dangerous position from which to handle a hose. To prevent an accident, he should lash himself to the ladder before the hose is passed down to him. At the first stage of the washing process, this man's head should be no farther down into the tank than a foot or so, just enough to enable him to see the interior. The blower should be rigged and a supply of fresh air should be supplied to him continuously. He will start washing the overhead and continue down the bulkheads and sides of the tank. At the beginning, the heat will be intense and the man should be relieved after not more than 10 min of washing. Gradually, the time for each man in the tank can be increased as the tank cools down. It is important to start the hand washing before the tank cools excessively, because as the tank cools the waxy elements in the oil tend to coagulate and become difficult to wash away. The steaming period should melt most of the wax and grease and make it easier to wash the tank. The hand hose cannot be handled at the high pressures and temperatures used with a machine. The temperature of the water when washing by hand should be 125°F and the pressure 125 lb/in.^{2}. This is about as much pressure and as high a temperature as a man clinging to a slippery ladder can handle. If started while the tank is still warm, this washing procedure will be effective. While the washing process is being carried out, the tanks should be slowly stripped by operating the pumps steadily. This is for the same reason as mentioned in the case of machine washing, to have a steady stream of water running toward the suction foot and to prevent a rise in the dirty water level in the tank.

As the washing continues, the tank will be cooling down and the man handling the hose in the tank will have to go farther and farther down. After he has done all the washing possible with the hose, the tanks should be well ventilated with a mechanical blower. Following the ventilation process, all hands should be sent into the tank for wiping down and mucking out. When the tank has been wiped down and mucked out, it should be clean, dry, and ready for the next cargo.

Whether washing a tank by hand or by machine, if the tank is not cleaned sufficiently well after the regular washing process there is no alternative except to repeat the process. However, if only certain spots are found to be dirty, the interior may be cleaned in those areas only.

If a tank has been used for dry cargo for some time and the principal defect in the tank's condition is heavy rust and dirt accumulation, the interior may have to be scraped and chipped before washing. This depends on the type of cargo that is to be loaded into the tank. For example, when loading latex in bulk in Malaya it is imperative to have every square inch of the tank free of rust and dirt. The interior surface of the tank is coated with wax before the latex is loaded, and there must be no trace of rust under the coating of wax. Luckily, in that part of the world labor is not costly, and sufficient workers can be engaged to cover the interior surface literally foot by foot. Each worker labors all day long on a few square feet, but in the end the tank is scaled clean without any further washing or preparation being necessary. The wax is applied after removing all the scaffolding and equipment, and then the latex is loaded.

Before any tank is loaded, the interior is inspected by a representative of the shipper, the shipowner, and one of the ship's officers. The shipowner will, for his own protection, generally have a surveyor certify the tank as being clean, the surveyor being a qualified representative of the cargo underwriters, probably from an organization such as the National Cargo Bureau, Inc.

CHECKING THE CONDITION OF THE CARGO HOLD

To the casual observer glancing down into a ship's cargo hold, it does not appear that there are many things that could go wrong with the space below. It has the appearance of a big steel container with spaced wooden lining against the frames and little else. Yet before an officer can make an entry in the ship's log relative to the condition of the cargo spaces, he must have checked in one way or another a long list of items. Some of these things must be checked each time the ship empties her holds; others, such as the striking plate under the bilge sounding pipe, need to be checked very infrequently. These items are listed and commented upon briefly in the following sections.

Permanent Dunnage

The term *permanent dunnage* refers to the sweat battens secured to the frames, the spar ceiling laid over the tank tops, the wooden sheathing sometimes secured to the transverse bulkheads separating boiler rooms or engine rooms from adjoining cargo holds, and any wooden casings around pipes or conduits going through the cargo spaces. These items can be inspected as the ship is being discharged. That is to say, there is no need to make a special inspection trip through the ship to keep informed about their condition. The policy should be set by the chief officer for all officers to take note of the condition of these items as they travel from hatch to hatch during their normal day's work of observing the cargo operations. If something is noted to be missing or broken, the chief officer should be given the information as soon as possible. In the case of sweat battens, the chief officer should replace the missing sweat batten during the lunch hour or whenever there will be no longshoremen in the hold. In this manner, the ship's sweat battens will always be in first-class condition. If casings or the spar ceiling is noted to be damaged, it may be repaired by the ship's force, but more than likely a requisition to the company's repair department will have to be made. The sweat battens can usually be taken care of by the ship's force in the manner described above, and the work and trouble are negligible. If the batten is missing above a height that can be easily reached from the tank tops, the cargo provides a convenient platform upon which to work.

Limber System

The term *limber system* refers to the drain pipes and drain holes that constitute the means for carrying off all liquids to the drain well in a hold. In other words, "limber" is a synonym for "drainage." The limber system or drainage system must be checked each time a ship has been discharged. If a ship is on a run where cargo is always coming in and going out, the check must be made whenever practicable.

In the 'tweendecks, the principal thing to be checked is the drain pipe leading down to the ship's drain well in the after end of the hold. This drain pipe can be checked by pouring water from a bucket down the pipe and observing whether the water flows through or not. Obviously, two men are needed to see if the flow is obstructed or not.

If the flow is obstructed, the pipe should be cleared with an ordinary wire snake or by using a rubber plunger. If the pipe cannot be cleared by these methods, it must be removed and either renewed or cleaned out. Cargo should not be stowed in a 'tweendeck space where the scupper drains are plugged up. There are cargoes that must be stowed so that leakage from them is prevented from reaching these drains, but the drains themselves should never be plugged. One reason for this is because of the tremendous effect on stability that the free surface in the upper levels of a ship has. The free surface effect itself is the same whether high or low in the ship, but the added weight of the liquid trapped in the upper levels adds to a poor situation. If it becomes necessary to use a fire hose in the 'tweendeck where there are no drains, the water will create a condition that endangers the ship. One of the contributing causes to the capsizing of the *Normandy* in 1941 at a dock during the fighting of a fire in her in New York City was the fact that the water from the fire hoses collected in the upper decks and *remained* there. If the water could be drained to the lower levels through freely running pipes, the ship might have been saved.

These drains from the upper decks to the drain well are located in the after end of the hatch. This fact must be kept in mind when stowing certain cargoes. Cargoes that are known to drain heavily should normally be loaded near these drains. There are exceptions; for example, if the cargo is of such a nature that it would obstruct drainage from other cargoes, then it should be stowed away from the drains. In the latter case, however, it would be highly important to make sure that the dunnage was laid to form an efficient drainage system under the cargo stowed near the drain pipes. Lack of attention to such details on the part of stevedores or ship's officers may be the cause of costly damage to cargo.

Side Bilge and Drain Wells

In the lower holds, two types of drainage systems may be found. On older vessels (those built prior to World War II), the side bilge system is almost always found. The Victory ship built during World War II also has side bilges in the no. 2 and no. 3 holds and in the deep tanks of the no. 4 hold. When using the deep tanks for dry cargo, they should be stowed as lower holds with side bilges. In side bilge construction, the double bottom stops a few feet inboard from the hull plating and then continues downward at an angle of about 45° from the horizontal to join the hull plating at about the middle of the turn of the bilge.

In most modern ships, the tank top plating continues outboard in a horizontal plane and connects with the hull plating. A drain well is provided for draining off any water or other liquids that might find their way into the ship's hold. The drain well is located between the last two frames and floors of the hold. It drops down below the level of the tank tops to within about 18 in. of the outer bottom, runs athwartship for several feet, and stops about 18 in. inboard from the turn of the bilge. Figure 6-4 is a midship section illustrating the side bilge construction; Fig. 6-5 illustrates the modern construction.

When the ship has side bilge construction, it is important to check all of the limber (drainage) holes through the bilge bracket where the hull and tank top intersect. These holes should be clear. This bilge system is not open to the hold. It is usually covered with boards 2 by 12 in. in dimension. These boards are known as the *limber boards* and their purpose is to keep dirt and cargo residue from entering the limber system. They are removable for inspection purposes. It is necessary, therefore, to open up the bilges by removing these boards each time one inspects them.

In holds equipped with the drain well system, there is a perforated plate that covers the drain well and must be kept clean. If the holes

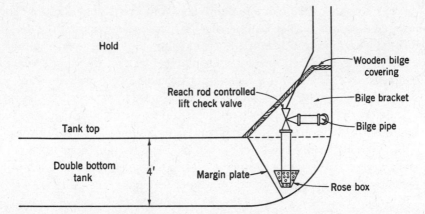

FIG. 6-4. Side bilge system. The rose box and bilge suction are located in the after end of the compartment. To seal off the side bilge system, it is necessary to cover the wooden bilge covering laid over the face of the bilge bracket so that the bulk commodity cannot leak through. This requires a good deal more care than sealing off the entrance to a drain well. More burlap is required, and it must be carefully secured in place by use of battens laid over the burlap and nailed down.

become plugged, no water or liquid cargo can drain into the well; hence, if the hull plating is pierced or water finds its way into the compartment in any way, there will be no way for it to be removed. It will back up into the hold and cause free surface, which will impair stability, or rise vertically and damage the cargo.

The liquid that finds its way into either the drain well or the after part of the side bilge is removed from the compartment through a bilge suction

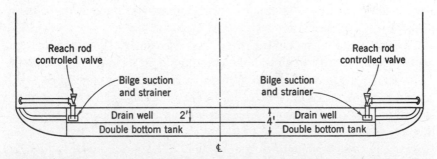

FIG. 6-5. Drain well. The drain well may run almost all the way across the ship as shown here or it may be only about 4 ft transversely. Its depth is about 2 ft, and it is located just forward of the after bulkhead of the compartment, extending forward about 2½-ft. The opening into the drain well from the hold is 2 to 4 ft long in a transverse direction. It is this opening that must be made dust tight but not watertight when preparing for stowage of bulk commodities.

pipe that is connected to a bilge pump manifold in the engine room. The pipe used to pump out the well or bilge is 3 in. in diameter and the end is not fitted with a bell. It terminates about $1\frac{1}{2}$ in. above the lowest point in the well or bilge. When checking the limber system, the ship's officer should always check the wire cage or perforated box covering the suction foot in the well or bilge. The wire cage or perforated box acts as a strainer of the liquid picked up by the suction, and it must be clean and clear. This strainer is known as the *rose box*. The rose box keeps rags, bits of wood, and other debris from being sucked into the bilge suction pipe. If the rose box is neglected, it will eventually become clogged; this will make it impossible to pump out the hold in case of leakage.

Stop-Lift-Check Valves

The bilge suction pipes have valves inserted in them just above the level of the suction foot. These valves are known as stop-lift-check valves. The action of this valve is as follows: If the valve stem is turned clockwise as far as it will go, the valve is closed. If the valve is opened one and a half turns, its action is the same as any lift-check valve. If the valve stem is turned counterclockwise for several turns, the valve is raised off its seat and acts as an open valve.

Ordinarily, this valve should be set to operate as a lift-check valve. That is, it should be backed off about one and a half turns. In this position, if suction is lost when pumping the bilges, the contents of the pipe will not be vomited back into the well. This fact makes it easier to regain suction and to retain suction when pumping liquids at low levels, which is almost always the case when pumping out the bilges and drain wells of cargo ships.

The valve should be closed when there is danger of flooding an adjacent compartment through the bilge system, as in the case of a collision. This may not be necessary unless the pipe is broken by the impact, but it should be done as a safety precaution. Part of every ship's collision drill should include the closing of these valves by crew members instructed in the procedure.

The valve will have to be opened wide if it becomes necessary to pump water into the hold through the bilge system. This may be the only way to extinguish some types of fires.

This valve should be checked to see that it turns freely each time that it becomes accessible after discharging cargo. It is important to reset it, however, as required by the chief engineer. It may be opened wide in some cases so that the lift-check feature is not utilized.

Reach Rods

Obviously, stop-lift-check valves cannot be reached when the hold is filled with cargo. To operate them from the deck, they are fitted with *reach rods*. These are simply extensions of the valve stem running up through the main deck through watertight glands and fitted with a wheel for turning. They must be checked to see that they are not broken and that they turn freely. If they are rusted and frozen, the ship may be placed in a dangerous condition in case of an emergency. It is recommended that the reach rods and bilge suction valves be checked at the same time. This can be accomplished by one man turning the reach rod from the weather deck level while another man observes the action in the hold.

The 'tweendeck drains, side bilges or drain wells, rose boxes, bilge suction valves, and reach rods make up an important group of items all related to keeping the drainage system of the ship functioning in the best manner. These items should be checked each voyage and if any one is not working correctly it should be repaired before cargo is loaded.

Bilge Sounding Pipe

To determine whether a given compartment is taking water, each one is equipped with a pipe running from an accessible upper deck down to the drain well or bilge called the *sounding pipe*. At regular periods, an iron or brass rod is lowered down this pipe until it strikes the bottom of the well or bilge. The rod is coated with chalk, and if any water or other liquid is in the bilge its presence and depth will be indicated on the rod. This chore is performed by the ship's carpenter or designated AB seaman and is called *sounding the bilge*. It should be done at least twice a day under ordinary circumstances, every half hour when taking on fuel oil or other liquids in the ship's tanks, and immediately after a period of heavy weather. There may be times when the bilges cannot be sounded because of heavy seas placing water on deck continually; during such times, orders should be given to the watch engineering officer to pump all the bilges at least once a watch. Such orders would generally be given by the master of the ship through the chief engineer.

The Striking Plate

After several years of daily sounding, the steel directly below the bilge sounding pipe will wear away. If a pinhole develops in this spot from wear and corrosion, oil or water from the tank below can enter the hold. The amount of oil that can enter the hold might be sufficient to cause damage to cargo in the lower tiers, especially if the tank were pressed up and the ship had a large trim by the stern. Another bad feature of such a hole is the possibility of contamination of any drinking water carried in the tank below.

To guard against the possibility of the steel directly below the bilge sounding pipe wearing out, this area is covered with a small doubling plate called the *striking plate*. This plate does not wear out quickly and hence need not be inspected every time the drainage system is checked. However, every officer joining a ship should satisfy himself that this plate is not worn through on both sides of every compartment. This can only be done by opening up the drain well and sighting the plate or running a finger over it to feel the amount of wear. If it is badly worn, it should be renewed at the first opportunity. It is not necessary to renew the entire plate; the worn spot can be filled with a bead of weld metal.

Electrical Conduits and Outlets

Conduits and outlets can also be checked as the cargo is being worked. Whenever a defect is noted, a repair requisition should be sent in immediately or the defect should be taken care of by the ship's force.

Ladders

As a safety measure, a broken rung in a ladder should be repaired as rapidly as possible. If a ladder is damaged during the cargo operation and cannot be repaired immediately, the longshoremen must be provided with a Jacobs ladder or other means of climbing into and out of the hold. Rungs can be replaced very easily by the ship's force. The uprights of the ladder can be bored with a $\frac{5}{8}$-in. hole and a $\frac{1}{2}$-in. steel bar, threaded at both ends, can be secured in place with nuts. This may be used as a temporary repair measure, providing the uprights are not bent out of line. Ladders are frequent casualties in ship's holds and their repair goes on continually.

Rivets, Seams, and Pipes

The rivets and seams of the hull plating, tank top plating, deck plating, and the plating of the transverse bulkheads forming one side of a tank should be surveyed constantly by the ship's officers as they make their frequent visits to the holds. Prior to loading cargo, a special inspection should be made for evidence of leaks. Leaks may be spotted by the presence of rust streaks or oil traces. Such measures also apply to pipes running through the compartment, especially the flanges of such pipes.

Manholes and Side Ports

These openings into the hold should be checked more frequently than any of the other items mentioned. Obviously, they should be checked when they are closed and made ready for sea. If any of the dogs are difficult to turn, they should be made free at the first opportunity, which may not be at the time of closing. The gaskets should be kept clean and intact. If gaskets are made of flax or woven fiber material of any kind, they should be kept oiled and free of paint. If they are made of rubber, they should be kept clean and free of oil, grease, and paint. The knife edges against which the gasket presses when the door of the opening is closed should be clean and free of broken or jagged places.

While at sea, side ports should be checked twice daily. This is not of great importance when the ship is light and the weather is fair, but if the ship is loaded and the seas are running up the hull and placing hydrostatic pressure over the side port as the ship works in the seaway, it is extremely important. If possible, some means of gaining access to the side port without having to open a manhole on the weather deck should be made. This can be accomplished by building *crawl holes* in the cargo in the shelter deck spaces leading from entrances other than the manhole openings on the weather deck, if such openings exist.

If a side port is found to be leaking while at sea, it should be caulked with oakum. Directly in the way of most side ports, a drain will lead to the drainage system of the hold. This drain should be kept clear so that in case of leakage the entering sea water can be drained off quickly. In such a case, it would be common sense to have the bilges serving the compartment pumped frequently by the engineer officer on watch. In very bad weather it may be impossible to have the bilges sounded periodically because of water on deck and generally unsafe conditions in the exposed area of the bilge sounding pipe opening on the weather deck.

Firefighting Equipment

The steam smothering or carbon dioxide outlets should be checked to see that they are not broken and are clear. The inlets to the smoke detecting system should be checked also to see that they are clear and in good working order.

Hatch Covering Equipment

It may not be necessary to place this item on the checkoff list; however, it is important to have all such equipment in good order. If tarpaulins are used, there should be at least three and they should be without holes or weak seams.

Ventilation Ducts

The most important thing that needs checking with respect to the ventilation ducts leading into the holds is that the system of closing them off from the outside atmosphere is in good working order. In the case of fire, it is imperative that all openings to the atmosphere be closed and kept closed. In some instances, it may be necessary to close these openings because of heavy seas coming on board. It should be seen to, of course, that these same openings into the hold are clear so that the hold can be ventilated correctly during the voyage.

THE STOWAGE OF BREAKBULK CARGO

This chapter is entitled "The Stowage of the Cargo," but it is impossible to deal with the stowage of all types of cargo in this single work. In the remaining pages of this chapter we will discuss the stowage of a number of important cargo types. We will also deal with the problem of calculating space consumption using the stowage factor. The discussion will be based on the principles already covered and should serve to illustrate their application. The cargoes selected for discussion are not necessarily any more important than hundreds of others that might have been chosen;

however, some selectivity had to be exercised and those covered do bring out a number of interesting points. A full discussion of a large number of commodities will have to await the writing of a book with that objective. Such a discussion cannot be an adjunct to a book dealing with fundamental principles.

The Stowage of Bags

The preparation of the hold for the stowage of bags commences with the laying of a dunnage platform upon which to stow the bottom tier. This platform is primarily for drainage purposes, although in some cases it is equally important to keep bagged cargo from contacting the steel deck. If the ship is equipped with side bilges, the first layer of dunnage should be laid down running athwartships or, preferably, at an angle to the keel. If laid at an angle to the keel, the forward end of the dunnage should be inboard.

If the ship is equipped with drain wells in the after end of the hold, the first tier of dunnage should be laid down running fore and aft. The dunnage should be spaced not more than 1 ft apart.

The second tier of dunnage should be laid at right angles to the first, except that if the first tier is laid diagonally the second tier may be laid either fore and aft or athwartships. For bagged cargoes, the dunnage of the second tier should be spaced not more than 1 in. apart. If the dunnage is too widely spaced, the pressure of the top tiers will cause the bags on the bottom tiers to split.

If ventilation is not a major problem with the commodity in the bags, the dunnage of the second tier should be placed flush together. This eliminates any possibility of the bottom bags splitting and requires very little additional dunnage. If sufficient dunnage is available, the drainage tiers of dunnage should be 2 or 3 in. high for commodities requiring maximum ventilation, such as rice and soy beans. This provides a series of ventilation ducts under the cargo.

The stowage of the bags may be in any one of three ways, depending on the nature of the commodity in the bags. If the commodity requires maximum ventilation, the best stowage is one bag directly on top of the one below, laid with the length fore and aft (see Figs. 4-21a and 6-6b). This stowage will provide small air spaces throughout the stowed cargo block and help in the circulation of air. It is difficult to get longshoremen to do this consistently as the stowage proceeds; it generally requires careful explanation of the need and desirability of care in placing the bags in the hold and the presence of an officer after stowage commences. Stowage in this manner is known as *bag on bag* stowage.

If the commodity does not require maximum ventilation, the best stowage is to place the upper bags between two lower ones, which results in a brickwork style of stowage. This stowage eliminates the spaces between rows and tiers and reduces the broken stowage percentage. If a full hatch is being stowed, it might mean that one complete additional tier can be stowed. This type of stowage is referred to as *half bag* stowage (see Fig. 6-6).

When stowing bagged cargo in small blocks in which the stability of the block is a factor, it may be advisable to stow one tier of bags fore and aft and the next tier athwartship. This stowage ties the tiers together so that the stowed block is less likely to fall over while working the cargo in port or when the ship is rolling and pitching in a heavy sea. Such blocks should not depend entirely on this cross tiering for their security; they should be shored up or blocked up by adjacent cargo as well. It may not be necessary to cross-tier the bags throughout the stowed block, but simply to do this on the last two rows of the block. Still another way to add to the security of a bulkhead of bagged goods is to place a strip of dunnage along the outside edge of every other tier as the loading progresses vertically. This will tend to tip the bulkhead toward the stowed block or at least to lessen the tendency for the face of the bulkhead to lean away from the stowed block.

FIG. 6-6.   (*Opposite*) (*a*) Bag stowage. This is a view of bag stowage extending out into the hatch square for a little more than one section. It is not the best stowage. The bulkhead has been carefully built vertically and looks fairly stable; however, by running every other tier of bags along the bulkhead face with their lengths athwartship and tying the bulkhead bags into the stow, a much more stable bulkhead could have been made. The effort to do this would have been no more than that of building the bulkhead as shown here. Almost all of the bags in this bulkhead are running longitudinally and are built up half-bag style. Thus, these bags are more or less a unit and can more easily fall away from the rest of the stowage in really heavy weather. Of course, if other cargo will always be bearing against this bulkhead or if no bad weather is encountered, no damage will be done. However, cargo should not be stowed so that a good out-turn depends on so many uncertain factors, especially when there is no difference in the amount of effort required. Courtesy of U.S. Maritime Commission. (*b*) Bag-on-bag method (top) for maximum ventilation and half-bag method (bottom) for maximum use of available cubic. Note dog ears for longshoremen to handle bags. Courtesy of J. Leeming, *Modern Ship Stowage,* U.S. Government Printing Office, Washington, D.C., 1942.

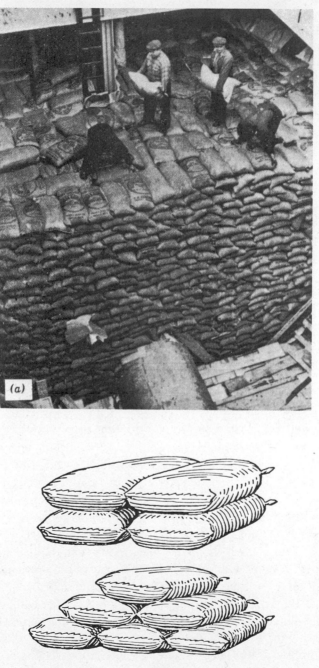

(a)

(b)

261

Although there are exceptions, generally no dunnage is used between tiers of bagged cargoes. Venetian vents are sometimes used *within* a stowed block of commodities needing heavy ventilation (see Fig. 4-22).

The only dunnage used besides the drainage floor is vertical dunnage laid against the transverse bulkheads, the sweat battens, and all vertical structural members, such as stanchions, ladders, and ventilator ducts. This vertical dunnage is used to keep the bags from contacting the steel members of the ship. For additional protection and cleanliness, the vertical members may be wrapped with heavy paper and tied with rope yarn.

If the bags contain coffee, cocoa beans, or any other commodity that may leak and must not touch the ship's deck, the bags should be stowed over separation cloths. Any leakage will be caught in the cloth and can be rebagged before being contaminated. The use of these separation cloths is also advisable when stowing bagged commodities over other cargoes. The most efficient covering for this purpose is made of burlap. The burlap comes in strips about 4 ft wide, which must be cut to reach the entire distance across the stowed block with enough slack to enable tucking them down the sides of the ship about 2 ft. These cloths should never be stretched tightly over the cargo, but rather should be left with some slack.

Hooks used on some bags will result in tears that allow the contents to leak out. On loosely woven bags a hook can be used without tearing the bag. The hook will slip through the fibers and out again without actually leaving a hole for the contents to leak through. The commodity must be large enough to allow the use of such bags, such as cocoa beans. If a bag is tightly packed and heavy (over 100 lb), there is little doubt that the longshoremen will use hooks on it. They cannot get a grip on such a bag with their hands and to handle it they are forced to use a hook. If the shipper packs the bags with corners tied to form a place for the longshoremen to grab, known as *dog ears* (see Fig. 6-6b), hooks will not be used as readily. Paper bags of cement or plaster are very vulnerable to the hook, but they are often small enough and light enough to allow the longshoremen to lift them easily. Every effort should be made to prevent hooks being used on such cargoes. A hook with several small slightly curved prongs known as a *bag hook* may be issued to longshoremen handling fiber bags if it is desired that they not use their larger cargo hooks. This will help reduce hook damage in some cases. The bag hook should not be used on paper bags.

In the outboard tiers, bags may be stowed athwartships or placed on end, providing the vertical dunnage cannot be used. The reason for placing the bags athwartships is to present only the relatively small surface at the end of the bag to possible damage from contact with the frames. Standing the bags on end presents a surface to the sweat battens that is too large to protrude through them and is a device that may be used in lieu of vertical dunnage. If the dunnage is available, it should be used.

The bags must not be allowed to rest on the edges of the sweat battens, stringers, upper ends of vertical dunnage, or other surfaces that might be present along the periphery of the hold. If they rest on these surfaces, they will be torn as the cargo settles during the voyage.

In spite of all precautions, some bags will be torn in almost every shipment; hence, it is important to use separation cloths as mentioned above to catch leakage. If the contents that leak out are caught in the separation cloths and rebagged, they are known as *ship fills*. If the contents are not caught but allowed to reach the deck of the ship or dock and are swept up and rebagged, they are known as *sweepings*. Ship fills or sweepings are distributed to the consignees having that type of cargo in the hold in proportion to the total amounts they had originally. Ship fills may be handled by the consignee in the same manner as the rest of the original shipment. Sweepings must be condemned as unfit for eating and either used for the oil content or re-exported.

The Stowage of Bales

In stowing bales, a drainage floor is prepared as described for bags. Vertical dunnage is not as important in the wings as with bags, but it should be used against transverse bulkheads and vertical structural members in the manner described for bags. Bales contain either unprocessed fibers or strands of a commodity or a flat material shipped in layers. Examples of the former are cotton, sisal, jute, wool, and manila. Examples of the latter are cotton piece goods, tobacco, skins, and dry goods.

Bales are generally parallelepiped in shape (Fig. 6-7). If they contain fibrous commodities, they may be placed in the ship on their flats, edges, or crowns and no harm will come to them. They stow more securely, however, if stowed on their flats in a fore and aft direction. All cargo should be stowed to conform with the shape of the ship unless there is a good reason for not doing so. One good reason for deviating from continuous flat stowage in the case of bales is to use up as much of the cubic as possible in the compartment. Therefore, it is important to measure the hold clearance vertically and stow the bales so that the top of the last tier

FIG. 6-7. Wool bales in stowage and on the sling. Wool bales do not need to be stowed carefully on their flats because the contents are fibers and also because this type is almost a cube and does not have a pronounced flat area. These bales are mostly stowed on their flats nevertheless. Note that the clearance between the top bales and the deckhead is very small. This is good, for the lost stowage has been reduced. Note the special type of sling consisting of eight snorters with two bale hooks in the eye. Each bale is picked up by two hooks, eight bales being hoisted at once. Courtesy of Boston Port Authority.

will come as close as possible to the underside of the beams. It may be that all the bales should be on their flats or edges on all but the last tier, which may have to be placed on their crowns.

If the contents of the bale are in layers, the bales in the first tier are laid down on the dunnage platform on their flats. The outboard bales in each tier should be placed on their edges so that their flats will be presented to the sweat battens. This precaution is necessary to lessen the damage in case of chafage. If the bales chafe or become wet on their flats, only one or two layers will be damaged. If the edges or ends are damaged, every item in the bale is affected. After the bales in the periphery are stowed, it is not necessary to stow the remainder on their flats. In fact, the remaining distance should be measured and the bales should be stowed so that the greatest amount of space is consumed by the cargo and there is a minimum of lost space over the last tier. The bales that bear against stanchions and vertical ventilation ducts should be placed with their flats against these members with dunnage between them and the structural members.

Some of the fibrous commodities may present a danger of fire. Two apparent causes of fires in such commodities are: (1) spontaneous heating until the kindling point is reached, resulting in *spontaneous combustion;* and (2) sparks caused by friction of metal bands used on bales against steel parts of the ship. The first cause of fire can be eliminated by making certain that all bales are dry when stowed and that they remain that way. They must also be free from oil. The second cause can be eliminated by making certain that dunnage always stands between a bale and the metal parts of the ship. In 1978 a cargo hold of cotton bales caught fire in this manner owing to the lack of dunnage between the bale bands and the ship's stanchion.

Carelessness in the use of matches and smoking in the hold is undoubtedly the cause of some fires. Sparks from electrical fixtures in the hold or carelessly placed cluster lights are also responsible for some fires. These causes, however, are simple safety violations and not a direct result of the stowage.

When baled commodities are received, exceptions should be taken if the bands holding them together are broken or missing, if the covering is ripped off and part of the contents is exposed, or if the covering is hanging from the bale in loose strands. The cotton bale, for example, is supposed to have at least three-quarters of its surface, including the ends, covered by burlap or other suitable cloth covering. The bands are spaced about 8 in. apart. During the handling process from the warehouse to the ship, some bands may be broken and the covering ripped and tattered. If this happens, the possibility of fire from all sources mentioned above is increased.

The Stowage of Cardboard Cartons

The dunnage platform on which the stowage of cardboard cartons commences is the same as for bags except that the spacing of the second tier may be up to 4 in. The stowage of cartons may start at the centerline and proceed outboard, or vice versa. Probably one of the most important points to make about the stowage of cartons is the care needed to maintain the level in the wings. As the stowage proceeds outboard in the lower holds, care must be taken not to stow a carton beyond the level portion of the tank tops. That is, a row should not be stowed upon reaching a point where the next carton, when placed in position, will rise above the level of the rest of the tier. In place of this carton, dunnage must be laid to fill in the triangular space on the outboard side. The next tier can then be stowed. This procedure must be used in the wings until all of the curvature has been stowed under.

The cartons are placed in the hold so that each carton rests on two cartons below it. This results in stowage that is known as *brick fashion* stowage. The dunnage in the wings should be placed against the sweat battens in a vertical position. Dunnage should also be used to separate the cargo from the transverse bulkheads and vertical structural members. If the vertical dunnage is not used in the wings, some of the cartons may become hung up on the sweat battens and as the ship works in a heavy seaway the side of the carton will be sheared off from chafing and superimposed weight from above.

The need for dunnage floors between cartons in deep lower holds is explained on pp. 195 and 196.

The tiers must be kept level. This requirement is important. Dunnaging in the wings is one of the ways that this is accomplished. Obstructions in the compartment require care in the use of dunnage to keep the tiers level.

The Stowage of Crates

The term *crates* as used here refers to wooden containers that are built as a framework with open sides and tops. These crates are generally stiffened by the use of diagonal pieces, but are, for the most part, open so that the contents are exposed. Crates without diagonal stiffeners are insufficient for ocean transportation. Refrigerators, stoves, and light machinery are shipped in crates. The bottoms are solid, with well-built foundations for supporting the contents. The crate is a poor container, because it lacks strength and lacks protection for the contents. Crates require a layer of dunnage between every tier. The dunnage may be spaced up to about 4 in. apart. If a shipment consists entirely of crates, the best place to stow it is in the 'tweendecks. Top tiers in the lower hold are suitable, providing, of course, that the contents of the crates are not liquid. If a shipment of crates is stowed in the lower hold and it is necessarily placed on the tank tops, the height to which it is stowed should be limited to 12 ft or about half the depth of the lower hold. Cargo stowed over these crates should be lightweight merchandise.

Because of the dunnage layer between each tier of crates, it is not necessary to stow them brick fashion. This is an advantage because it means that in the wings the crates will rise vertically without the stepping in and out resulting from brick fashion stowage.

The Stowage of Cases

The word *case* as used here refers to a wooden box that is sheathed to form a tightly closed container. A wide variety of commodities are shipped in cases. Not only refrigerators, stoves, automobiles, and machinery, but also paints, drugs, beverages, and foodstuffs are shipped in cases. Uniform-sized cases may be stowed brick fashion in the same way as cardboard cartons without the dunnage floors between tiers in the bottom of lower holds. The wooden case depends on its contents for part of its strength, as do all containers, but unlike the cardboard carton, dampness will not weaken it.

Depending entirely on the contents, cases and cartons may at times be placed on their ends or edges if this procedure will facilitate the stowage without endangering the cargo. Another important point to make with respect to the stowage of cases and cartons is that when their length is twice as long or approximately twice as long as their width, security of the stowed block is obtained by placing one tier with the length fore and aft and the next tier with the length athwartships. The pattern of the stowage would be as in brick fashion stowage, but the stepping in and out of the outboard edges would be avoided. This is the manner in which such items are usually stowed, because most cargoes are not constructed with a square base. In some cases the direction of the cargoes in any one tier will be changed so that the division of the commodity is staggered.

Most of the comments above refer to shipments of crates or cases of uniform size. If a cargo lot consists of a number of variable sizes, the stowage precautions include some additional points. First of all, those cargoes with the heaviest items should be stowed in the lower tiers. Secondly, a smaller crate or case must never be stowed within the perimeter of a larger one below it without dunnage to support the upper crate. Such dunnage should extend from at least one side of the lower case to the other, but preferably the dunnage should extend over two or more cases. The upper crate or case is best stowed over two lower ones.

As the stowage of variable-sized crates and cases proceeds, the longshoremen should note the variations in height of the top of the stow and use dunnage to level off small or large sections, whichever is possible.

The use of dunnage in the stowage of such cargo should be extensive. Piling additional cargo on top of irregular platforms of cargo in the lower tiers invites heavy crushing damage. The crates and cases may be fitted together in much the same manner as a set of odd-size blocks, but in so doing the probability that uneven pressures will cause failure of containers of cargo and damage to contents is greatly increased. The use of ample amounts of dunnage to fill in slight steps and to floor off large sections is mandatory for the safest stowage. The lack of such dunnage is improper stowage.

The Stowage of Casks

The word *cask* refers to a type of container. The words *barrel, puncheon, pipe, butt,* and *tun* refer to sizes of the cask type of container. The term "barrel" is often used to mean all containers of the cask type. The actual amount of liquid contained in the various types of casks varies widely depending on the type of liquid being carried and the trade in which it is being handled. To give some idea of the relative sizes of a few of the commoner casks and the names used, the following table has been prepared:

Cask	Capacity in U.S. Gallons
1 barrel	$31\frac{1}{2}$
1 hogshead	63
1 pipe	126
1 butt	126
1 tun	252

It must be emphasized that these volumes are not universal. For example, the English barrel of beer has a volume of 36 U.S. gal whereas the $31\frac{1}{2}$ gal shown in the table is the volume of a barrel of wine. A butt of beer in England has a volume of 108 gal, whereas a butt of wine has 126 gal. The volumes of the various casks are so variable that there is little point in trying to remember any of them. When concerned with the stowage of such items, the officer should be certain he determines correctly the volume of the cask listed on the manifest or bill of lading before calculating the space required to stow it.

The U.S. gallon is the equivalent of the English wine gallon, 231 in.^{3}. The Imperial gallon consists of 277.418 in.^{3} or 1.2 U.S. gal. One U.S. gallon is equal to 0.8331 Imperial gal.

Because of the frequency with which the term "barrel" is used when speaking of the capacity of tank ships, it should be pointed out that the barrel used in the petroleum industry has a volume of 42 U.S. gal.

The stowage of cask type containers is the same whether they are barrels or tuns, except that the height limitation is less as the container grows larger. There are also two broad categories of casks. There is the wet or tight cask made of oak. The construction of this type of cask must be done with great care. The scarfed joints of the staves must be precisely cut and the staves must narrow from the bilge out to the head with great precision (see Fig. 6-8). This cask is used to ship beer, wine, spirits, and olives. The aluminum cask is now utilized much more extensively than the tight wooden cask.

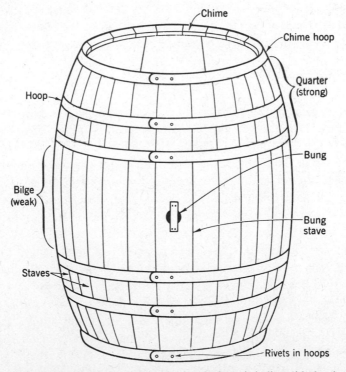

FIG. 6-8. The nomenclature of a cask. Note that the bung is in line with the rivets on the hoops and that the head staves are vertical when the bung is up.

The other type of cask is the dry or slack type cask. This type is usually made of softer wood than the former, fir being widely used. The joints and other construction details do not require the care and precision of the tight cask. This type of container is used to ship flour, cement, and other dry products.

The weakest part of the cask, whether wooden or of aluminum construction, is the bilge. The weight should never be placed on this part of the cask when it is stowed. The strongest part of the cask is the quarter. The stowage of wet or tight casks, if in small lots and when there is no need to overstow them with heavy cargo, may be on end with a layer of dunnage between the tiers. The same stowage should always be given to the dry or slack cask. When stowing casks on end, the most important things to remember are to protect the chimes of the lower tiers by using strips of dunnage between tiers and to prevent excessive weights by limiting the weight of cargoes stowed over the casks. In most lower holds it is possible to stow barrels about seven tiers high. This height is too great if barrels are stowed on end; six tiers is the maximum. This type of stowage may open up the seams of the casks, beginning along the bilges. The lower tiers of casks will leak first because they will be supporting the most weight.

When carrying large numbers of the wet or tight casks, or when the best possible stowage is desired, the casks should be stowed as follows. The cask is laid in the ship with the length running fore and aft and in regular athwartship rows. The bilge of the cask is raised off the deck by dunnage at least 2 in. thick laid under the quarters. The cask is secured in position to some extent by the use of small wedges placed on either side of each barrel at both ends. If dunnage and wedges are not used, a more elaborate and safer method is to use *quoins* (Fig. 6-9). These are chocks or supports cut to fit the curvature of the cask at the quarters and raise the cask vertically to clear the bilge from the deck. This type of stowage results in what is sometimes referred to as *bilge free* stowage.

In the wings, great care must be given to wedge the outboard cask securely into position with dunnage and bracing. Assuming that we are starting the stowage in the after part of the hold, the next cask forward must be precisely in line with the adjacent cask. Their chimes should fit together. One must not be inboard or outboard of the other nor higher or lower. As the stowage proceeds longitudinally, the curvature of the hull in the entrance or run of the ship will make it necessary to drop a cask from

FIG. 6-9. The use of quoins. The quoins are placed under the first tier at the quarters of each cask. Courtesy of U.S. Department of Commerce.

the outboard longitudinal rows and fill in with dunnage. The longitudinal rows should be precisely in line fore and aft.

The second tier should be stepped one half the length of the cask away from the bulkhead against which the bottom tier was stowed and a half diameter athwartship. No dunnage need be used between the first and second tiers. The movement of the second tier forward and athwartships will place the bilge of the second cask over the intersection of four lower casks and the quarters of the upper cask will rest on the quarters of the lower cask. Hence, the bilges of the upper casks will be free without the use of dunnage (see Fig. 6-10).

Before stowing the third tier, dunnage should be placed in the space between the bulkhead and the chime of the second cask. This dunnage should be installed to support the quarter of one end of the third cask while the other quarter is supported by the second cask. Thus the stowage proceeds upward. It is of great importance to stow the bottom tier securely and to pay careful attention to dunnaging the wings and ends of the hold.

The curved intersection of two casks is known as the cantline of the stowed cargo. Hence, this system of stowage is referred to as *bilge and cantline*. However, the casks are stepped one half of their lengths fore and aft, which gives the more descriptive phrase *bilge and cantline half-cask stowage*. If the casks are stowed bilge and cantline *full cask,* the

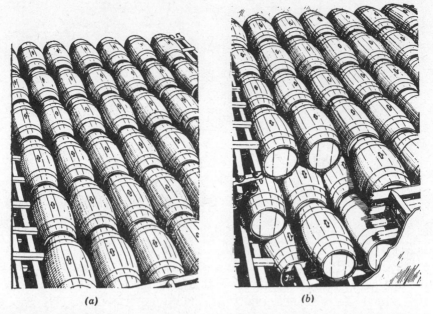

(a)   (b)

FIG. 6-10. Stowage of casks. (*a*) The stowage of the first tier with dunnage on the sides and one end as necessary, with all bungs up. The safety of the entire stow depends on this first tier. These casks are resting on athwartship pieces with bilges raised off the deck and quoins to keep them secure. (*b*) The stowage after the second and third tiers have been stowed, with dunnage being used to chock and fill in. All of the upper casks rest in the cantlines of the lower casks, with four lower casks supporting an upper cask. Courtesy of U.S. Navy.

bilges would not be free without additional dunnaging. Full-cask stowage is easier, however, because it obviates the need for the dunnage at the ends of every other tier. It may be used safely if the casks are empty.

The bung on a cask is placed at the bilge in line with all the rivets holding the hoops together. The rivets and bung are always along a side stave in line with the direction of the head staves. Hence, if the end staves are vertical, the bung is pointing either directly up or directly down. A glance to see if the rivets are on top will tell whether the bung is down or up. The bung should always be up in stowage.

If the casks are properly stowed in this bilge and cantline half-cask style with the bung up, the stowage results in what may be described as *bung up and bilge free* stowage. However, this phrase tells only *the result* of proper stowage, not the method.

Occasionally, shipments of casks will be in two or three sizes. If it is desired to give these shipments the best possible stowage, they should be stowed bilge and cantline half-cask style with necessary modifications. It may be necessary to use dunnage and wedges between tiers to give proper support and security when smaller casks are stowed on top of larger ones. Also, instead stepping the third tier back against the bulkhead, it may be expedient to continue upward forming a pyramid, with dunnage and bracing or suitable filler cargo being used to fill in the space left between the casks and the bulkhead. It is recommended that pyramiding be avoided by providing a platform of dunnage for stepping the third tier back against the bulkhead.

The Stowage of Cylinders

Flammable and nonflammable compressed gases are shipped in steel bottles commonly referred to as *cylinders*. These containers are under pressures of 2000 lb/in.^{2} and over. They must be handled carefully and stowed securely to preclude disastrous accidents.

Cylinders may be stowed on deck or under deck, but the distance from the side of the ship is regulated. Also, certain cargoes must not be in the same hatch with flammable gases (see Table 6-3). In general, such gases should not be stowed with cargoes that are liable to give off heat.

Cylinders should be stowed on their bilges in a fore and aft position; the second tier should rest in the cantlines of the lower tier, reversing the direction of the ends in each tier. They should be placed over dunnage and, inasmuch as the cargo is fore and aft, the dunnage should be laid diagonally if the cylinders are stowed on deck with only one tier being used. When stowed below deck, two tiers of dunnage should be laid, with the first tier fore and aft or diagonal and the second tier athwartships. For on-deck stowage, 2 by 4 dunnage spaced every 4 in. is suitable.

Cylinders are shipped in three forms with respect to the way the inlet valves are protected. They are equipped with a screw cap cover or a dished head or they are boxed. The inlet valves must be protected during shipment by one of these three forms.

Under certain conditions, it may be more convenient to stow cylinders standing up instead of lying down bilge and cantline style, for example, when clear deck run is lacking or is at a premium and needed for other deck cargoes. Standing cylinders on end is permissible only when extremely sturdy racks are built out of 2-in. lumber and thoroughly lashed down with wire rope, chain, or steel strapping with turnbuckles on every turn over the stowed block. The clear deck run can be saved by building these racks just forward of or abaft the hatches and well inboard to clear the deck run. If it can be prevented, however, cylinders should not be stowed on end. They are long, narrow, and extremely unstable. They stow very naturally in a fore and aft bilge and cantline fashion, and that is the way they should be placed if possible.

When stowed in the bilge and cantline manner, they must be securely lashed, braced, and tommed. Tomming, of course, is only possible when they are stowed in the 'tweendecks. Tomming is not necessary if the cylinders are extremely well lashed, but if there is any doubt as to their security, it is recommended that plans be made for setting in some toms. Tomming is also not necessary if sufficient heavy cargo is stowed over the cylinders and this cargo is well secured.

When handling cylinders, it is important that the valves be protected as much as possible in addition to the protection afforded by caps or construction. At a pier in New York in 1952, a sling load of cylinders was carelessly discharged from a hold. One cylinder's cap was extending beyond the edge of the platform sling and was struck a blow sufficiently strong to break the cap and the valve. The cylinder immediately took off in flight, propelled like a rocket by the expelling of the gas under pressure. Fortunately it went up instead of down. It spent itself rapidly, but not before it had gained enough velocity to carry it down the waterfront and land it on the roof of a pier warehouse two piers from where it took off. This gives an idea of the power and potential danger in these charged cylinders.

As mentioned before, these cylinders are under pressures of 2000 lb/in.^{2} and over at normal temperatures. If the temperature rises alarmingly because of fire, direct sunlight on a tarpaulin lying next to them, heating copper concentrates, or other possible causes incident to water transportation, the pressures will be increased. It is quite possible for the pressure to be raised so high that the cylinder will blow a safety plug or, if that fails, that the entire cylinder will blow up. This emphasizes the need for preventing an undue rise in temperature. If carried on deck, cylinders should have dunnage between them and any tarpaulins over them to provide circulation of air. They should never be stowed near cargoes likely to heat in stowage. If fire breaks out, these cylinders should be jettisoned if they are in danger of being affected by the heat. The latter action would be especially important if the cylinders contained hydrogen or other flammable gases. Such an action would be for the safety of the ship and all the other cargo on board.

The Stowage of Drums

Drums may be stowed directly on the deck of a ship if stowed below the weather deck; however, a single layer of dunnage to provide better drainage and to offer more friction in case of heavy rolling is strongly recommended. Drums are stowed on end with the end bung up and packed as closely together as possible. If the distance athwartships allows stowage of an integral number of drums plus a fraction of one, they should be spaced at regular intervals instead of being close together, except for the end drum. The next row should be set in the oversized cantlines formed by the above action. The third row should again be set tightly into the cantlines (vertical) of the second row, and so on. This eliminates the need for any bracing or additional dunnaging in the wings, providing the wings are vertical themselves. If the wings have curvature, the triangular space in this part of the stowage should be filled in with filler cargo, bracing, or dunnage (see Fig. 4-23). Dunnage stripping should be placed between every tier. It is not necessary to use solid dunnage floors. Two strips of dunnage over a single row of drums is sufficient. These two strips will tie the stowed cargo block together, spread the weights satisfactorily, and prevent bending of the drum chimes.

In the wings, plenty of dunnage should be used to fill in the space left between the last outboard drum and the ship's side if there is curvature such as is found in the entrance and run of the vessel. A level must be provided for laying the dunnage strips for the next tier of drums. The drums are usually the 55-gal size, and the height to which they can be stowed depends on their contents and their construction specifications. On most ships, this height limitation is naturally enforced by the limit of vertical drift in the lower holds. Four tiers is the maximum height that such drums can possibly be stowed in most 'tweendeck compartments. This particular drum is usually stamped with the letters "S.T.C." on the head, meaning "single-trip container" and usually "single trip with flammables." Such containers are often used a second time for the shipment of vegetable oils.

The stowage position for drums on the ship is often limited because of the contents. There are also a number of commodities that cannot be in the same hold. These will be discussed later.

The Stowage of Reels

What is mentioned hereafter with respect to reels also applies to any heavy cylindrical object, such as a millstone. The stowage of such items should always be on their sides unless they are light enough to be capsized easily by one or two men. The type of reel envisaged here is the extremely heavy reel with some type of cable on it, sheathed and generally marked to be rolled in one direction only. Such a reel should always be stowed with its axis running athwartships. Thus, the curvature is fore and aft. This requirement is simply to lessen the possibility of the reel exerting sufficient force to break its lashings during heavy and continuous rolling in a seaway. A ship will roll, pitch, and yaw in a seaway, but the rolling motion will be the most violent. The reel could work loose with long periods of continual stress due to extended bad weather. It might chafe lashings in two or pulverize blocking cargo. This may happen when the reel is stowed with the axis in either direction, but the probability is less with the axis athwartships.

The safest level at which to stow reels is in the lower holds. This is because the rolling motion is not as violent here as in the upper decks. The reels should be blocked in place with 8- by 8-in. balks on both sides. It is preferable to have the timbers cut to fit the curvature of the reel. The timbers should be tied together by nailing 2- by 6-in. battens between them. Besides this chocking, the reel should be braced, shored, and lashed into position. If suitable cargo is available, the reel may be blocked in with cargo. Some examples of suitable cargoes are bales of wood pulp, rags, hay, or certain types of lumber. Cargoes that cannot stand some chafing without being damaged should never be used. If suitable cargo for stowing around the reel is not available, a bulkhead of dunnage should be provided against which other cargoes can be stowed adjacent to but safely protected from the potentially dangerous reel. Reels should be given about 6 in. clearance by such bulkheads.

Reels generally are well marked with precautionary measures that the shippers want taken to protect the contents. One of the most important of these, often seen on sheathed reels, says to be careful when using nails to build bracing or shoring around the reel. Some of the cable contained on these reels is enclosed in a tube that is filled with a gas to cut down corrosion and general deterioration. If the cable exterior is punctured by a nail, the entire reel is ruined, and this will result in a very costly claim. This type of damage would almost certainly be adjudged improper care and custody. Arrows are stenciled on the hub of the reel to indicate which way it should be rolled.

The Stowage of Uncased Automobiles

Automobiles that are shipped in the assembled form may be divided into two broad categories: the new car and the secondhand car. There is little difference in the actual stowage of the two types, although there is more preparation required and possibly more potential danger when carrying old cars.

One of the points needing emphasis when discussing the handling and stowage of this commodity is the importance of checking very carefully for exceptions. Here again, the older car will give the most trouble, because there are more dents, scratches, and broken parts to take exceptions on. Many companies provide forms with various views of an automobile on which marks can be made to point out accurately where deficiencies are found. In addition, written statements are made regarding the extent of the deficiencies.

Cars may be stowed in any of the 'tweendecks, facing in any direction that allows the greatest use of the available cubic. If carried in the lower holds on the tank tops, they preclude the use of all the clearance over them. Unless the space is not needed, they should never be stowed on the tank tops.

If cargo is stowed to within 7 ft of the beams in the lower hold, leveled off, and covered with at least two layers of crisscrossed 1-in. dunnage without spacing and a third dunnage layer of 2-in. deals, automobiles may be stowed over this platform. The lost space is practically eliminated.

Automobiles are sometimes safely carried on deck. If carried on deck, they should be lashed securely on top of the hatches and provided with protective light canvas covers. Even if the weather is fair during the voyage, paint from the sailors working aloft might ruin the finish of a car. Stowage on deck should be avoided if the ship is a full-scantling type and loaded to her marks. Such a ship is likely to be extremely wet on deck.

The best method of securing automobiles is by the use of the Peck & Hale type of adjustable cable, as shown in Fig. 4-41. As shown in Fig. 4-43, it is possible to use these fittings without special deck fittings. It is not as convenient, but they may be attached to chains stretched athwartships and secured by the use of frame clamps and turnbuckles. This would be the only possible way to use these cables to secure automobiles that are stowed over cargo in the lower holds on a platform of crisscrossed dunnage.

Without these adjustable fittings, a large amount of 4- by 4-in and 2- by 4-in. lumber must be on hand to secure the automobiles. Two 4- by 4-in. balks are cut to fit in front of and behind the front and rear tires. The balks should be long enough to extend beyond both sides of the car about 6 in. Two 2- by 4-in. battens are then cut to run the length of the car on each side from the forward balk to the after balk. The 2 by 4's are nailed down securely to all the 4 by 4's. This forms a chock in which the automobile will ride. In addition, some lashings should be attached for greater security. The lashings are needed if the car is not new because the springs may not be compressed. The spring action during heavy weather may cause the auto to move out of its cribbing unless it is lashed. The brakes should be set and the car placed in park.

Automobiles should not be stowed closer together than 6 in., nor should they be stored closer than 6 in. to any obstruction in the hold. The clearance overhead should also be 6 in. The possibility of a heavy sea causing the car to rise above its stowed level is greater, of course, if it does not have its springs compressed. It is advisable when setting up any lashings that tend to place a downward strain on the car's frame to have several men stand on the bumpers. The weight of the men will compress the springs and the lashings will be tighter and hold the car down much better.

Prior to loading the automobiles, they should be prepared for stowage. Preparation includes disconnecting the battery terminals and draining the gasoline tank and radiator. All removable items should be taken off and locked in the trunk of the car; this includes hub caps, outside rear view mirrors, and search lights.

After stowage, each car should be secured by one of the methods mentioned above. The windows should be closed and the doors locked. The keys for all the cars should be collected and placed in the custody of one officer. The officer accepting the keys will be required to sign a receipt for them, and when he releases them to a member of the agent's office in the port of discharge, he should obtain a signed receipt. Each key is identified by a key number, the make and model of the car, and a bill of lading number.

Under older cars, it is advisable to spread a few thicknesses of heavy paper to catch any oil drippings from the crankcase or transmission. Finally, cars should be covered with lightweight canvas covers if there is any possibility of heavy dust from other cargoes during any stage of the voyage. These covers fit over the entire automobile and reach the deck. They are coveted items and should be issued and checked very carefully. When the cars are discharged, these covers should be gathered up, folded, and stored away in a locked compartment to be discharged at the ship's home port. The chief officer may have to sign a receipt for them when they are provided at the loading port, and he should be sure to obtain a receipt when delivering them after the voyage is completed.

The Stowage of Deck Loads in General

During World War II there was so much cargo waiting on every dock that ships began carrying more and more on deck. On occasion, the deck loads were so large that the amount of cargo was almost equal to the cargo that would have been stowed if the ship had been provided with an additional 'tweendeck space. Besides the tremendous call for deadweight, some of the built-up units that were shipped had to go on deck because they would not fit anywhere else on the ship. Units that are shipped on deck are those that are not permitted below because of their dangerous characteristics, those that will not fit below, those that if placed below consume a tremendous amount of cubic and can be safely exposed to the weather, and those that are difficult to stow below and can be carried safely on deck. Large deck loads are less frequent when cargo offerings are light.

The two principal concerns of the ship's officers when deck cargo is carried are: (1) the security of the cargo, and (2) the accessibility of the equipment needed to operate the ship safely.

Providing security for the cargo begins with making thorough plans about how the cargo is to be lashed, braced, and shored. A scale drawing of the ship's deck and a scale model of the cargo enable greater accuracy when making such plans. The details of such plans should include the sizes and number of all braces, the under-deck shoring, the number and positions of all pad eyes, and the cribbing required between the load and the deck. If pad eyes are not numerous enough or are placed so that they afford poor leads for the lashings, it is an easy matter to have them relocated. All this should be done long before the cargo is scheduled to be loaded. After the cargo is placed on the ship as per the plan, the lashings and bracings should be installed as directed by the chief officer. The chief officer should not hesitate to insist on a different arrangement of additional lashings if he is dissatisfied. Lashings and bracing should be applied with the assumption that during the first night at sea the ship will pass through a full hurricane. Using this philosophy, the ship will generally be stowed safely.

Keeping the equipment accessible is a matter of marking off with chalk everything that should be left clear for working the ship. The things to be left clear are the bilge sounding pipe openings, cleats, fair-leads, reach rod valve control wheels, bitts, fire hydrants, and fire hose racks. Large clearances should be indicated and the words "KEEP CLEAR" printed on deck. To insure that these notices are respected, the ship's officers should be on deck when preparations commence. Furthermore, at least one officer should be on deck throughout the actual process.

The type of lashing should be either chain with pear links, pelican hooks, and turnbuckles; wire rope with shackles, clamps, and turnbuckles; or steel strapping with turnbuckles. Manila or other fiber rope lashings should not be used except for frapping purposes during heavy weather or emergency use in general. Fiber rope will stretch and loosen in time, and so many turns are required for strength that setting up such lashings is a long and arduous task.

The place of stowage of deck cargoes is dictated by the nature of the container and the cargo. For example, uncased automobiles should be stowed on top of the hatch and preferably on the hatch just abaft the midship house for greatest protection. Carboys of acid should be placed on the after deck as far inboard as they can be bunched. As mentioned before, drums of flammable liquids must all be placed on the same side of the ship. A 65-ft motor launch may fit in only one or two spots on the ship's deck; hence the choice of where to stow it is automatically dictated.

Shoring up a Deck

When the deck load is excessive, the main deck may have to be shored up. Shoring will be required whenever the deck load is greater than 350 lb/ft^2 on the average ship. The shoring up of the main deck may not have to go farther than the 'tweendeck; however, if the load is excessive, the shoring will have to extend all the way down to the tank tops. It is a mistake to load a ship so that shoring is required in all hatches. This stiffens the hull girder and places stresses where the naval architect that designed the ship did not intend them to be concentrated. The possible result in extremely heavy weather is structural failures in the deck or bottom plating.

The size of the shoring depends on the total strength needed. For loads up to 120 lb/ft², 12- by 12-in. balks spaced at 6-ft intervals should be used. This means that such shores would be under every other beam on the average ship and that about two rows would be needed between the hatch carling and the side of the ship. The spacing between the hatch carling and the side of the ship should be such that the three divisions thus formed are approximately equal. If the load is actually equal to 1200 lb/ft² over the area it rests on, shores would be required all the way to the lower hold. This is because the 'tweendeck load capacity is seldom over 700 lb/ft². It should be obvious that smaller shores can be used if the spacing between them is reduced or the load is less than 1200 lb/ft².

The shores are placed in position so that the lower end rests on a 6- by 12-in. *stringer* running longitudinally. The upper end bears against a *header* that runs longitudinally under all the beams. The shores are cut slightly shorter than the distance between the header and the stringer and any slack is removed by driving soft wood wedges between the lower end of the shore and the stringer. Fishplates are used on both sides of each shore at the upper and lower ends. The fishplates are made of 2- by 12-in. planks cut about 24 in. long and securely nailed to the shores and the

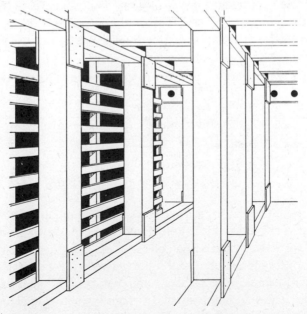

FIG. 6-11.   Shoring up the main deck from the 'tweendeck. Courtesy of U.S. Navy.

stringer and header. The space above the header, between it, and the deck above should be filled with 4-in. timbers cut to fit. This means that the width of this beam filler piece will be equal to the outside measurement of the beam. The length will be equal to the distance between the beams. These pieces offer additional support the the deck plating under the load and prevent the beams from tipping. Figure 6-11 shows this arrangement of shores in a 'tweendeck.

THE STOWAGE OF LUMBER

Lumber in sawed form is shipped in full loads from several parts of the world, the largest exporting volume being from the Pacific northwest coast of America. The lumber shipped falls into two broad categories: (1) deals, battens, and boards, which are the smaller pieces, such as the 2- by 4-in. batten and the 1- by 8-in. board, used for light construction work; (2) balks, flitches, and squares, which are the larger timbers, such as the 6- by 10-in. flitch and the 16- by 16-in. square, used for heavy construction work or intended for resawing at the country of destination for use in light construction work.

In the European trades timber is measured in standards of 165 ft^3, while in the North American trades 192 ft^3 is the average. In other trades the amount can be 216 ft^3, that is, a cube 6 ft by 6 ft by 6 ft.

Some timber shipping terms include:

A shipping ton	42 ft^3
A load	50 ft^3
A load (untrimmed)	60 ft^3
A stack	108 ft^3
A St. Petersburg and Baltic standard	165 ft^3

Lumber is shipped in loose board lots and also in packaged lots. Packaged lumber is not a new idea. Full shiploads of lumber in standard-sized units were carried on the Pacific coast before 1930. Packaged shipments are now carried on the intercoastal route between the west and east coasts of the United States. The package unit has been standardized by one major lumber carrying company at 2 ft high by 4 ft wide. The lengths in any one package are all the same (see Fig. 6-12).

FIG. 6-12. The Weyerhaeuser full unit lot 4 ft high and 4 ft wide being hoisted on board. The standard package size is 2 ft high by 4 ft wide. The lot shown here is not strapped because it is planned to stow the lumber by hand. Wire snorters with sliding hooks comprise the sling. The bolsters upon which the load was built and transported to the pier apron can be seen lying on the dock under the load. The straddle truck to the right is standing by with another load. The empty bolsters are stacked as shown on the left. When enough empties have accumulated, a straddle truck will pick them all up and take them to the place where the loads are built up. Courtesy of Weyerhaeuser Co.

283

The two most important problems in the stowage of lumber are the protection of the higher grades of lumber and stowage so as to combat tenderness in the ship. The first of these requisites is met by keeping the surfaced and kiln-dried lumbers in the wings, using fiber rope slings, being careful not to stain the lumber, using no hooks on milled or surfaced lumber, and taking other common-sense precautions.

The means used to combat tenderness in the lumber carrier, not including the obvious method of eliminating or drastically limiting deck loads (which would cut into the ship's earning power) are three in number: (1) Place the heaviest woods in the lower holds. (2) Distribute lumber types so that those stowing most compactly per unit of weight are in the lower holds. (3) Pack the ship's frame spaces.

The lumber below decks, if in loose lots, is stowed piece by piece and packed in carefully by the longshoremen. Uneven surfaces are dunnaged

FIG. 6-13. Packaged lumber in the hold of the ship. The units on the right are 4 by 4 ft, but this company has changed its standard size package to a unit 2 ft high by 4 ft wide. Two- by two-inch battens separate the lots; this affords room for passing slings and to insert and remove the forks of the truck used to stow the packages in the hold. Note the large voids that unavoidably occur when working with packaged lots. Loose lumber of various lengths has to be brought in and stowed by hand to fill up such broken stowage spaces. Courtesy of Weyerhaeuser Co.

FIG. 6-14. Longshoremen filling in the voids and butting the loose lumber together very carefully so that a firm and smooth floor is provided on which a forklift truck can work to stow additional tiers of packaged lumber. Note the peaveys in the background, an indispensable tool for the handling of lumber. The last few loads in the square of the hatch are stowed with the slings remaining on them. This is necessary in order that the first few key loads may be removed at the port of discharge and to give the men working room in which to pass slings around the remaining loads. Courtesy of Weyerhaeuser Co.

with lath, and the broken stowage is reduced by using *ends* as filler cargo. Ends are pieces of lumber less than 6 ft in length with the same cross section as the rest of the shipment. Once a lot is stowed, it is marked off with rope yarn or by painting stripes across the top of the block. Arrowheads are used to indicate a change in vertical direction of the stowed cargo.

A lumber deck load is always found on ships carrying full or large amounts of lumber. While the deck load is being started, an officer should be on deck continuously, and the building of the deck load should be watched all during the process. The sides of the load and the ends must be shaped to afford the least resistance to heavy seas. The deck load must also be stowed in accordance with the basic requirements for deck loads in general, the provisions of the International Maritime Organization (IMO) Code of Practice for Ships Carrying Timber Deck Cargoes, and the regulations of the International Convention of Load Lines.

The deck load is secured by use of $\frac{3}{4}$-in. chain spaced not more than every 10 ft. The lower ends of the lashings are shackled to pad eyes; the upper ends are brought together approximately amidships and secured by

(a)

(b)

286

(c)

FIG. 6-15. Three views showing the care and planning that must go into the preparation of the loads of packaged lumber to prevent an excessive amount of lost cubic. (a) The load on the right is ready to be placed in the vacant space on top of the two previously stowed 2- by 4-ft units. Note that the top has been stepped so as to fit the load next to the beam knees with minimum clearance. (b) The fork truck placing the load in place. The man in the center is shoving a 2- by 2-in. batten under the load before the truck operator sets the load in place. (c) The stowage of this unit as seen from the ship's centerline. Note the minimum clearance from the overhead, which can be obtained only by correct flooring off before starting to handle packaged lumber. Courtesy of Weyerhaeuser Co.

the use of turnbuckles, pear links, and slip hooks. For maximum security, the deck load should be solid from side to side where it is possible to build it like this. The height should not impede navigational visibility in any way. Safe access to crew members with guard lines or rails and life lines should be provided in accordance with the IMO Code. As a general rule the quantity stowed on deck should not be more than one-third of the weight of timber carried.

Lumber shipped in packaged lots is packed into the ship with surprisingly little lost space by filling the bigger voids with loose pieces as necessary. The technique of stowing packaged lumber in the hold with the use of the forklift truck is clearly depicted in Figs. 6-13 to 6-15.

The trend in stowage of the increasing tonnages of timber being traded internationally is toward unitization of timber handling.

Since deck timber cargoes are mostly in packaged form, there is a need to ensure that: (1) the packages are securely bound such that they are in "solid" form; (2) a level surface of dunnage provides a secure foundation; (3) in general, stowage is in a fore and aft direction to assist proper and adequate athwartship lashing arrangements; and (4) provisions are made for the tightening of the lashings.

The calculations relative to the amount of lumber that will go into a given space and the inverse of this problem are presented in the section dealing with stowage factors.

THE STOWAGE OF BULK GRAIN

Grain in bags offers no special problems other than those of other bagged products requiring stowage to prevent damage from sweat and loss of contents through damage to the container itself, namely the bag. Grain in bulk, on the other hand, has proved to be dangerous during transportation, due to the possibility of its shifting and thereby rendering the ship unseaworthy. Bulk grain is also interesting in that it is handled by methods that exploit the use of gravity (see Fig. 6-16).

The Carriage of Bulk Grain

Due to the fact that shifting grain may cause an unstable condition, all vessels loading bulk grain in the United States must comply with the USCG regulations for the carriage of bulk grain. Compliance with these regulations is enforced by the National Cargo Bureau, which issues a Certificate of Loading on successful completion of the loading operation. This certificate is one of the documents required for vessel clearance.

In 1976 the United States adopted as regulations the IMO amendments to Chapter VI of the 1960 SOLAS (Safety Of Life At Sea) regulations as they apply to the carriage of bulk grain. These current regulations address the individual stability characteristics of each vessel, rather than providing a single strict requirement for securing the surface of the grain, as in the previous regulations. The new regulations require the master to calculate the effects of an assumed shift of the grain surface on the stability of his vessel. If the planned stowage is such that the effect of the assumed

FIG. 6-16.   A ship being loaded with bulk grain at a grain gallery. Four spouts lead into the ship. The grain is brought to the spouts from the elevator by conveyor belt. Courtesy of Port of New York Authority.

shift is within the parameters of the regulations, then the vessel may sail without further expensive securing of the grain. If the effect of the assumed shift is to put the vessel in an unsafe condition, then the regulations provide for approved methods to limit the effect of such a shift or to secure the surface of the grain.

To understand the requirements of the bulk grain regulations, the ship's officer must be familiar with the following definitions:

ANGLE OF REPOSE: The angle to which a horizontal may be rotated before the grain will start to flow. The angle of repose of wheat is 23°.

ASSUMED SHIFT: Assumption that the grain surface has assumed an angle to the horizontal. This assumed shift is different for different classes of vessels.

GRAIN HEELING MOMENT (G.H.M.) The upsetting moment of the assumed shift of the grain surface in ft-tons or m-tons. When the G.H.M. is divided by the displacement, the result will be the off-center shift in G.

VOLUMETRIC HEELING MOMENT (V.H.M.): The potential volume shift for the assumed shift of the grain surface. When the V.H.M. is divided by the stowage factor of the grain to be loaded, the result is the G.H.M. This information would be provided to the master in the form of tables.

LOADING CERTIFICATE: A document issued by the National Cargo Bureau upon loading in compliance with the appropriate USCG regulations.

CERTIFICATE OF READINESS: A document issued by the National Cargo Bureau upon preparation of the vessel for loading in compliance with the appropriate USCG regulations.

DOCUMENT OF AUTHORIZATION: A document issued by the authority recognized by the vessel's administration. This document indicates that the master has the information to determine the effects of the assumed shift on the stability of his vessel. If there is not a valid Document of Authorization on board, the vessel must load under the older, more stringent regulations.

RESERVE DYNAMIC STABILITY: The area on the vessel's statical stability curve between the righting arm curve and the heeling arm curve.

Preparation for Loading of Bulk Grain

The compartment into which the grain is to be loaded must be prepared to the satisfaction of a cargo surveyor from the National Cargo Bureau. In general, this preparation is to provide a clean, dry, odor-free, and vermin-free compartment. Additionally, the bilges must be sealed so as to be graintight but not watertight, and the sounding tubes must be free. Preparation of tank vessels requires the boxing of tank suctions and the blanking off of all cargo lines in the pump room. Also, sounding tubes have to be constructed.

Once the surveyor has inspected the compartments and bilges, and found them satisfactory, a Certificate of Readiness is issued. To load under the amended Chapter VI of the SOLAS regulations the vessel also requires a valid Document of Authorization.

Intact Stability Requirements for Carriage of Grain

The current regulations recognize three general classes of vessels: *all vessels, specially suitable vessels, and tank vessels*. Each class of vessel has different stability requirements under the SOLAS regulations.

All Vessels

The stability requirements for vessels loading bulk grain are as follows:

    a.   A minimum *GM* after correction for free surface of 0.9 m or the minimum *GM* required by the Trim and Stability Book, whichever is greater.

    b.   A maximum angle of heel of not more than 12° for the assumed shift for these vessels shall be:

    1.   15° in filled compartments.
    2.   25° in partly filled compartments.

    c.   A minimum reserve dynamic stability of 0.075 m-radians (14.1 ft-degrees). This is the area on the vessel's statical stability curve between the righting arm curve and the heeling arm curve, or between the angle of list and (1) the angle of flooding, (2) the angle of the maximum righting arm (*GZ*), or (3) 40°, whichever is least.

Specially Suitable Vessels

Specially suitable vessels are those vessels with two or more longitudinal graintight divisions so disposed as to limit the effect of a grain shift. To limit this effect, the divisions shall have an angle to the horizontal of not less than 30°. Since these are the original requirements of the 1960 SOLAS regulations, the assumption of 2% grain settling is still used to determine the volume of the void above filled compartments. The only intact stability requirement limits the angle of list to 5° for an assumed shift of 12° in both filled and partly filled compartments.

Tank Vessels

Tank vessels with two or more longitudinal bulkheads may load in any manner, provided that in the most unfavorable loading conditions the vessel will not list to an angle greater than 5°.

Trimming of the Grain Surface

All grain surfaces should be trimmed level, and filled compartments should be trimmed to fill void spaces under the decks and hatch covers to the maximum extent possible.

For more specific information the ship's officer should refer to the National Cargo Bureau booklet entitled *General Information for Grain Loading*.

Shifting of Grain

A reasonable demonstration that the grain loading regulations have proved their worth appeared in the August 1983 *Proceedings of the Marine Safety Council* as follows:

On September 23, 1979, the SS PILGRIM departed Port Arthur, Texas, with a cargo of 12,478 tons of grain for discharge in Durban, South Africa. This was scheduled to the PILGRIM's last trip, for after discharging the grain the vessel was to proceed to Taiwan for scrapping. As events unfolded, however, the voyage almost ended prematurely off the coast of Africa with near-tragic consequences for the crew.

With the exception of a detour to Tampa, Florida, for repairs to the engineering plant, the voyage to Durban proceeded routinely until the morning of October 18, when a series of unsettling events began. On that morning, the vessel was hove to in high winds and rough seas while the crew secured a cargo boom which had broken free. When the vessel resumed course, a 4° port list was detected. The ship's officers suspected a cargo shift as the cause of the list. Later that same day, water began accumulating in an athwartship passageway as well as in other below-deck spaces. Water could be seen entering around the port side-port door, the garbage chute, and through below-deck drains which penetrated the hull. As the vessel rolled routinely 5° starboard to 45° port, the cargo continued to shift and the vessel continued to take in water.

On October 19, with the PILGRIM listing and laboring in deep swells, the master directed that a distress signal be transmitted. A salvage tug arrived on scene the following day and transferred portable pumps to the

vessel. Two days later, the PILGRIM, listing 17° to port, entered Capetown, South Africa, under its own power. There, efforts to right the vessel
were successful, and the PILGRIM later proceeded to East London to
discharge its cargo.

Investigation of this incident revealed several elements which either led to
or aggravated the situation. First, rough seas most likely caused the initial
grain shift. The possibility that a cargo might shift was taken into account
when the International Maritime Organization (IMO) Grain Regulations
were written. The maximum allowable vessel list resulting from such a
shifting is 12°; this is based on the assumption that the surface of the grain
within the cargo holds would shift no more than 15° from the horizontal.
Efforts were made to limit shifting of cargo when the PILGRIM was loaded.
The grain was trimmed so that all grain surfaces were level and all spaces
below 'tweendecks and hatch covers were filled. "Bundles" were installed
in certain cargo hatches to restrict the shifting of the grain. These bundles
were large, tightly secured bags of grain placed in the square of the cargo
hatch after the hold was loaded. Six of them were used on the PILGRIM.
Also, hatch covers were used where required by the National Cargo Bureau.

The PILGRIM was fitted with "feeder holes" six inches in diameter in
the 'tweendeck hatch-side girders; these provided a means of feeding additional grain into any void spaces which might form outboard of the hatch-
side girders after the hold was filled. The feeder holes were not used in this
loading, however.

After the vessel arrived at Capetown, the crew discovered that these precautions had not accomplished their purpose and that shifting in excess of
that allowed in the grain regulations had occurred. The average angle of
grain surfaces within the cargo holds was about 17.5° from the horizontal.
The greatest shift was measured at 19°, while the smallest shift was 13°.
Vertical shifting of the grain occurred in every hold in which the hatch
cover was closed and grain loaded in the hold above it. In each of these
holds, grain, varying in depth from $1\frac{1}{2}$ to 2 feet, was found on top of the
bundles. Grain was found to have sifted through the feeder holes in every
hold except No. 6 upper 'tweendeck, which was not fitted with feeder holes.
Of the six bundles installed at Port Arthur, two were torn and had shifted to
the port side. A third bundle, while remaining intact, also had shifted.

Rough seas, heavy rolls, and shifting grain led to the vessel's taking on
water. The ingress of water into the vessel was allowed by numerous
breaches in the vessel's watertight integrity. In particular, deterioration in
the main deck in way of the No. 5 cargo hold, inoperative sideshell check
valves, deteriorated gaskets in the side ports, wasted deep-tank ventilation
piping at the main deck level, and wasted refuse chute closures were noted.
In all, it was estimated that a total of 560 tons of water were being shipped
on board the vessel prior to its arrival in Capetown.

The Coast Guard investigating officer for this incident recommended that the Coast Guard and the National Cargo Bureau jointly pursue efforts to improve methods for preventing shifts of grain cargoes. Among the points covered in the recommendations were: making sure hatch covers were graintight, preventing grain from sifting through feeder holes, and possibly securing bundles to the vessel's structure to prevent them from shifting. Also emphasized was the importance of a thorough examination of vessels for the type of specific hull defects which existed in this incident, namely, deterioration of the main deck between hatch coamings, deterioration of vent piping on the weather deck, deterioration of closures for drains and other openings which penetrate the hull, and deteriorated gaskets on side port openings.

HANDLING AND STOWAGE OF BULK ORE

The stowage of ore on the ship built especially for the bulk trade is accomplished by placing the ship under a series of chutes that may be lowered into the ship's hatches and allowing the ore to flow into the holds. The ship has a large number of hatches, but usually only four compartments. A typical arrangement provides four hatches in two end compartments and five hatches each in two middle compartments. The engine room always is aft. The ore weight is distributed 30% in each end compartment and 20% in each center compartment.

The average loading time for 12,000 tons of ore is about 4 hours. The loading can be accomplished in less time, but there is danger of causing structural damage to the ship. A test case was made on September 7, 1921, to demonstrate how quickly a ship could be loaded by these methods. The *S.S. D. G. Kerr* arrived at the Two Harbors, Michigan, ore docks at $1623\frac{1}{2}$ on this day and cleared at 1700 the same day. Thus, from arrival to departure the time was $36\frac{1}{2}$ min. Sixteen and a half minutes of this was for loading 12,506 tons of iron ore at a rate of 758 tons/min.

Discharging Operations

Discharging is accomplished by either a transporter type crane with a cantilever runway that supports a trolley on which a clamshell bucket is mounted or by use of the giant Hulett unloaders. These two crane types are illustrated in Figs. 6-17, 6-18, and 6-19.

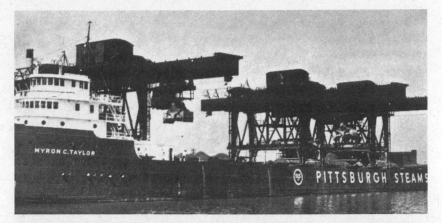

FIG. 6-17.   The transporter type cranes used for unloading iron ore at Lorain, Ohio. This type of crane uses the offshore and inshore cantilever boom to support a runaway for the man-trolley and bucket. The entire structure can move up and down the length of the dock. Courtesy of Pittsburgh Steamship Co.

Bulk Ore on the General Carrier

When carrying bulk ore on the general cargo ship, the precautions with respect to the weight distribution as discussed in Chapter 4 and to preparation of the drainage system discussed on p. 244 must be given. If a general cargo carrier is used to carry a full load of bulk ore, the distribution in any given hold should be in proportion to the ratio of the given hold's cubic to the total cubic of the ship including deep tanks and refrig-

FIG. 6-18.   The *S.S. Governor Miller* lying at the Pittsburgh and Conneaut Dock of Conneaut, Ohio, with five hulett unloaders working. Courtesy of Pittsburgh Steamship Co.

FIG. 6-19.   A closer view of hulett unloaders at work. The arrow points to the operator of the unloader. The operator rides with the bucket as it oscillates between ship and dock and has an excellent view of the operation. Courtesy of Pittsburgh Steamship Co.

erated spaces if they remain empty. If these latter spaces are not empty, the weight of the material in them should be subtracted from the weight found by the proportional method.

The cargo in each hold should be trimmed so that the bulk of it lies toward the after bulkheads in the forward holds and the forward bulkheads in the after holds.

The above is a general rule for use only on ships with engine room spaces amidships. It should be used with caution. A check of what the ship is actually doing should be made continually as described in Chapter 4 under the discussion of hull deflection. The concept of loading by proportionality should not be used for ships with their engine rooms aft. On these latter ships, the cargo spaces should be divided into four sections and the two end sections should receive about 30% of the cargo while the two middle sections should each receive about 20% of the cargo. The problem on the ship with the engine room spaces amidships is to prevent hogging, whereas the ship with engine room spaces aft tends to sag with full loads.

Shifting boards as were mentioned for bulk grain are not needed with bulk iron ore. The angle of repose of iron ore is about 45°. In some trades the iron ore is allowed to remain in a cone, much as it appears when it has finished running into the hold through the chutes. This is done for two reasons: (1) to raise the weights somewhat and prevent a stiff ship; and (2) to minimize the cost of handling by eliminating the cost of trimming the ore. This practice may produce a dangerous condition if the ship runs into heavy weather and the violent working of the ship in a seaway causes the cone to fall off to one side. It is much better management always to trim down this cone and if it is necessary to raise the weights to do it by placing some ore in the 'tweendecks. Even in fairly recent times, ships have been listed dangerously from this very cause, and a few ships have been lost at sea with many seamen. The cause, it must be emphasized, is poor judgment in loading the cargo.

Carriage of Ore and Ore Concentrates

Ore is a very commonly carried commodity on both general cargo vessels and specialized carriers. Ore and ore concentrates are normally loaded by means of gravity using chutes or crane grabs. Cranes may be either shoreside or the vessel's own cargo gear. There has been some success with loading certain ore cargoes by means of slurries. In a slurry transfer system the cargo is mixed with water and pumped through a pipeline. Discharge of ore cargoes is usually done by crane grabs, with a bulldozer used to move cargo into the square of the hatch.

The hazards associated with bulk ores are structural damage, loss of stability during the voyage, and chemical reaction. Structural damage may occur during the loading operation if the rate of loading is not controlled to protect the structure of the vessel. Additionally, damage may be caused by improper distribution of the weight within the vessel. General cargo vessels must be loaded in a manner that will not overstress the hull.

Loss of stability may be caused by the shifting of improperly trimmed cargo or the liquefying of certain cargoes due to vibration. Cargoes with low angles of repose should be trimmed level, and all high-density cargoes should be loaded in the lower hold whenever possible. Cargoes that may liquefy should be tested for water content prior to loading. Additional information on these procedures may be obtained from the IMO Code of Safe Practice for Solid Bulk Cargoes.

Carriage of Bulk Coal

Coal in bulk is subject to two recognized hazards: spontaneous heating and hazardous-gas generation. In 1981 there were several ship fires involving the spontaneous heating of bulk coal. These incidents have led to a great deal of study on methods to control these hazards. At this time there are no hard and fast rules governing the carriage of coal. There is some agreement that not all coal is potentially hazardous. The ship's officer should work with the shipper to determine if the cargo to be carried has a history of heating or gas generation.

STOWAGE OF REFRIGERATED CARGOES

Preparation of the Refrigerated Space

Preparing the refrigerated cargo space to receive cargo includes some special steps in addition to the steps required with regular general cargo spaces. The specialized nature of this preparation is evident from the discussion that follows.

Certification of Spaces

When loading any cargo on a ship, in any compartment, the shipowners may call in a surveyor to have the space certified as being in all respects in good condition and ready to receive the cargo. This survey is frequently made in the case of refrigerated spaces. Whether any cargo survey is made is entirely at the option of the shipowner; most large companies do have surveys made before and after stowage and during the stowage of cargoes. These surveys may not be costly, depending on the type of survey being made.

The survey of the spaces of a refrigerated ship includes the parts of the refrigeration system plus the insulation and equipment in the actual compartment. The machinery survey includes tests to insure that:

1. All controls are reacting properly.
2. Correct pressures can be maintained on both sides.
3. The system is fully charged.

The survey of the insulation of the compartment in an inspection of this type is superficial. Little more than a visual check to determine whether the insulation has any breaks in it can be made; a check of insulation plugs and their condition is also made. The coils of the evaporator should be carefully tested for leaks. The interior of the reefer space must be adequately equipped with gratings and battens to protect the inside surface of the insulation sheathing. The space must be exceptionally well cleaned and all drains clear; a reefer compartment will be equipped with two to four drains instead of only one. Surveys of machinery and insulation are made by the American Bureau of Shipping and similar organizations. Surveys of the actual cargo stowage are made, in the United States by the National Cargo Bureau.

Cleanliness of the Spaces

Although the cleanliness of the space is mentioned as a requisite for stowage for practically every item that may be stowed on a ship, nothing requires a greater degree of cleanliness than the reefer space. Improperly cleaned spaces may cause heavy mold to develop on commodities despite optimum conditions of temperature, humidity, and air motion. Fungus growth and rotting may be caused on fruits and vegetables infected by imperfectly cleaned spaces; however, it should be understood that the appearance of such damage on chilled or cooled products is not proof that the space was improperly cleaned. The product can be infected during its handling prior to loading in such a way that out-turn will show the same effects that it would if infected by unclean conditions in the reefer compartment. In fact, the most frequent cause of infected fruit is improper handling prior to stowage on the ship. The point to be made here is that damage can be caused by an improperly prepared space, damage that can be eliminated only by effective cleaning. If the fruit shows evidence of being unclean or of already having developed mold or disease spots of any kind, thorough exceptions should be taken and written up for submittal to the claims department. To make this cleaning procedure completely practicable is a problem for the shipowner. A recommended procedure is as follows:

    a.   Remove all gratings and sweep up the compartment.

    b.   Any residue from past cargoes that cannot be removed by sweeping must be scraped off or washed off the interior surface.

    c.   Spray the interior with a light antiseptic solution that will kill mold spores and bacteria or at least tend to retard their growth. Such a spray must not be so powerful that it will leave odors behind that might taint the cargoes to be loaded. There are many such solutions; three are given here:

1.   A solution of $\frac{1}{2}$ lb of chloride of lime to 3 gal of water.
2.   One-quarter teaspoon of high test hypochloride (HTH) to 3 gal of water. HTH is very powerful and the solution must be made carefully and tested by smelling the mixed solution. If the odor of chlorine is too strong, the solution is too strong. Do not allow the powder to lie around; if the powder in full strength comes in contact with dunnage, the dunnage should be condemned.
3.   A solution of sodium hypochlorite. This solution is made by adding sodium hypochlorite to fresh water, but it should also be very weak. The recommended strength is 0.8%, made by adding 0.8 lb of sodium hypochlorite to 99.2 lb of water.

If spraying or washing thoroughly is not possible, the next best step is to wipe down the interior, including the gratings, with rags or hand swabs that have been dampened in the solution.

    d.   Replace the gratings after washing them with one of the above solutions and allowing them to dry in the sun. If they cannot be allowed to dry, wipe them off as much as possible.

The interiors of reefer spaces are usually sheathed with surfaced tongue and groove spruce, pine, or fir plywood in such a manner that the cleaning is made easier than that of general cargo spaces. All frame spaces are eliminated, corners are coved, edges are chamfered, and the entire surface is painted with several coats of a high-grade varnish. The finish on the inside should not be allowed to deteriorate; revarnish the interior at least annually.

It should be understood that a thorough cleaning may not be necessary every time the spaces are to be prepared for a new reefer cargo. However, if in the judgment of the master or mate of the vessel the space is dirty and needs such a cleaning, the recommended procedure is as described. The exact procedure to be followed should be carefully outlined by the ship-owner's operating department and incorporated in the company manual. With some cargoes, such as chilled meats, the cleaning should be frequent and thorough. The final decision rests with the ship's master or mate, a company shoreside refrigeration expert, or with the cargo surveyor retained to survey the cargo space as ready for loading a given reefer cargo.

Precooling

Besides the above preparations, there is also the matter of *precooling* the compartment before loading commences. There are many procedures that may be followed to insure that the space is ready when the cargo is ready. The important thing is to have the ship's chief engineer notified well in advance of the need for the space to be at the temperature specified. Such notification should be given to the chief engineer at least 24 hours prior to the actual loading of the commodity. This officer will then place the refrigeration machinery in operation and commence to pull the temperature down. During this period, all of the dunnage that will be needed to stow the cargo securely should be stacked in the compartment in such a way that its heat load can be picked up easily along with that of the interior of the space. Considerable damage may be caused by using warm dunnage between the containers of refrigerated cargoes. Precooling of cargo spaces is universally required; precooling of the cargoes is not. Bananas are never precooled before stowage.

Temperature Records

Hourly recordings of the temperatures maintained in all reefer spaces are made and logged in the engineering log book. These temperatures are read at a remote station. The refrigeration engineer simply has to close an electrical circuit including a galvanometer set to read degrees of temperature instead of ohms of resistance. The electrical circuit is closed by pressing a button at the centrally located recording station. On other installations the refrigeration engineer must make periodic rounds of all spaces and read a recording thermograph on the outside of each refrigerated compartment. On modern ships, there is nearly always a recording thermograph regardless of the method used to obtain periodic temperature readings during the voyage. This thermograph chart can be used by the shipowner to ascertain whether the ship's operating personnel actually do maintain temperatures as requested. In the case of claims, these same charts can be used to prove that the temperatures required were, in fact, maintained.

A portable type of recording thermograph is frequently used. These can be secured in a centrally located position in the space at the beginning of the voyage and removed by properly designated personnel just prior to discharging.

Receipt and Care

The stowage of refrigerated cargoes is a task that requires careful attention to many details on the part of the ship's officers. The officers must be with the cargo from the time the cargo is received on the dock until it is safely stowed on the ship. Refrigerated cargo, commonly referred to as reefer cargo, requires the time of two officers during some phases of its receipt and care.

Inspections

All cargoes should be observed by the ship's officers as they are received by the dock force and as they are loaded on the vessel. In the case of refrigerated cargoes, this observation should take the form of a continuous inspection of the cargo to ascertain whether there is any evidence that would make it doubtful that the cargo could stand the voyage. The form of this evidence will be presented in the context of this and the following sections; it is important to note, however, that the officer who is not aware of what to look for cannot protect the shipowner's interest or his ship's interest regardless of how conscientious his inspections may be. The ship's officer's check should be a double check following inspection by the receiving clerk's staff or a specially appointed company expert. Regardless of the number and type of inspections made by dock force personnel, the master of the vessel should require a thorough inspection by at least one of his officers, who should submit a written report to him. The written report is an excellent educational device for the junior officers and it will tend to make the officer pay greater attention to what he is actually doing. The need for writing about what he finds and evaluating it will also build up in him an accurate vocabulary concerning the damaged conditions found.

The inspection commences when the refrigerated cargo is received on the pier. At this time a ship's officer must carefully inspect the containers as they appear in the connecting carrier. The connecting carrier will be either a refrigerated railroad car or truck. Infrequently some leafy vegetables are carried to a ship that have only been iced before leaving the hinterland and have been transported in an unrefrigerated truck. Such facts should be noted and included in a report to the master and claims department. If there are outward signs of improper handling or evidence of any type that might cause the cargo to be suspect, exceptions should be taken. If conditions warrant, the cargo may be rejected. Inspection by simply observing the containers in their connecting carrier is not sufficient. Individual units of the commodity, such as an apple or a pear, should be picked out of the lot and sampled. They should be cut in half to inspect for brown heart or other signs of internal decay. They should be tasted and smelled carefully. Many troubles, such as soggy decay in apples, manifest themselves in the condition of the flesh only, without any external symptoms. The appearance of the skin should be checked for signs of mold, freezer burn, scald spots, or other diseases that may be common to the particular type of fruit in question. The types of diseases and their symptoms are so numerous that they cannot be covered completely here. It is strongly recommended that just prior to receiving a given type of refrigerated commodity, a reference should be consulted to ascertain possible sources and manifestations of troubles. Finally, if the product is supposed to be precooled, an appropriate thermometer should be inserted into the pulp of the commodity and its temperature taken. Such thermometers are ordinary in their construction except for a strong steel tip that makes it possible to insert the instrument into a chilled quarter of beef or a green and firm apple or pear. It is important to note this temperature and either reject the shipment or take exceptions if the temperature is not down. The need for this is twofold: (1) If not precooled, the commodity will have ripened much faster than it should have since being harvested, and it may not be able to last the voyage. Most fruits will not keep well under cold storage conditions unless they are precooled within 24 hours of being picked. (2) The requirement of cooling the commodity after stowing on the ship may be more than the refrigeration system of the ship can handle in the time allowed. In this case, even if the commodity could stand the voyage if brought down to proper temperature, the trouble lies in the fact that the temperature cannot be reached and maintained by the facilities on hand. In this instance also, there is danger of losing the entire shipment. The fact that pulp temperatures were taken and a record of the temperatures should be included in a log book entry. These temperatures are required for any refrigerated product that is supposed to be at some specific low temperature when arriving at the ship.

The operating departments of some companies require that refrigerated products be inspected by a competent surveyor and a report made to the company claims department. Fresh fruit shipments out of the United States should be inspected by a representative of the U.S. Department of Agriculture. If the shipment is found suitable, the inspector will issue an inspection certificate that certifies that the shipment is clean and contains no disease. In many foreign ports such a certificate must be on hand before the shipment can be landed.

When inspecting refrigerated cargoes as received, frozen or otherwise damaged fruit should be set aside. This fruit must not be mixed with fruit in good condition. Damaged fruit mixed with sound and healthy fruit may be the cause of the entire shipment going bad.

Stowage

The basic principle to be followed when stowing refrigerated products is to build suitable air channels throughout the stowed cargo block. There are a few exceptions to this general requirement, which will be mentioned later. Baskets and bulge-top fruit cases can be stowed so as to provide excellent air passageways without using dunnage expressly for this purpose. Dunnage used with such containers serves the purpose of binding the load together for more secure stowage. The size of the strips of dunnage used varies considerably from one ship or one company to another. One inch by 2 in., 1 in. by 1 in., $\frac{1}{2}$ in. by 1 in., and the common building lath of $\frac{3}{8}$ in. by $1\frac{1}{2}$ in. by 48 in. have all been used extensively. In placing the containers in stowage, take care to see that the contents of the crates do not take the weight of the succeeding tiers, but that the crate supports this weight. Crates that have cleats at the ends only should be stacked in the reefer compartment one on top of the other with dunnage running at right angles to the cleats between every tier. If these crates are staggered in the recognized brick fashion used on many ordinary cargoes, the load will be taken up by the contents, which will become bruised and damaged.

Crates that have a supporting transverse member dividing the crate into two equal parts and are provided with a center cleat can be safely loaded in the brick fashion. Brick fashion stowage, of course, makes for a more securely stowed cargo block. Vertical air channels can be provided by placing pieces of dunnage between the rows on their forward and after sides. Crates of citrus fruits, melons, apples, and similar products are packaged in containers with bulged tops. Such containers can be stowed on their ends with the bulged tops together. This arrangement provides ample air channels, and the crate, rather than the contents, will carry the weight of top tiers. It is recommended that stacking no more than five tiers high be allowed with this type of container, with possibly a sixth tier lying longitudinally with the bulge up; this arrangement will aid in tying the entire stowed block together.

Refrigerated Cargo Types

The commodities received for refrigerated stowage are numerous, and increasing every year. Most of these are well known and include meats, fish, poultry, dairy products, fruits, and vegetables. It is impossible, in this limited discussion, to set forth formally precise methods of handling and stowing each commodity type that may be classified as refrigerated. Not only is our space limited, but the methods used in any one instance may vary from nation to nation, trade route to trade route, company to company, and even from ship to ship. Thus far there has been no effort on the part of any group, whether specifically interested in the welfare of the shipping industry or otherwise, to investigate thoroughly and determine the true merits and problems of extant practices found within the marine industry. Shipper, consignee, shipowner, consumer, and underwriter alike would benefit from research leading toward a determination of the best practices.

Apples, grapes, and pears are the fruits shipped in the greatest amounts in partially refrigerated vessels and offer some interesting problems. Bananas are carried almost exclusively on one type of ship and on one trade route. The methods used in handling and stowing bananas are interesting and reflect how efficient the process can become when the efforts of all interested parties are coordinated. Meat carriage is done on a large scale only in the merchant marines of nations other than the United States, principally on British ships.

Types of Refrigerated Ships

There are four classes of refrigerated ships: (1) *the all-refrigerated chilled meat carriers;* (2) *the all-refrigerated frozen meat carriers;* (3) *the all-refrigerated cool air fruit carriers; and* (4) *the general refrigerated carrier* designed to accommodate many types of refrigerated commodities at temperatures from below 0 to 55°F. To be most efficient, each of these types should be specially designed for the purpose to which it is going to be put. In practice, however, compromises must be made because refrigerated cargoes are carried on these carriers on only one leg of the voyage. Such compromises are made so as to accommodate general cargoes. For the purposes of discussion, an outline of the major points of each type will be sufficient.

All-Refrigerated Chilled Meat Carrier

The all-refrigerated chilled meat carrier was the first type of refrigerated carrier developed. A consignment of chilled meat stowed in a compartment cooled with ice was shipped from the United States to Great Britain in 1875.[1] Beef can be carried in chilled condition for not much more than 35 days. The temperature is generally between 28.5°F and 29.5°F. Higher temperatures can be used for short voyages, and in fact authorities on the subject recommend that the highest temperature possible for a good outturn be maintained.[2] The shipowner does not specify the temperature to be maintained; this is the responsibility of the shipper of the commodity regardless of the class of refrigerated cargo involved. The relative humidity in a compartment of chilled beef should be between 88 and 92%. Air velocities within the compartment should be maintained at approximately 32 ft/minute. Chilled beef is hung from meat rails in such ships in 'tween-decks that are approximately 6 ft 6 in. in depth or greater, at increments of 3 ft. If the depth of the hold is greater than 9 ft 6 in., more than two tiers of beef quarters will have to be carried. The space on older ships is cooled by brine coils that are suspended between the meat rails on the overhead and coils secured to the sides and ends of the compartment. In such compartments deck gratings are unnecessary. The brine piping should be exten-

[1] A. Woods, "Transport of Refrigerated Cargoes Under Modern Marine Practices," *Transactions of Institute of Naval Architects (British)*, 1935, p. 32.
[2] L. Williams, "The Design and Construction of Refrigerated Ships," Transactions of SNAME, 1924, Vol. 32, p. 131.

sive so that higher brine temperatures can be maintained and the average temperature within the compartment still kept at the desired level with only slight variation. The temperature of the compartment should not be allowed to fluctuate more than 0.5°F below and above the specified carrying temperature.

In compartments containing chilled beef, the weight of the cargo is carried by the overhead. If the deck above is used to carry commodities stowed on the deck in the ordinary manner, the deck must be able to handle the greater deck load. The meat carried in this manner packs in at approximately 125 ft³/ton. The meat must not hang too freely, nor must it be packed in too tightly. If it is too loose, it will suffer some damage from striking adjacent structural members or other meat during violent rolling or pitching in a seaway. If it is packed too tightly, it will be damaged by crushing stresses exerted throughout the stowed block during the voyage, and it may also suffer because of the reduced air flow around the meat. (A good air flow is necessary for a good out-turn.) Because of the necessity of maintaining an even temperature level, and since it is easier to maintain uniform temperature levels in small compartments, an ideal chilled meat carrier would have a number of smaller compartments rather than a few large compartments. Leeming states that in the trade between Buenos Aires and Great Britain some of the ships had as many as 160 separate refrigerated cargo spaces.[3] These spaces were served with separate brine piping, but the brine was obtained from a common brine cooling tank. Such ships would not lend themselves very conveniently to the carriage of general cargoes; the owners sacrificed general cargo handling ability to obtain a more perfect refrigerated type of ship.

All-Refrigerated Frozen Meat Carrier

The all-refrigerated frozen meat ship does not require the limited height in the 'tweendeck spaces and lower holds that is necessary in the chilled meat carrier. An ordinary cargo vessel can be converted into a frozen meat carrier by properly insulating the spaces and installing coils within the compartments or constructing cold air bunker rooms and ducts. This is because the meat can be stowed to a height limited only by the depth of the compartment. On a ship designed and built for the purpose of carrying only frozen meats, the simplest and most economical installation is wall

[3] J. Leeming, Modern Ship Stowage, U.S. Government Printing Office, Washington, D. C., 1942, p. 99.

coils. However, for more versatility the use of cooled recirculated air is called for.

Because of the extremely low temperatures maintained in the frozen meat compartments (0 to 10°F), the insulation must be more effective and is generally thicker than on a ship constructed for the carriage of chilled or cooled cargoes only. No outside air is allowed to enter the frozen meat compartment once the cargo has been stowed. Any air brought in from the outside would deposit its moisture content on the cooling coils and necessitate frequent defrosting or decrease the general efficiency of the cooling system. For the purpose of maintaining a constant temperature throughout the compartment, the air within is agitated with blowers, but the insulation plugs are caulked and an effort is made to prevent the entrance of outside air. The temperatures maintained in a frozen meat compartment should be 0°F or less. The reason for this is that although the freezing point of meat is 27°F, pork will turn rancid after 2 months at 15°F, and beef, lamb, and veal will turn rancid after 3 months.[4] The needs for some degree of safety when transferring the product from one carrier to another and for prevention of thawing in high-temperature pockets are why the maximum temperature should be below 0°F.

All-Refrigerated Cool Air Carrier

The all-refrigerated cool air carriers differ greatly from both of the other all-refrigerated types mentioned above. For one thing, the cooling is done entirely by recirculating cool air through the cargo spaces after it has been passed over a bank of coils with cold brine in them. In general the refrigeration machinery, such as the compressor, evaporator coils, brine tanks, condenser, and receiver, is located amidships in or near the machinery spaces. The cold brine is piped to a compartment where large fans capable of handling 15,000 to 25,000 cubic feet of air per minute (cfm) blow air over the coils in which the brine is contained. The air is cooled to the desired dry bulb temperature level and sent through ducts to be discharged in the refrigerated spaces. There are numerous variations in the arrangement of the units and methods used in recirculating the cool air type of refrigeration installation. The sizes of the units vary; in smaller installations, direct expansion coils may be used instead of the brine coils; and the air may be recirculated by means of a cold air diffuser or by use of a periphery duct.

[4] *Refrigerating Data Book, Application Volume,* 2nd ed., A.S.R.E., New York, 1946, p. 373.

General Refrigerated Carrier

The refrigerated general cargo carrier is a ship that has some part of her total cubic capacity devoted to refrigerated cargo. There is a great variance of opinion as to how much refrigerated space a general carrier should have to make her a versatile carrier. Of course, the run in which the ship operates would have a major influence in fixing this figure. For a carrier to be truly a *general* carrier she must have *some* provision for the carriage of refrigerated products. The amount of partial refrigeration space runs all the way from the 1000- or 2000-ft^3 portable, wall-coil reefer boxes up to 90,000 ft^3. Vessels with 70,000 ft^3 of reefer space are examples of adequate provision for refrigerated cargoes if and when they are available.

In the partially refrigerated general carrier the reefer boxes are located in the 'tweendeck areas. Thus, when not in use as reefer spaces, they can be used for general dry cargo of the packaged type without much loss in efficiency. If ingress to the hatchway around the boxes can be obtained via a side port opening, the inconvenience of loading dry cargo into the confined spaces of a reefer box is reduced to a minimum. The lower hold can be worked continuously with loading through the side ports using roller conveyors.

All of the completely refrigerated ships in the U.S. Merchant Marine are cooled by recirculation of cool air through insulated spaces. Of the partially refrigerated vessels, the vast majority are cooled by recirculated cool air; the others are cooled by inserting wall coils within the space and cooling the air directly in the box.

With respect to the type of refrigerant used, it can definitely be said that in all ships built for the U.S. Merchant Marine since 1940 the cargo refrigeration system utilizes dichlorodifluoromethane (R-12), commonly referred to as Freon-12. In recent times ships have been equipped with centrifugal compressors, which operate more efficiently with trichloromonofluoromethane (R-11), commonly referred to as Freon-11.

Factors Involved in the Safe Handling and Stowage of Low-Temperature Cargoes

Dry Bulb Temperature

The proper maintenance of the dry bulb temperature level within a reefer space is not the only factor affecting the condition of refrigerated cargoes at the discharging port. But the temperature level is, of course, of primary importance when carrying hard-frozen cargoes. The exact temperature to be maintained for a good out-turn varies with the type of commodity, the species of the commodity, the length of the voyage, and the authority establishing the temperature range. The first three factors are fairly obvious; the fourth is placed on the list because the authors noted a wide variance in suggested temperatures at which to carry a given commodity when consulting various lists published by different authorities. In many cases the recommended temperatures were the same, but the point is that such lists must be used as guides only and not as hard and fast rules. In addition to the three important variables mentioned above in regards to the temperature levels to be maintained, there is one more. *This is the stage of ripeness of the fruit when picked.* The shipper of the commodity specifies the carrying temperature and the limits of temperature variation during the voyage; however, all ship's officers should have a general knowledge of the factors involved in deciding the carrying temperature and should check the specified temperature against the temperature listed in some authoritative text.

Condition of Atmosphere

The condition of the atmosphere in the reefer space is another factor affecting the condition of the cargo when discharged. Controlling the atmosphere means regulating the amount of carbon dioxide (CO_2) and oxygen in the reefer space. The technique is successful when the total amount of these two gases equals 21% of the total atmosphere in the space. All fruits continue to respire (ripen) after being picked, and the end products of this process are CO_2, water vapor, and heat. It has been established that respiration is diminished not only by lowering the temperature, but also by reducing the amount of oxygen and allowing the CO_2 to accumulate. The percentage of CO_2 can be controlled by ventilating when the concentration becomes too great, but this technique is possible only when the total amount of these two gases is 21%. To obtain a concentration of oxygen less than that which can be obtained by ventilation, it is necessary to absorb excessive amounts of CO_2 by passing the atmosphere of the space through a solution of strong sodium hydroxide or potassium hydroxide. Control by ventilation then becomes possible when the amounts of CO_2 and oxygen have been brought to a total of 21%, such as 8% CO_2 and 13% oxygen. To determine the amount of each it is necessary only to obtain the percentage of CO_2 and subtract from 21. If the initial concentration was 4% oxygen and 6% CO_2, it would be impossible to obtain this ratio except by the absorption technique as mentioned above. To understand this it is necessary to understand that CO_2 is produced in a volume equal to that of the oxygen consumed.

The principal advantage of using a controlled atmosphere is the ability to keep the fruit in storage for longer periods with less deterioration or for the same periods as before the use of controlled atmospheres but at higher temperatures. Another advantage is the reduction in the rate of mold growth in atmospheres high in CO_2 concentration.

Marine cargo refrigeration practice does not yet include the controlling of storage atmospheres beyond the introduction of air by ventilation to prevent the concentration of CO_2 from becoming too high. With respect to how much changing of the air is required in a given reefer space loaded with a certain cargo, no exact answer can be given. The air change rate may vary from a complete air change every minute, or a change of 2 or 3% of the total capacity of the space (unloaded) every minute. This applies only to those products that are carried above their freezing points and are therefore producing CO_2.

Installations for the storage of identical products ashore in many cases call for treatment completely different from that for storage on board a ship. For example, apples in storage ashore may not be ventilated at all, although the apple may suffocate and suffer a diffuse browning of the skin known as "apple scald" if the CO_2 content becomes too great. Overaccumulation of CO_2 is avoided by circulating the storage atmosphere through a washer or "scrubber" filled with a 30% solution of sodium hydroxide. In such an installation, a gastight compartment must be available and all the commodities must respond in the same way to the concentration of CO_2 maintained. On board ship it is impractical to consider such an expensive installation. It would be difficult to obtain shipments that would entirely fill a compartment and much space would thus be lost. Probably the deciding factor is that the technique of controlling the atmosphere to such a fine degree is primarily for the purpose of retaining products in storage for many weeks or months, whereas on the ship the prime objective is to retain them in good condition for a relatively short period and deliver them in that condition to the consignee of the shipment at the port of discharge. Three weeks on board the ship would be a long time. Ship operators plan the loading of refrigerated products on the last day in port and make it a point to arrange for the receipt of such cargoes the day the ship arrives in the port of discharge; hence the time in storage on aboard the ship is usually only the length of the voyage. Under some conditions of operation, however, it may be considerably longer.

Relative Humidity

The relative humidity of the storage atmosphere must be maintained at an optimum level in a reefer space to obtain a good out-turn of the cargo. This factor is closely allied to the problem of controlling the condition of the atmosphere, but it is important enough to deserve a separate entry on our list of factors involved in the safe handling of reefer cargoes. The relative humidities recommended are high for practically all commodities; the exceptions are chocolate candies and dried egg products. The humidity level in a reefer space carrying fruit above its freezing temperature is of the greatest importance. Before and after fruit is picked it is ripening, and this ripening process causes the fruit to give off water vapor, as mentioned before. But before being picked, additional water is available to the fruit to replace that given up in respiration during ripening. When this source of water is removed by picking, the ripening process, with its accompanying release of water, must be slowed down. The loss of water vapor is largely a function of the vapor pressure of the water in the ambient air in which the fruit is stowed. Hence, there is the need for a large amount of absolute humidity, which results also in a high percentage of relative humidity, inasmuch as the vapor pressure of the ambient air varies directly as the absolute humidity. Theoretically then, it appears that 100% relative humidity would be the most desirable. This, however, is not so, because at this humidity the water vapor released from the fruit would condense on the skin of the fruit as free moisture and cause conditions that would promote other forms of disease and mold growth. On the other hand, many products are not carried at humidities that are the best for prevention of moisture loss, because that relative humidity level within the compartment in which the commodity is stowed would be too high for the control of mold growth on the containers and on the fruit itself. Actually, the relative humidity used is a compromise between that level which would allow such a rapid discharge of moisture that the skin of the fruit would shrivel and that level which would entirely control moisture loss. If the relative humidity is maintained at too high a level, mold will develop, the skin of the fruit will become slimy, bacterial growth will be promoted, and in general the out-turn will be poor. If the relative humidity is too low, the moisture loss will be excessive, resulting in loss of weight, shriveling and discoloration of the skin, and again a poor out-turn.

Control of relative humidity. The control of the relative humidity in refrigerated spaces on board a ship is not precise. Under shipboard conditions it is especially difficult to maintain during the period when the sensible heat load of the product is being stabilized. During this period, generally referred to as the *pull-down period,* the difference between the temperature of the circulating air as it enters the reefer space and as it leaves is at a maximum. This period should not be more than 72 hours. Conditions of the storage atmosphere should be stabilized within that time, which is short enough that little damage should occur as a result of low average relative humidities. That low relative humidities would exist is evident when it is noted that if the incoming air is at 40°F, at nearly 100% relative humidity and the outgoing air is at 45°F, with a drop in relative humidity due to the rise in temperature of about 17%, or down to 83%. This relative humidity would be increased, of course, by some water vapor picked up by the air from the cargo in storage. Whether the amount picked up would be sufficient to bring the relative humidity up to a suitable level is not certain. On the other hand, once the reefer space temperature level has been stabilized and the incoming air is at 40°F at nearly 100% relative humidity and the outgoing air is at 42°F, the relative humidity at the outlet end will be 93%, which means that the average relative humidity in the reefer space will be very high, and especially so if additional water vapor is picked up from the ripening fruit. In the case of frozen products the relative humidity would be even higher, for the temperature rise of the circulating air would be less. It is evident that with frozen products a two-temperature brine coil system is necessary to prevent the relative humidity from being constantly too high. With products that give off some vital heat and continue to ripen within the refrigerated compartment, the best control of the relative humidity is obtained by operating with a low mean effective temperature differential. The dry bulb temperature within the compartment can be maintained at correct levels under a low mean effective temperature differential by having a correspondingly large area of coils and operating with low air velocities. (The mean effective temperature differential referred to is that between the temperatures of the refrigerant in the coils and of the medium to be cooled.)

Air Motion

The rate of air movement within the reefer space also has its effect on the condition of the refrigerated cargo at the out-turn. Air velocities are stated in feet per minute (fpm). The recommended exact rate of air flow within the reefer space depends on the commodities stowed in it. The velocity may range from 30 to 100 fpm. Products requiring precooling to the proper storage level in the reefer space require a greater rate of air flow. Products that have high vital heats need greater velocities than those with no or low vital heats. Although products can be damage by too high or too low an air velocity, it is difficult to obtain exact data concerning the optimum rate for a given type of cargo. The effects of too great a velocity are similar to those of too low a relative humidity: discoloration, excessive moisture loss, and shriveling of skin on fruits; and, of course, a loss of weight. Too low a velocity will produce effects similar to those of a high relative humidity, such as mold growth, slime on surfaces, and some diseases in fruits that cause a loss of flavor or texture or both.

The rate of air flow within a space can be calculated very easily if the volume output of the fans moving the air and the cross-sectional area of the path along which the air must pass are known. Knowing the two values mentioned, we immediately have:

$$V = R \cdot A \qquad R = \frac{V}{A}$$

where R = the velocity of the air in fpm.
V = the volume of the fan output (cfm).
A = the cross-sectional area of the air path.

Types of Refrigerated Spaces

Some Early Systems

The earliest refrigerated compartment constructed on a ship for the carriage of chilled beef allowed convection air currents to flow and maintain the interior of the compartment at a temperature close to 32°F. With the addition of salt, a slightly lower temperature could be reached. This system was used on a ship between New York and London in 1875.[5] For this short voyage, a temperature of 30 to 32°F was sufficiently low for safe carriage of the chilled beef. If the system was not such a tremendous space-consumer, it could be considered as effective as the modern installations of today, except, of course, for the labor involved and the lack of flexibility. On December 6, 1879, the S.S. Strathlevan sailed from Mel-

[5] A. Woods, "Transport of Refrigerated Cargoes Under Modern Marine Practices," *Transactions of Institute of Naval Architects (British)*, 1935, p. 32.

bourne for London with a cargo of frozen beef. She arrived 70 days later, on February 15, 1880, and discharged her cargo in good condition. This was perhaps one of the first shipments of frozen cargo.

The above shipment of frozen beef could have been accomplished only by the use of mechanical refrigeration. All mechanically cooled spaces in which cargo is carried on ships today are constructed with insulated boundaries ranging in thickness from approximately 10 to 15 in. The necessary low temperatures are obtained by circulating a cooling agent through coils mounted directly in the compartment or by circulating a precooled air stream through the compartment. The exact thickness of insulation required to reduce the rate of heat flow into a refrigerated compartment most economically is not easily calculated. A great deal of experimental work has been and still is being done to gain greater knowledge concerning this specialized branch of refrigeration engineering.

Wall Coil Spaces

Spaces cooled by wall coils have the coils secured to the sides and ends of the refrigerated spaces, and sometimes to the overhead. These coils are approximately 1½ to 2 in. in diameter and are generally thin-walled galvanized steel coils. They are generally found in layers of two or three. The reason for the multiple layers is to provide a greater surface area for heat transfer, which allows the cooling agent inside the coils to be maintained at a higher temperature than would be possible otherwise. Maintaining the refrigerant at a higher temperature results in more economical operating costs for the refrigeration machinery. The cooling agent may be either a type of refrigerant that is actually evaporating within the coils or a brine solution that has been cooled in a centrally located brine tank and then pumped into the coils. The former type is known as a *direct expansion system,* while the latter is known as a *secondary* or *indirect system.* Coils in spaces of this type are not of the finned type. Finned coils have the advantage of giving a greater surface area per linear foot of coil length, but they frost up very rapidly between the fins if their temperature is below the freezing point of water. Once a coating of frost has formed on any coil, the frost acts as an insulator and cuts down on the heat-transferring efficiency of the coil.

When a compartment cooled with wall coils has been stowed with cargo and the doors have been shut in preparation for stabilizing the temperature for the voyage, there will initially be a considerable amount of condensation. There will also be heavy condensation within the compartment if any system of air renewal is used, as in the storage of certain fruits. As a consequence, it is imperative that these spaces be equipped with several drains leading from the compartment to the drain well for that section of the ship. These spaces are most suitable for the carriage of hard-frozen commodities; however, they can be used for the carriage of chilled or cooled cargoes. Chilled cargo is that which is carried at a temperature of 1 or 2°F above the commodity's freezing point. For example, chilled beef is carried at 28.5 to 29.5°F; the freezing point of beef is 27°F. Cooled cargoes are those that are carried at relatively high temperatures for the purpose of retarding the growth of bacteria or slowing down the ripening process without causing any damage to the commodity. The best example of a cooled cargo is bananas, which are carried at 53°F. The freezing temperature of green bananas is 30°F, but if they are brought down to temperatures near 30°F they rapidly turn black and the fruit is not suitable for eating after being removed from refrigeration. They will suffer from exposure to temperatures only a few degrees below 53°F and when removed from refrigeration will develop spots on their skin and rot very rapidly.

When stowing hard-frozen products in wall coil compartments it is necessary to use the same security precautions that are standard for any cargo. This means leaving no voids within the stowed cargo block and stowing so that the containers can take, without damage, all pressures likely to be placed upon them. Sufficient air space should be left above and on the sides and underneath the stowed cargo to allow the agitated air to flow around the cargo. With hard frozen cargoes, it is not mandatory to build air passageways between each and every tier. The air is agitated in such compartments to prevent some local spot in the periphery of the compartment from becoming warm enough to cause damage to the cargo.

It is necessary to keep most chilled cargoes and cooled cargoes separated from the coils by at least 8 in., and 12 in. is preferable. This is necessary because the coils are several degrees cooler than the average temperature maintained in the compartment, and if the commodity is too near the coil it will be frozen or cooled below the safe carrying limit. This separation is maintained by stowing the containers away from the coils and then using light dunnage to shore them up and to act as distance pieces. In addition, air channels must be built into the cargo block between every tier, as in most cases of stowing chilled or cooled cargoes.

From the above, it can be seen that the wall coil compartment is best suited for the stowage of hard-frozen beef and other products.

Cool Air Systems

The cool air system consists of an insulated space into which is blown a current of cool air. As this low-temperature air passes through the cargo space, it picks up heat from the cargo, and eventually it is discharged from the space at a slightly higher temperature than it had when it entered. The heat picked up by the air is in turn given up as it passes over a bank of coils kept suitably cool. The air is then recirculated back through the refrigerated space. In some cases, it is necessary to introduce fresh outside air into the recirculated air stream. This new air may be added intermittently or continuously. This air renewal is necessary when carrying fruits at relatively high temperatures, for example, lemons at 55°F or bananas at 53°F. The air taken from outside must always be adjusted thermally before being introduced into the refrigerated space. Although it is true that the temperature adjustment is downward in almost all cases, there are times when it is necessary to adjust the temperature upward. There is a need for some means of raising the temperature of new air when carrying cargoes maintained at temperatures above the winter temperatures along the vessel's route. All ships equipped to carry bananas into northern waters during the wintertime have heating units as well as cooling units in their recirculating systems.

Cool air bunker room.   The component parts of one cool air system are essentially the same as those of any other, but there is a difference that leads to the use of two terms in describing them. In one type the air stream passes over a coil, containing the cooling agent, that is installed in a compartment which may be adjacent to or remote from the space being cooled but is in any case definitely separated from such a space. Such an arrangement is known as a cold air bunker room system; it is common in all-refrigerated vessels.

Cool air diffusers.   Another type of installation, known as a cool air diffuser, consists of cooling coils and fans with appropriate controls installed as a unit within the space to be refrigerated. Cool air diffuser units may be so small that they can be secured to the overhead of the compartment and the air current maintained by simply agitating the air within the space, not using air ducts. The larger cool air diffuser units may take up approximately 10% of the available cubic. These units may utilize air ducts for the distribution of the cool air within the space. All cool air bunker room systems use air ducts, generally secured to the upper periphery of the compartment.

Basic Considerations When Stowing Cargo in These Spaces

When stowing refrigerated cargoes in any one of the above spaces, the principal precaution is to allow adequate space between the tiers and rows. Adequate space is provided by using common building lath for dunnage. The possibility of improper stowage is greater when stowing a partial load in such spaces. Partial loads in compartments using periphery ducts for air distribution should be spread out over the deck area of the compartment, with good spacing between the vertical piles, that is, good horizontal spacing. The air flow here is out from the duct slit and downward to the deck, flowing under the gratings to the outlet. If the cargo is piled up in a block anywhere in the space, the air flow passes over the cargo and then downward and to the outlet. Thus, the air by-passes the cargo block and the interior of the block is not properly cooled, which may result in a poor out-turn.

In a space fitted out with a cool air diffuser unit but without ducts, the best way to stow a partial load is to build it up as high as feasible across the compartment at the end where the outlet louver is located. The cargo must not be at the opposite end from the outlet louver. If it is stowed here, the air flow will naturally follow the course of least resistance, and after leaving the outlet blower the air will drop and return directly to the outlet louver, having penetrated the cargo hardly at all.

The general principle to be followed in all partial loadings of refrigerated spaces is to avoid stowing cargo so that the air flow will by-pass the cargo and flow directly to the outlet, rather than being forced through the cargo block. Obviously, there must be no deviation from general good stowage principles when meeting this special requirement.

STOWAGE OF EXPLOSIVES AND OTHER DANGEROUS ARTICLES

The quantity and diversity of hazardous materials transported on cargo vessels has increased significantly in recent years. Industry is continually developing different compounds and chemicals as our technologies demand them. As these commodities come on the market, they appear on our bookings for transportation, necessitating the development of safe methods of transporting them.

Regulations for the transport of hazardous cargoes are increasing apace with the above-mentioned trends, with various regulatory bodies drawing up and enforcing them. In the United States, the Materials Transportation Bureau of the Department of Transportation (DOT) formulates the regulations and the Coast Guard is responsible for their enforcement. The only exception is bulk cargoes, for which the Coast Guard is solely responsible.

In January 1977, a new set of regulations went into effect. Many of them are unchanged, but there is much new material as well. The increase in hazardous cargoes makes it more necessary than ever for the ship's officer to be thoroughly familiar with the hazards involved in and the regulations governing their transportation on his vessel.

These regulations are contained in 49 Code of Federal Regulations (CFR) 171-177, "Hazardous Materials Regulations." This is a lengthy and complicated publication. It contains regulations for transportation by air, rail, and highway as well as by water. Some sections apply to all modes, whereas others are limited to one. Many sections contain a great deal of material that does not concern the ship's officer and a few items that are very important to him. The only way to make oneself really familiar with all the regulations necessary is to go through the whole book, picking out those items that apply.

With this in mind, the following summary is not intended as a substitute for the CFR publication, but only as a guide to its use.

Part 171—General Information, Regulations, and Definitions

Definitions and Abbreviations

This section should be read over for applicable items, then referred to as needed when reading other parts.

Incidents

These sections require that immediate notice be given the Department of Transportation of any incident involving hazardous materials that involves any of the following: a person killed, injuries requiring hospitalization, damage exceeding $50,000, damage to or suspected contamination involving radioactive material or etiologic agents, or any other situation that in your judgment should be reported. This notice must be given by telephone. The notice must contain the name of the reporter; name and address of owner; phone number of reporter; date, time, and place of incident; extent of injuries; class, name, and quantity of hazardous material involved; type of incident; nature of hazardous material involvement; and whether a continuing danger to life exists. A written report must then be made on DOT Form 5800.1 within 15 days of the incident.

Part 172—Hazardous Materials Table and Hazardous Materials Communications Regulations

Part 172 consists of the hazardous materials table and the instructions for its use. Each officer should be familiar with the use of this table and should refer to it when any hazardous material is to be loaded. All sections referred to in the table should be read when looking up an item, because additional information and requirements may be found in them.

Subpart C: Shipping Papers

The preparation of these is not the responsibility of the ship's officer, but improper manifests and the like can cause problems, particularly on homebound loadings. The book should be referred to if you are in doubt about the documentation of any shipment.

Subpart D: Marking

You should be familiar with Sections 172.300 through 172.326. They give detailed requirements for the markings that must appear on packages containing hazardous materials. Violations of these could lead to confusion in emergencies and are readily apparent to Coast Guard boarding officers.

Subpart E: Labeling

On July 1, 1977, the new labeling regulations took effect (see Table 6-1). Sections 172.400 through 172.406 give requirements for label placement, number of labels, and so forth. Ship's officers should read these sections carefully and refer to them if in doubt about any package.

Part 173—Shipper's General Requirements for Shipments and Packagings

Subparts A and B

These subparts give general requirements, many of which are not applicable to carriage on ships. However, the following sections at least should be read:

173.2   Classification of a material having more than one hazard

173.21   Prohibited packing

173.26   Quantity limitations

173.29   Empty packages, portable tanks

173.32   Qualification maintenance and use of portable tanks

Subparts C through N

These subparts contain detailed requirements for specific commodities listed in the hazardous materials table and define the various classes. The material should at least be skimmed over; it could be useful if you encounter commodities with which you are not familiar or that appear to be hazardous but are not classed as such on your loading list.

Part 176—Carriage by Vessel

This part applies directly to seagoing cargo vessels. Every deck officer should be thoroughly familiar with the entire part, as the ship is responsible for compliance. Certain especially important sections are quoted or referred to here, but cannot take the place of a thorough study of this part by each officer.

Exceptions

Note that although shipments may comply with IMO regulations for marking, listing, and so forth, they must still be loaded and stowed in accordance with the regulations in this section.

TABLE 6-1

BASIC COMPARISON OF **IMO** AND **DOT** HAZARD CLASSES[a]

IMO/UN Class No.	International Code (IMO/IMDG)	U.S. Regulations (DOT/49 CFR)
	IMO Class Name	*DOT Class Name*
1	*Explosives*	*Explosives*
	1.1	Class A
	1.2	" A or B
	1.3	" B
	1.4	" C
	1.5	Blasting Agent
2	*Gases (Compressed, Liquified, or Dissolved Under Pressure)*	
	2.1 for Flammable Gases (or Inflammable Gases)[b]	*Flammable Gas*
	2.2 for NON-Flammable Gases	*Nonflammable Gas*
	2.3 for Poison Gases	*Poison A*—Poison Gas (May also apply to volatile *liquid* class A poisons.)
3	*Inflammable Liquids*[b]	*Flammable Liquids* =
	3.1 = Flash Point below 0°F (−18°C).	Flash Point *Under* 100°F (38°C).
	3.2 = Flash Point 0°F to Less than 73°F (23°C).	*Combustible Liquid* = Flash Point 100°F to Less than 200°F (93°C).
	3.3 = Flash Point 73°F to *141*°F (61°C) INCLUSIVE.	Note: Flash points by "closed cup" method.
	Note: Flash points by "closed cup" method.	
4.1	*Inflammable Solid*[b]	*Flammable Solid*
4.2	*"Spontaneously Combustible"*	*No* equivalent class (but note Pyrophoric liquids are classed as Flammable Liquids per 49 CFR). (Note 49 CFR sect. 173.150.)
4.3	*"Dangerous When Wet"*	*No* equivalent class (but "Dangerous When Wet" label also required with some Flammable Solids, per 49 CFR 172.101). (Note 49 CFR sect. 173.150).

TABLE 6-1 (*Continued*)

IMO/UN Class No.	International Code (IMO/IMDG) *IMO* Class Name	U.S. Regulations (DOT/49 CFR) *DOT* Class Name
5.1	*Oxidizing Agent*	*Oxidizer*
5.2	*Organic Peroxide* (Caution: some organic peroxides require secondary "explosive" labels.)	*Organic Peroxide*
6.1	*Poisons* (Liquids & Solids) (Note: Poison *Gas* is Class *2.3* (amend. 17). Package grp. III poisons take "Harmful " label ("St. Andrew's Cross")	*Poison B* (Liquids & Solids) Note: Former Poison C is now classed as "Irritating Materials" (Poison A is usually a *gas*.)
7	*Radioactive Substances*	*Radioactive Material*
8	*Corrosives* (Liquids or Solids)	*Corrosive Material* (Liquids or Solids)
9	*Miscellaneous Dangerous Substances* (All Class 9's are "No Label Required")	*No* General Equivalent in 49 CFR
IMO/UN CLASS: (NONE)	IMO has an equivalent class for ORM's, Hazardous Substances or Hazardous Wastes.	*Other Regulated Materials* (*ORM*): ORM-A, ORM-B, ORM-C (see 49 CFR (173.500 for all ORM designations) ORM-D (Consumer Commodity) ORM-E (Hazardous Substance & Waste—see 49 CFR sections 173.500(b)(5) & 171.8)

[a] For nearest DOT equivalent class (for multimodal documentation use) see nearest "corresponding" class in 49 CFR, section 172.102(h).

[b] "Inflammable" has *same* meaning as "Flammable" (Ref IMO IMDG Code Vol. II, pg. 3002, footnote.)

Note: This comparison chart is only a guide for the user's convenience and must not be used without simultaneous application of *all* appropriate laws, regulations and codes.

Dangerous Cargo Manifest

a. The master of a vessel transporting hazardous materials or his authorized representative shall prepare a dangerous cargo manifest, list, or stowage plan. This document must be kept in a designated holder on or near the bridge. It must contain the following information:

1. Name of vessel and official number.
2. Nationality of vessel.
3. Shipping name of each hazardous material on board, as given in the hazardous materials table, or the "correct technical name," as given in the International Dangerous Goods Code published by IMO. When the shipping name of a material is an "N.O.S." entry, this entry must be qualified by the technical name of the commodity in parentheses, for example, "Corrosive Material, N.O.S. (Caprylyl Chloride)."
4. The number and description of packages (barrels, drums, cylinders, boxes, etc.) and the gross weight for each type of packaging.
5. Classification of the hazardous material in accordance with either (i) the hazardous materials table or (ii) the IMO Dangerous Goods Code.
6. Any additional description required by section 172.203 of this subchapter.
7. Stowage location of the hazardous material on board the vessel.

b. The hazardous material information on the dangerous cargo manifest must be the same as the information furnished by the shipper on the shipping order or other shipping paper, except that the IMO "correct technical name" and the IMO class may be indicated on the manifest as provided in paragraphs (a)(3) and (a)(5) of this section. The person who supervises the preparation of the manifest shall insure that the information is correctly transcribed, and shall certify the truth and accuracy of this information to the best of his knowledge and belief by his signature and notation of the date prepared.

c. The master or licensed deck officer designated by the master and attached to the vessel shall acknowledge the correctness of the dangerous cargo manifest, list, or stowage plan by his signature.

d. Each carrier who transports or stores hazardous materials on a vessel shall retain, for one year thereafter, a copy of the dangerous cargo manifest, list, or stowage plan, and shall make that manifest or list available for inspection in accordance with Section 176.36(b) (see Table 6-2).

TABLE 6-2   DANGEROUS CARGO MANIFEST

Form TE-11 1/TE 12J/TE20J (Rev. 6/78)

MOORE McCORMACK LINES INCORPORATED—2 BROADWAY, NEW YORK, N.Y. 10004
DANGEROUS CARGO MANIFEST

American S.S. MORMACSAGA   Voy. No. 26/0   Official No. 289547   Net Tons 5486

Wherefore _____ is Master.   Loaded at DUNDALK MARINE TERMINAL   Date 22 APRIL 1982   Bound for DURBAN

Name of Local Agent MOORE MCCORMACK LINES, INC.   Address BALTIMORE, MARYLAND 21202

B/L or D/R	MARKS	QUANTITY	COMMODITY (TRUE SHIPPING NAME)	GR. WGHT.	CLASS	LABEL (or state "None Req.")	STOWAGE
	ADVANCED CHEMICALS VIA DURBAN	15 DRUMS	CEMENT, LIQUID, N.O.S. (CONTAINS XYLENES) F/P 83°F, IMCO PAGE #3123, UN #1133, CLASS 3.3	8235#	INFLAMMABLE LIQUIDS	INFLAMMABLE LIQUID	#4 ON DECK FWD STB SIDE
	ADVANCED CHEMICALS VIA DURBAN	15 DRUMS	CEMENT, LIQUID, N.O.S. (CONTAINS METHYL ISOBUTYL KETONE) F/P 66°F, IMCO PAGE #3064. UN #1133, CLASS 3.2	7395#	INFLAMMABLE LIQUIDS	INFLAMMABLE LIQUID	#4 ON DECK FWD STB SIDE
	BRITISH INDUSTRIAL PLASTICS DURBAN	3 DRUMS	LITHIUM HYDROXIDE MONOHYDRATE IMCO PAGE #8105-3, UN #2680 CLASS 8	1254#	CORROSIVES	CORROSIVE	#6 T/DC AFT PORT SIDE
	ABBOTT C/O RENNIES SEAFRT JOHANNESBURG VIA DURBAN	5 BOXES SAID TO CONTAIN 7-7½ OUNCE TUBES EACH BOX	CEMENT, ADHESIVE, N.O.S. (CONTAINS KARAYA PASTE) F/P 58°F, IMCO PAGE #3064. UN #1133, CLASS 3.2	178#	INFLAMMABLE LIQUIDS	INFLAMMABLE LIQUID	#4 ON DECK FWD STB SIDE

This manifest shall be prepared* whenever dangerous cargo is loaded in accordance with the DEPARTMENT OF TRANSPORT'S HAZARDOUS MATERIALS REGULATIONS.
* When no dangerous cargo is carried, the Master shall post a properly headed and signed Dangerous Cargo Manifest bearing the remarks "no dangerous cargo carried." No further distribution required.

DISTRIBUTION:
Two (2) copies to the Master before vessel sails.
One (1) copy to Captain of Port—U.S.C.G., Governors Island, N.Y.C. 10004 (U.S. Ports Only).
One (1) copy to Captain of Port—U.S.C.G., Port of Loading (U.S. Ports Only).
One (1) copy to Stevedoring Dept., 23rd Street Terminal, Brooklyn, N.Y. 11232.
One (1) copy to Operations, 23rd Street Terminal, Brooklyn, N.Y. 11232.
One (1) copy to Local Agent.

PREPARED BY: DATE   4/21/82

AUTHORIZED SIGNATURE

PLEASE PRINT NAME

ON BOARD COPIES CHECKED BY:

MASTER/CHIEF OFFICER SIGNATURE

PLEASE PRINT NAME

Exemptions

"If a hazardous material is being transported by a vessel under the authority of an exemption and a copy of the exemption is required to be on board the vessel, it must be kept with the dangerous cargo manifest." Since in most cases a copy of an exemption must be on the vessel, copies of all such papers should be put with the dangerous cargo manifest in its proper location.

Inspections

Section 176.39 requires that an inspection of spaces containing hazardous materials be made after loading and every 24 hours thereafter. The fire detecting system need not be inspected daily, but must still be inspected after heavy weather. In addition, an inspection is required immediately before entering a U.S. port. The section does not restrict this to entering from foreign waters, so an inspection is required before each port on the coast. All of these inspections must be entered in the deck log.

Notification

Section 176.48 requires that the nearest District Commander USCG be notified when a "fire or other hazardous condition" exists and there are hazardous materials on board. Notification is also required when any hazardous material is lost or jettisoned. These notifications are separate from the ones to the DOT required in 171.15 and 171.16.

Supervision

Section 176.57, paragraphs (a) and (c), requires that the handling and stowage of hazardous materials be supervised by a ship's officer. This makes it necessary for a mate to be *on deck* at all times when this cargo is being worked.

General Stowage Requirements for Hazardous Materials

Hazardous materials must be stowed in a manner that will facilitate their inspection during the voyage and their removal from a potentially dangerous situation, including fire. Each package marked "this side up" must be stowed so as to remain in the position indicated during transportation.

On-Deck Stowage of Breakbulk Hazardous Materials
(See Figs. 6-20 and 6-21)

a.   Packages containing hazardous materials must be secured by enclosure in boxes, cribs, or cradles and by proper lashing with wire rope, strapping, or other means, including shoring, bracing or both. Lashing of deck cargo is permitted if pad eyes are used to attach the lashings. Lashings may not be secured to guard rails. Bulky articles must be shored.

b.   Packaging susceptible to weather or water damage must be protected so that it will not be exposed to the weather or to sea water.

c.   Not more than 50% of the total open deck area may be used for the stowage of hazardous materials.

d.   Fireplugs, hoses, sounding pipes, and access to these must be free and clear of all cargo.

e.   Crew and passenger spaces and areas set aside for the crew's use may not be used to stow any hazardous material.

f.   A hazardous material may not be stowed within a horizontal distance of 25 ft of an operating or embarkation point of a lifeboat.

g.   Hazardous materials must be stowed so as to permit safe access to the crew's quarters and to all parts of the deck required in navigation and necessary working of the vessel.

h.   When runways for use of the crew are built over stowed hazardous materials, they must be constructed and fitted with rails and life lines so as to afford complete protection to the crew when in use.

Other Sections of Concern to the Ship's Officer

Section 176.78 gives comprehensive and detailed requirements for the use of forklifts. Their strict enforcement would require the ship's officers to observe forklift use closely. Note that paragraphs (g)(8) and (9) require that each truck be equipped with a fire extinguisher or that one must be kept in the space the truck is working. The ship's firefighting system must also be kept "ready for immediate use."

Section 176.83 consists of two tables, explanations, and definitions that together set forth segregation requirements. This whole section should be understood thoroughly and the tables used whenever hazardous material is to be loaded. These requirements are in addition to any others (see Tables 6-3 and 6-4).

Subparts G through O give detailed requirements for the handling and stowage of each class of hazardous material. Some of these, such as the

FIG. 6-20. Four drums of lube oil stowed on deck at the forward starboard side of the no. 2 hatch are classified as a flammable liquid, UN Class 3.1. Requirements for stowage from Title 49 172.101 state that these drums must be stowed on deck separated by 10 ft from other hazardous materials or below deck away from heat. Courtesy of M/N S. Willett, Class of 1985, USMMA.

FIG. 6-21. On the port side of the no. 2 hatch on deck are two cylinders of tri-*n*-hexyl aluminum, also known as isopentane. This commodity is a UN flammable liquid, Class 3.1, and thus must be stowed either on deck or below deck away from other flammable materials. Courtesy of M/N S. Willett, Class of 1985, USMMA.

TABLE 6-3
SEGREGATION TABLE [a]

Material Class / Material Class		1.1 1.5	1.2	1.3	1.4	2.1	2.2 2.3	3.1 3.2	3.3	4.1	4.2	4.3	5.1	5.2	6.1	7	8
Explosives	1.1 1.5	*	*	*	*	4	2	4	4	4	4	4	4	4	2	2	4
Explosives	1.2	*	*	*	*	4	2	4	4	4	4	4	4	4	2	2	4
Explosives	1.3	*	*	*	*	4	2	4	4	3	3	4	4	4	2	2	2
Explosives	1.4	*	*	*	*	2	1	2	2	2	2	2	2	2	X	2	2
Inflammable gases	2.1	4	4	4	2		X	2	2	1	2	1	2	4	X	2	1
Other than inflammable gases	2.2 2.3	2	2	2	1	X		2	2	X	1	X	X	2	X	1	X
Inflammable liquids	3.1 3.2	4	4	4	2	2	2			2	2	2	2	3	X	2	1
Inflammable liquids	3.3	4	4	4	2	2	2			1	2	2	2	3	X	2	1
Inflammable solids	4.1	4	4	3	2	1	X	2	1		1	1	1	2	X	2	1
Spontaneously combustible substances	4.2	4	4	3	2	2	1	2	2	1		1	2	2	X	2	1
Substances that are dangerous when wet	4.3	4	4	4	2	1	X	2	2	1	1		2	2	X	2	1
Oxidizing substances	5.1	4	4	4	2	2	X	2	2	1	2	2		2	1	1	2
Organic peroxides	5.2	4	4	4	2	4	2	3	3	2	2	2	2		1	2	2
Poisons	6.1	2	2	2	X	X	X	X	X	X	X	X	1	1		X	X
Radioactive substances	7	2	2	2	2	2	1	2	2	2	2	2	1	2	X		2
Corrosives	8	4	4	2	2	1	X	1	1	1	1	1	2	2	X	2	
Miscellaneous dangerous substances	9	No general segregation recommended; individual schedules should be consulted.															

[a] This table shows the general requirements for segregation between classes, but since the properties of substances or articles within each class may vary greatly, the individual schedules and relevant introduction to that class should always be consulted for individual requirements for stowage and segregation.

Numbers relate to the following terms as defined in sub-section 15.8:
1. away from
2. separated from
3. separated by a complete compartment or hold from
4. separated longitudinally by an intervening complete compartment or hold from
X no general segregation recommended; individual schedules should be consulted
* see subsection 5.4 of the introduction to Class 1 for separation of goods of Class 1 of different compatibility groups from one another.

TABLE 6-4
SEGREGATION REQUIREMENTS

Segregation No.	Stowage Definition	Underdeck Fore & Aft	Underdeck Athwartships	On Deck Fore & Aft	On Deck Athwartships	Vertical Stow
1	Away from	None	None	None	None	Separated by a lid. A solid may be stowed over a liquid if at least one van (8 ft) intervenes
2	Separate from	Minimum 20 ft or a bulkhead	Minimum 16 ft	Minimum 20 ft	Minimum 16 ft	Separated by a lid
3	Separate by a complete hold from	Separated by a bulkhead	Not permitted	Minimum 20 ft	Minimum 24 ft	May never be stowed in the same vertical line at all
4	Separated longitudinally by an intervening complete hold from	Two bulkheads or one bulkhead and 80 ft	Not permitted	Minimum 80 ft	Not permitted	Not applicable

330

construction of magazines, will rarely apply to your vessel. Others will be commonly encountered and ship's officers must be conversant with them. Some of the more important sections are mentioned here, but the material should be read over.

Section 176.105(a) requires that class A or B material be loaded last in any one port and that other explosives not be worked at the same time as other cargo.

Section 176.115(a)(2) states: "An explosive may not be stowed nearer than 25 feet in a horizontal plane to the crew quarters."

Section 176.125 states: "A deck load over which explosives must be passed may not exceed the height of the hatch coaming, bulwark or 3 feet, whichever is greater."

Section 176.205(7) requires fire screens on the weather ends of all vent ducts from compartments containing compressed gases, and paragraph (8) prohibits compressed gases from compartments with gooseneck type vent heads.

Section 176.210 requires that poisons be stowed away from quarters and any vents serving them. It also states that a package having both poison gas and flammable gas labels must be segregated as a flammable compressed gas.

Section 176.800 says that corrosive materials must be stowed so as to be readily observable, and prohibits their stowage above any combustible substance. It also prevents the stowage of corrosive material above a compartment containing cotton or fibers.

Section 176.900 requires that a 'tweendeck containing cotton or fibers be closed with "hatch covers, tarpaulins and dunnage", and requires that hatches containing cotton or fibers be closed when not being worked unless a fire watch is kept in the hold. It also requires that cotton or fibers be segregated as flammable solids that may not be stowed with other flammable solids or with any combustible liquid. If cotton or fibers are stowed in the same hold or compartment with rosin or pitch, they must be separated by dunnage or noncombustible cargo. Where large amounts are involved, the rosin or pitch must be floored off with two layers of 1-in. dunnage and the cotton or fiber stowed above them.

Fire Protection Requirements

Fire protection requirements are involved with shipment of any hazardous material. Whenever such material is to be worked, "No Smoking" signs must be posted in the vicinity, the fire hose at each hold must be led out and fitted with an all-purpose nozzle and pressure maintained on the fire line, and two 15-lb foam or dry-chemical fire extinguishers must be placed in the vicinity of each hatch.

Segregation

Segregation requirements for a hazardous material and any incompatible package can be determined from Tables 6-3 and 6-4.

THE STOWAGE FACTOR

Stowage Factor Defined

The *stowage factor* is defined as the number of cubic feet required to stow 1 ton of a given cargo. It is a value that is used to answer two very important questions: (1) When given a certain amount of cargo, what is the amount of space that will be consumed in stowing it? (2) When given a certain volume of space, what is the number of tons, units, or pieces that will go into the space? These two questions or ones related to them are continually coming up during the voyage of every ship, especially when the plans are being made for stowing the ship, whether with a full load or simply a few hundred tons.

These questions cannot be answered precisely, because the actual amount depends on the broken stowage resulting when the cargo is stowed. This latter value varies greatly, as pointed out on p. 207. We will use the following equation when calculating broken stowage:

$$L = \frac{V - v}{V} \times 100$$

where L = the percentage of broken stowage.
V = the volume consumed in stowing the cargo.
v = the volume of the cargo stowed in V.

Lists of stowage factors should not be used without great care. If the list fully describes the containers and gives other particulars so that there is no doubt that the listed item is the same as the item being considered for stowage, there is little danger of a serious mistake. It is recommended, however, that unless the list is one made personally by the officer using it and relates to a particular trade route and commodity type, current data should be obtained for the commodity in question. Then, a trustworthy stowage factor should be calculated prior to any attempts to estimate space or tons as in the above two typical questions.

Calculating the Stowage Factor

The stowage factor is simply the specific volume of the commodity expressed in units of cubic feet per ton. The stowage factor is equal to the cubic feet per 2240 lb; hence, if we divide 2240 by the density of the commodity we obtain the stowage factor. Density is defined as the pounds in 1 ft³ for our purposes. We can express these facts as follows:

$$f = \frac{2240}{D}$$

where f = stowage factor.
D = density of the commodity in pounds per cubic foot.
2240 = the number of pounds in 1 long ton.

However, data concerning cargo are not often received in terms of its density. More often the measurements of the container and its gross weight are available. Of course, we can calculate the density from these data, but why not express the equation for the stowage factor as one operation instead of two separated operations. This equation would be:

$$f = \frac{2240 \times v}{w}$$

where v = volume of the container.
w = gross weight of the container in pounds.

When handling grain, information concerning the cargo is often given in terms of pounds per U.S. bushel. Since the volume of the U.S. bushel is 1.2445 ft³, the equation for the stowage factor of grain is:

$$f = \frac{2240 \times 1.2445}{w}$$

$$f = \frac{2787.6}{w}$$

where w = weight in pounds per U.S. bushel.

When handling other bulk commodities such as ores, sulfur, and sugar, the only way that the stowage factor can be obtained is by weighing a known volume of the substance. A level bucketful can be used, or a box built with a volume of exactly 1 ft^3.

Obviously, it is not necessary to have the weight of a single container with its volume. The volume and gross weight of the entire shipment or of any part of the shipment can be used to calculate the stowage factor. Such data are given on the ship's manifest and on the bills of lading. Volume and weight information are readily available during all the phases of the cargo operation. The use of such data obviates the need of looking elsewhere for the information.

Using the Stowage Factor

The stowage factor is used to answer the questions mentioned above, and always with some estimated value for broken stowage percentage. The best way to illustrate the use of the stowage factor in conjunction with the broken stowage percentage is to present some general equations and then some numerical examples.

When given a certain volume of space (V), a cargo with a stowage factor (f), and an estimated broken stowage percentage (L), we find the tons, (T), that will fit into the space as follows:

$$T = \frac{V \cdot (1 - L)}{f} \tag{18}$$

When given the same data and instead of the tons (T) we want to know the number of pieces (P) that will fit into the given space, we change the denominator of the right-hand side of the equation to equal the volume of a single container (v); hence:

$$P = \frac{V \cdot (1 - L)}{v} \tag{19}$$

When a given number of tons of a certain type of cargo are to be stowed, we find the space that they will occupy by the following equation:

$$V = \frac{T \cdot f}{(1 - L)} \tag{20}$$

If the number of pieces is given instead of the tons, the space required would be:

$$V = \frac{P \cdot v}{(1 - L)} \tag{21}$$

There are times when it is convenient to combine the stowage factor with an estimate of broken stowage and express the result as a known value. We will define such a value F as the ratio of f to $(1 - L)$; thus:

$$F = \frac{f}{(1 - L)} \tag{22}$$

Note then that the following statements are true:

$$T = \frac{V}{F} \quad \text{and} \quad V = T \cdot F$$

NUMERICAL EXAMPLES

EXAMPLE 1

Given: A hold with a bale cubic of 60,000 ft³. A cargo consisting of cases weighing 400 lb and measuring 2.5 ft by 2 ft by 2 ft to be stowed. Estimated broken stowage: 10%.

Required: The number of tons that can be stowed in the hold.

Solution: Solving for the stowage factor of this cargo:

$$f = \frac{2240 \times 10^*}{400} \quad *(2.5 \times 2 \times 2)$$

$$f = 56$$

After finding the stowage factor, we use Eq. 18 to find the answer to our problem:

$$T = \frac{60,000 \times 0.9}{56}$$

$$T = 964 \text{ tons}$$

EXAMPLE 2

Given: The same data as for Example 1.
Required: The number of cases that could be stowed in the hold.
Solution: Using Eq. 19:

$$P = \frac{60,000 \times 0.9}{10}$$

$$P = 5400 \text{ cases}$$

EXAMPLE 3

Given: 500 tons of a cargo with a stowage factor of 50. Estimated broken stowage: 25%.
Required: The amount of space required to stow this cargo.
Solution: Using Eq. 20:

$$V = \frac{500 \times 50}{0.75}$$

$$V = 33,333 \text{ ft}^3$$

Note that the accuracy of such solutions depends entirely on the officer's guess as to the broken stowage that will result from the actual stowage operation. An officer in a given trade route can increase the accuracy of his estimations of broken stowage by maintaining records of the broken stowage that resulted in a number of specific cases. These records should be incorporated into a tabulation of average broken stowage percentages resulting with the most common cargoes carried on the route in question and in particular sections of the ship. Where longshore gangs are maintained as units, this fact should be noted also, because longshoremen's operations constitute a variable and some interesting facts may come to light. Without such records, the estimates of a given officer will not increase notably in accuracy. With such records, the estimates are based on facts and should become very accurate as time passes. If the estimate does prove highly inaccurate, the officer at least knows that there must be some good reason and can start checking into the facts. One observation gives some basis for judgment, but the basis becomes more trustworthy as the observations increase in number and are averaged.

Loading Full and Down

Once an officer fully understands the above stowage factor ideas, he can solve for the amounts of two different kinds of cargoes that must be loaded to obtain a full and down condition. Variations of the basic problem are more practical than the type of problem presented in most books on cargo stowage. The basic problem deals with choosing the exact amounts of two cargo types that will place the ship down to her maximum legal draft and at the same time fill up her internal volume. A good example of where this problem is met in the industry is at certain ports of Portugal where cork and pyrites are loaded. One of these is definitely a measurement cargo and the other a weight cargo. A measurement cargo is one that stows at or above a stowage factor of 40. Generally the freight charges for such cargoes will be based on measurement tons equaling 40 ft^3 each.

Without using specific values, let us look at the elements of the problem. We would always know four things: (1) the total free space available for stowing the cargo; (2) the cargo deadweight of the ship as she goes on the loading berth; (3) the stowage factor of the heavy cargo; and (4) the stowage factor of the light cargo. When solving these problems we will use a stowage factor that has been combined with a broken stowage factor; the resulting value is what we have called F. What we are trying to determine is how many tons of light cargo and heavy cargo must be loaded to put the ship full and down.

For a general solution let us write down the following factors: Let V equal the free space available. Let T equal the cargo deadweight available. Let s equal the stowage factor (broken stowage combined) of the light cargo. Let b equal the stowage factor (broken stowage combined) of the heavy cargo. These are all known factors. Let X equal the tons of light cargo and Y equal the tons of heavy cargo that must be loaded to put the ship full and down. The latter values are unknown.

It is obvious that to put the ship down, the total of X and Y cannot be more or less than T. Hence we write:

$$X + Y = T \tag{23}$$

We also note that the product sX equals the space that will be occupied by X and that bY equals the space occupied by Y. The sum of sX and bY must be equal to V to have the ship full. Hence we write:

$$sX + bY = V \qquad (24)$$

Equations 23 and 24 constitute a set of simultaneous equations; that is, we have two equations with two unknowns. We can solve for one of the unknowns and then insert the known value in either equation and solve for the second unknown. There are several ways of going about the first solution. The easiest method follows. Start with Eqs. 23 and 24. Multiply Eq. 23 by one of the coefficients of Eq. 24 and rewrite the set. We do this using the coefficient s:

$$sX + bY = V \qquad (24)$$

$$sX + sY = sT \qquad (25)$$

Subtracting Eq. 25 from Eq. 24 we obtain:

$$bY - sY = V - sT$$

Factor out Y on the left-hand side and solve for Y:

$$(b - s)Y = V - sT$$

$$Y = \frac{V - sT}{b - s} \qquad (26)$$

After solving for Y the value of X can be found easily.

The general solution of the problem given above is presented for purposes of providing a thorough explanation. It is not recommended that the reader try to memorize the formula derived, but rather the *method* used. Any given problem can be reasoned out more easily if the general idea is clear.

NUMERICAL EXAMPLE

Given: A ship is loading pyrites with $F = 14$ and cork with $F = 254$. She has 453,000-ft³ bale capacity and 8000-ton cargo deadweight available.

Required: How many tons of pyrites and how many tons of cork must be loaded to put the ship full and down?

Solution: Always write down the meaning of the symbols used. Let X equal the tons of cork, and Y the tons of pyrites. Then we have:

$$X + \quad Y = 8000$$

$$254X + 14Y = 453,000$$

Multiply the top equation by 14 and subtract the result from the bottom equation:

$$
\begin{aligned}
254X + 14Y &= \quad 453,000 \\
-14X - 14Y &= -112,000 \\
\hline
240X \qquad\quad &= \quad 341,000
\end{aligned}
$$

$$X = 1421 \text{ tons}$$

$$Y = 8000 - 1421$$

$$Y = 6579 \text{ tons}$$

Loading Full and Down with Required Trim[6]

A problem related to the above deals with the question of how much of two kinds of cargo must be loaded in the end hatches of a ship during the final stages of loading to get the ship full and down with a required trim. This problem is best illustrated by an example.

NUMERICAL EXAMPLE

Given: A Mariner class ship with free space in the no. 1 main deck space equal to 16,000 bale cubic. This space has its center of gravity 60 ft from the forward perpendicular. This is expressed as L.C.G.–F.P. = 60 ft. Free space is also available in the no. 7 second deck equal to 29,000 bale cubic. This space has L.C.G.–F.P. = 416 ft. The ship's tipping center is 264 ft from the F.P. The deadweight remaining is equal to 900 tons. The trim is presently 4 in. by the head. The $MT1$ equals 1522.

Two cargoes remain to be loaded: (1) wolfram concentrates in double cloth bags with $F = 14$; and (2) India tea in packages with $F = 86$.

---

[6] It is recommended that the reader review the material on pp. 133–147 as he studies this section.

Required: The number of tons of wolfram and tea that must be loaded in the no. 1 and no. 7 spaces to put the ship full and down with a trim of 6 in. by the stern.

Solution: The most direct approach is to solve first for the number of tons of the combined cargoes that must go forward and aft. Thus, let X = tons aft; let Y = tons forward. Obviously, then:

$$X + Y = 900$$

This is one equation of a set of two needed to solve for one of the unknowns. The second equation is obtained by working with the trimming moments needed to obtain the desired trim. The forward space is $(264 - 60)$ or 204 ft forward of the tipping center. Hence, the forward trimming moment is equal to $204\,Y$. The after space is $(416 - 264)$ or 152 ft aft of the tipping center. Hence, the after trimming moment is equal to $152X$. Note that the trim in the beginning is 4 in. by the head and that the desired trim is 6 in. by the stern. This means that we must change the trim, through our loading of the cargo, a total of 10 in. Since the ship is now down by the head and we want it to be down by the stern, we note that the after trimming moment must be in *excess* of the forward trimming moment. It must be in excess a sufficient amount that the difference yields a net trimming moment aft that will give the ship a total change of trim of 10 in. This can be expressed mathematically as follows:

$$\frac{\text{Excess of moment aft}}{MT1} = 10 \text{ in.}$$

Or:

$$\text{Moment aft} - \text{Moment fwd.} = MT1 \times 10$$

$$152X \quad - \quad 204\,Y \quad = 15{,}220$$

This is our second equation with the two unknowns. Multiply the first equation by 204 and then *add* the result to the second equation to eliminate Y. Solve for X, then for Y.

$$
\begin{aligned}
152X - 204\,Y &= 15{,}220 \\
204X + 204\,Y &= 183{,}600 \\
\hline
356X &= 198{,}820
\end{aligned}
$$

$$X = 558 \text{ tons aft}$$

$$Y = 900 - 558$$

$$Y = 342 \text{ tons fwd.}$$

With this information we proceed to solve for the exact amount of wolfram and tea forward and aft by using the same technique as was employed in the problem involving the loading of pyrites and cork. This time we know the same elements. Let us work with the forward end first. Note that we know that a total of 342 tons will go in the forward space and that the space is equal to 16,000 ft³. Hence we can write: Let A = tons of wolfram fwd.; let B = tons of tea fwd. Obviously, then:

$$A + \quad B = 342 \text{ tons}$$

$$14A + 86B = 16,000 \text{ ft}^3$$

Multiply the first equation by 14, and subtract the result from the second equation:

$$
\begin{aligned}
14A + 86B &= 16,000 \\
-14A - 14B &= -4,788 \\
\hline
72B &= 11,212
\end{aligned}
$$

$$B = 156 \text{ tons of tea fwd.}$$

$$A = 186 \text{ tons of wolfram fwd.}$$

In a similar manner, we solve for the distribution in the after space. Let W = the tons of wolfram aft. Let T = the tons of tea aft. Then:

$$
\begin{aligned}
W + \quad T &= \quad 558 \\
\hline
14W + 86T &= 29,000 \\
-14W - 14T &= -7,812 \\
\hline
72T &= 21,188
\end{aligned}
$$

By multiplying the first equation by 14:

$$T = 294 \text{ tons of tea aft}$$

$$W = 264 \text{ tons of wolfram aft}$$

This particular problem may arise during the last day of loading. If the problem does not entail both full and down conditions, only a partial solution with respect to the weight distribution for trim will be required. The holds involved do not necessarily have to be the end holds, but if a specific change in trim is to be obtained, the holds must be on opposite sides of the ship's tipping center.

Stowage Factors Used with Lumber

The stowage factor of general cargo has been defined as the number of cubic feet required to stow one ton of cargo. It was also pointed out that the stowage factor never varied and should always be used in conjunction with an estimated broken stowage value. It must be emphasized that these foregoing remarks were in reference to the stowage of ordinary dry cargo. These ideas are changed somewhat when dealing with lumber.

The stowage factor for lumber is defined as the number of cubic feet required to stow 1000 board ft of lumber. The symbol for 1000 board ft is M.

One thousand board feet will occupy $83\frac{1}{3}$ ft^3 if there is no broken stowage. But there is always some broken stowage as the lumber is stowed in the ship. There will be more broken stowage in the end holds than in the midship holds. With shipments that have large lots of uniform lengths, such as those found on the intercoastal run of the United States, the average broken stowage will be between 20 and 25%. Poor stowage might raise this percentage to 30%, whereas excellent stowage will lower it to below 20%. On shipments of lumber of varied lengths and many sizes, such as are found in foreign trade, the broken stowage will probably run between 25% and 30%.

When considering stowage factors for lumber, the estimate of broken stowage is included. As a general average figure in the problems discussed in this book, we will assume the stowage factor of lumber to be 110 ft^3/M.

In practice, the ship's officer should keep a record of the stowage in various compartments of his ship. From these data, he will be able to judge the broken stowage with accuracy. With a reliable stowage factor, we can estimate the space that is required to stow any given shipment of lumber. If given the size of the compartment, we can calculate the number of pieces of any given type of lumber required to fill the space.

Board Feet in a Shipment

To use the stowage factor for lumber, it is necessary to know the number of board feet of lumber in a shipment. A board foot is defined as the amount of lumber in a piece 1 in. thick and 1 ft square. The number of board feet in any given piece of lumber may be found by this equation:

$$\text{Board feet} = \frac{T \times W \times L}{12}$$

where T = thickness in inches.
  W = width in inches.
  L = length in feet.

To find the number of board feet in a shipment all we have to do is multiply the number so obtained by the number of pieces in the shipment.

The Decimal Conversion Factor

When lumber is surfaced, the thickness and widths are reduced. Therefore, when a 2- by 4-in. piece is received by the ship, it may be only $1\frac{3}{4}$ by $3\frac{3}{4}$ in. This fact introduces an error in stowage calculations unless it is taken into account. The number of board feet in any given shipment is reduced when the rough sizes are reduced. To find the true amount of lumber in such a shipment, multiply the gross amount of the lumber by the ratio of the finished cross-sectional area to the rough cross-sectional area. This ratio is called the *decimal conversion factor* (dcf). For a 2 by 4 finished to $1\frac{5}{8}$ by $3\frac{5}{8}$ in. the decimal conversion factor would be:

$$dcf = \frac{1\frac{5}{8} \times 3\frac{5}{8}}{2 \times 4} = 0.7363 = 0.74$$

To use this factor, always round off the figure to the nearest two decimal places. If it reads exactly 50 in the third and fourth places, then raise it to the next highest figure.

With use of the conversion factor, the calculation of the amount of lumber is first done based on the rough sizes; the amount thus obtained is called the *gross measurement*. Multiplying the gross measurement by the conversion factor, we obtain the *net measurement*. We always consider the net measurement when calculating stowage requirements.

Two General Equations: Space and Pieces

Two equations are used continually by the supercargo when checking and planning the stowage of lumber. One equation, for the space requirements of stowing a given shipment, is:

$$\text{Space} = \frac{T \times W \times L \times P \times 110 \times \text{dcf}}{12,000}$$

where Space = the space required to stow the shipment.
T = thickness in inches.
W = width in inches.
L = length of the lumber in feet.
P = total pieces in the shipment.
dcf = decimal conversion factor if there is a difference between rough (nominal) and finished (standard) size.
110 = the assumed stowage factor of sawed lumber.
12,000 = the product of 12 and 1000, the 1000 being used to convert board feet into M's.

The second equation, derived from the first, is used to solve for the number of pieces that can be stowed in a given compartment, knowing the type of lumber available:

$$P = \frac{\text{Space} \times 12,000}{T \times W \times L \times P \times 110 \times \text{dcf}}$$

It is recommended that the reader learn the reasoning involved in deriving the equation rather than memorize the equation itself.

Stowage of Logs

Solving for the space for stowage of logs and the inverse problem is approached in the same way as for sawed lumber except that the number of board feet is found by using a *mean* diameter to find the cross-sectional area and a different stowage factor is used. If the logs are all the same length, use 135 as the stowage factor; if they are of various lengths, use 150.

CHAPTER 7

THE SHIP'S LOADING AND DISCHARGING EQUIPMENT

The purpose of this chapter is to describe the methods and equipment used to load and unload ships today. Included with the presentation of extant practices is a discussion of the limitations of each, precautions on their use, and proposals for the future.

SHIP'S RIGGING

The Married Fall System

A common rig still found on merchant ships for the purpose of loading and discharging cargo is the *married fall* system. This rig is sometimes re-

345

ferred to as the *yard and stay* or *burton* system. The last two names are used most often among seafarers. In this rig, one of the ship's booms is guyed so that its outer or upper end is over the hatch and the other boom so that it is over the dock. In the days of sail, discharging and loading was performed by using two blocks with runners through them in a manner similar to the system in use today. The block plumbing the hatch was sometimes secured to the mast's stay and the block over the dock was frequently secured to one of the ship's yards. Thus the dock boom is still called the yard boom and the hatch boom is called the stay boom. The falls rove through the blocks attached to these booms are given the same names respectively. Hence, the designation *yard and stay* system.

After steam replaced sail, and also on many sailing vessels, cargo was handled on swinging booms for many years. The practice of using two booms, as seen in Fig. 7-1, in which both booms are *fixed,* originated on the west coast of the United States during the late 19th century. The swinging boom is used today only when handling loads considered beyond the capacity of the fixed rig. On most ships it is not safe to burton a load on fixed rigging if it exceeds 3 tons; however, if the guys are carefully positioned, the parts are large enough, or a *safe* rig such as the Ebel rig (discussed later) is used, then heavier loads may be burtoned. (To *burton* a load is to move it athwartship.)

The stay fall is an up-and-down fall, the yard fall pulls the load across the ship. Actually, both falls carry the load across the ship, and it is only after the athwartship movement is complete that either the yard or the stay fall takes the entire load, depending on which way the load is moving. The fall on the dock boom often is referred to as the burton, whereas the fall on the hatch boom is known as the hatch or up-and-down fall. This is so despite the fact that during loading the hatch fall also carries the load across the ship. It is evident, then, that a confusing array of terms exists for speaking about this rig. The fall over the dock may be termed the dock fall, the burton fall, or the yard fall; likewise, the fall over the hatch may be referred to as the hatch fall, the up-and-down fall, or the stay fall. Inasmuch as the use of the dock and hatch terminology results in the least possible confusion, these are the terms that will be used in this discussion.

The booms of Fig. 7-1 are able to pivot in two planes. The vertical movement of each boom is controlled by the topping lift. The transverse movement of the boom is controlled by the outboard or working guy and the midship or schooner guy. The midship guy is also known as the spanner guy, schooner guy or lazy guy. At the lower end of the boom, a

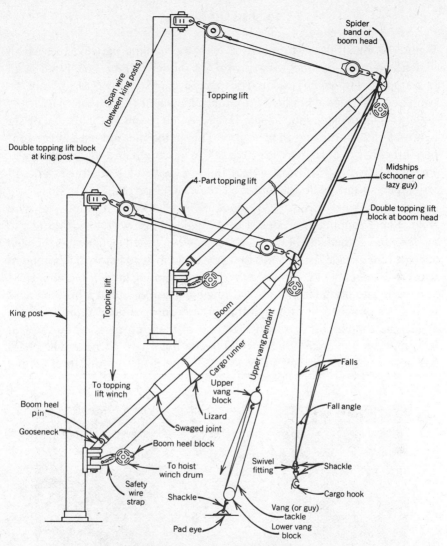

FIG. 7-1. The married fall system.

fitting known as the gooseneck allows movement in the two directions mentioned.

The cargo fall leads from the winch drum directly to the heel block mounted below the gooseneck fitting. The gooseneck and heel block generally make up a single assembly. The fall then leads up the boom through lizards or fair-leads to the head block and then to the cargo hook.

Topping Lift Rigging

Figure 7-2 illustrates a topping lift with the hauling part led to a small winch mounted on the king post. All modern ships are fitted with this *tipping lift winch*; however, there are old ships that are not so equipped. In ships without the topping lift winch the hauling part leads down to a large cleat and is made fast with one round turn plus five or six figure-of-eight turns and then seized. This is the method used if the topping lift consists of a wire rope rove off on a twofold or double luff tackle. The wire used in this case would be ¾-in. wire with 6 strands and 19 wires per strand made of improved plow steel or plain plow steel. Improved plow steel wire rope is about 13% stronger than plow steel wire rope. Some topping lifts consist of a single heavy wire shackled to the *spider band* at the head of the boom. This wire, called a bale, runs through a single block made secure aloft that acts as a fair-lead before it runs downward to the deck. On the lower end of such a topping lift a *flounder plate* is attached. The flounder plate is a triangular plate of about ¾-in. thickness with holes punched or drilled at the three corners about 1 in. in from the edge. The topping lift is attached to one of these corners, while the *bull rope* and the *bull chain* are attached to the other two corners. The chain has links of about 1-in. diameter and the rope is about 4-in. three-strand

FIG. 7-2. Cargo rigging: electric topping lift winch.

FIG. 7-3. Single-part topping lift. This shows the topping lift lead block secured to the crosstrees, the bale, the bull line, and the bull chain. Courtesy of U.S. Maritime Commission.

manila. The topping lift would be made of $1\frac{1}{4}$- or $1\frac{1}{2}$-in. improved plow steel wire rope, 6 strands and 19 wires per strand (see Fig. 7-3).

When the boom has been topped up with the wire purchase rig, it is simply stopped off and secured to the cleat. When rigged with a bale, the boom is raised or lowered by first taking the weight of the boom on the bull line and then unshackling the bull chain. The bull rope is led to a winch gypsy head for this purpose. When at the desired level, the bull chain's lowest link is shackled to a pad eye on deck and the bull line is slacked off until the chain has all the weight.

Obviously, the use of either the wire purchase and cleat or bull line and chain requires considerable time and entails several operations requiring good judgment and a certain amount of seamanship to be executed safely. The use of a topping lift winch obviates the need for any stopping off or shifting from cleat to gypsy head and back again or heaving with the hauling part on the winch, all of which can be disastrous if a mistake is made by the personnel. All that is necessary is to push a control button and the winch either tops up or lowers the boom. Some of these winches are made so that they can lower or top with the boom loaded; others cannot handle a loaded boom. If the topping lift winch is not designed to handle a loaded boom but an attempt to have it do so is made, the excessive torque required of the winch will cause a large rise in current through the motor's armature; this will result in a safety device breaking the circuit and an automatic braking mechanism being applied. This prevents the excessive current from burning out the armature. If the circuit breakers fail to operate, the armature will be burned out and a great deal of expense will be incurred on account of repairs and lost time.

Guying Systems

As can be seen from Fig. 7-1, there are two distinct systems for guying the fixed booms of the yard and stay rig. It is necessary to find a place down on deck to secure the inboard guy. This brings additional gear to a location already overly crowded, impairing the safety of the working area. The load on the cargo hook is always between the heads of the two booms or directly under one of them; therefore, there is little or no stress on the inboard or midship guys. Thus the lightweight midship guy is quite sufficient, and it is placed aloft out of the way.

The outboard guys are often referred to as the working guys because they are the guys that are under the greatest stress. The stresses on the guys appear when the load is being transferred athwartships (burtoned) or when it is being supported anywhere between the two boom heads. An analysis of the stresses on the falls and guys is made later. It is important for all officers to know how to keep these stresses at a minimum.

Preventers

In addition to the regular outboard guy on the fixed boom, an additional wire may be attached to the head of each boom and led to the deck to act as a *preventer guy*. This preventer guy is made of $\frac{5}{8}$- or $\frac{3}{4}$-in. wire rope. It may be completely of wire or have a tail spliced onto the lower end so that the part used to secure the preventer will have more flexibility than the wire itself. These tails are made of 4-in., three-strand manila rope or $\frac{3}{4}$- to 1-in. chain. If they are made of manila, they are susceptible to being cut by plates or other items being handled; if made of chain, they are less vulnerable to damage. Quite frequently the regular guy and preventer are set up in such a fashion that the regular guy takes all the load and the preventer is supposed to take a stress only in case the regular guy parts. *This is not a safe practice*; the net result is that when the regular guy parts the preventer parts also. The ship's officers must make certain that the load is being carried equally by the regular guy and the preventer. The safest procedure is to make both the regular and preventer guy secure as close together as possible and to equalize the stress on all parts. It follows also that it would be more sensible to have one guy rigged so as to have the strength of the ordinary guy plus the preventer.

The preventer is a constant source of trouble, and in many cases fails when the regular guy fails. In the records of the Accident Prevention Bureau of the Waterfront Employers of San Francisco, only one case is recorded in which the preventer held after the regular guy parted. In this particular case, the preventer was not made secure properly, and when the boom started to swing the preventer took up the strain gradually until the wire became jammed on the cleat and the boom's swing was checked. Because of this problem, the functions of the regular guy and the preventer have been combined in the vang guy.

Spotting the Guys

When securing the outboard guys to the deck after topping the booms, consideration should be given to the location of the lower end of the guy with respect to the stress that will be placed on it when under a load and with respect to the possibility of jackknifing.

Guy failures account for the great majority of cargo gear breakdowns. Bending of the boom due to too much compression caused by the component of thrust from fall, guy, and topping lift is the second most important cause of breakdowns. A number of very practical points relative to the task of positioning guys for greater safety are presented in the following section. These factors should be common knowledge to all deck officers and longshoremen.

P.M.A. PRACTICAL TESTS

Most of the facts presented below with respect to strains on guys under various conditions were obtained through experimentation carried out on a model closely approximating the C-3 type. The work was carried out by the Pacific Maritime Association's (P.M.A.) Accident Prevention Bureau, and results were published in its *Bulletin* 16, August 1948. Special mention of it is made because it represents a contribution toward safety and good management on all ships.

Conditions of the Tests

The model was adjusted to a beam of 70 ft, a hatch 24 ft wide, and the heel of the boom 10 ft aft of the hatch. The heels of the booms were placed 8 ft off the deck and outboard from the centerline a distance of 16 ft in one test and 6 ft in another test. The booms were made to scale at 55 ft of length. The guys were made fast at the ship's side at various positions fore and aft as discussed below.

Horizontal Stress

The greatest *horizontal* stress between the heads of the boom for any given height of load above the deck occurs when the falls make equal angles with the horizontal. Therefore, when a fall angle of 120° is referred to it means that each fall makes an angle of 30° with the horizontal and the two falls make an angle of 120° with each other. All tests in this experiment were made with a fall angle of 120°.

As usually rigged, the hatch boom's working guy carries much more strain than does that of the dock boom. Because of this, the hatch boom guy has been considered in the discussion that follows.

The load was considered to be 1 ton. This makes it possible to note the strains on the guys in terms of the load being hoisted. For instance, if a strain on a guy is read as 3 tons with our 1-ton load, it would be 15 tons if we had a 5-ton load on the hook.

Position of the Heel

The athwartship location of the heel of the boom has an important effect on the strain on the guys. In Fig. 7-4, the heels of the booms are 16 ft outboard from the ship's centerline. With the working guy of the hatch boom secured at a point in line with the two boom heads (indicated by the dashed line), the stress on the guy is three times the load being hoisted when the angle between the falls is 120°. In the experiment the force on the working guy would show up as a 3-ton stress because we are using a

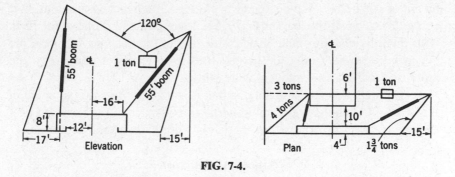

FIG. 7-4.

1-ton load. Note what happens, however, when the guy is made secure at a point about 4 ft aft of the heel of the boom. The stress on this guy (as indicated by the solid line) is four times the weight of the load. In this case the load would then be 4 tons, which is an increase of $33\frac{1}{3}\%$.

Now let us see what happens when we shift the heels of the booms inboard until they are only 6 ft from the centerline. In Fig. 7-5, the heels are arranged as mentioned above; nothing is changed from Fig. 7-4 except the shift in the positions of the heels. At a fall angle of 120° and with the guy made secure 4 ft aft of the heel, the stress on the working guy of the hatch boom is reduced from 4 tons to $2\frac{1}{4}$ tons, or $2\frac{1}{4}$ times the load on the hook. This represents a stress reduction of 42% caused by shifting the heels of the booms inboard.

As the heels are moved outboard, greater drift for the load is obtained. This is a definite advantage when handling certain types of cargo, such as long lengths. However, in Fig. 7-4 the fall angle of 120° is reached at a

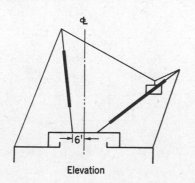

Elevation

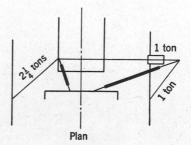

Plan

FIG. 7-5.

height of 40 ft above the deck and in Fig. 7-5 it is reached at about 32 ft above the deck. If the load in Fig. 7-4 is lowered to 32 ft above the deck with the falls making equal angles with the horizontal, the stress on the guy will be reduced only to $3\frac{1}{4}$ tons. Hence, in terms of guy stress, the increased drift has not proven an advantage.

It is evident from the above facts that it is not wise to base judgment of the strength of the gear on a ship with widely spaced king posts on past experience with ships in which the heels were close to the centerline.

Positions of the Guys

It is generally considered that the best position for the guy is at right angles to the boom when viewed from above. Going back to the conditions illustrated in Fig. 7-4, we will approximate this condition by placing the guy in line with the fall as indicated by the dotted line. As pointed out above, the stress on the guy here is less than when the guy is secured 4 ft aft of the heels of the booms.

If we swing the hatch boom outboard 8 ft from the coaming as shown in Fig. 7-6, with the guy in line with the fall, the guy stress is increased from 3 tons to $4\frac{1}{2}$ tons. This is indicated by the dashed line. Thus, it is seen that with the guy in line with the falls, a comparatively small movement of the head of the boom makes a difference of 50% in the stress on the guy. With the boom in this outboard position and the guy secured at a point 4 ft aft of the heels of the booms, the guy stress drops from $4\frac{1}{2}$ tons to $3\frac{1}{2}$ tons, a reduction of 22%.

It therefore appears that if a boom is angled outboard from a fore and aft line through its heel, less stress is put on the guy by leading it back to or behind the heel. If the boom angles inboard from a fore and aft line

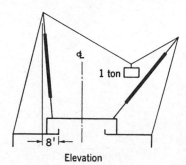

Elevation

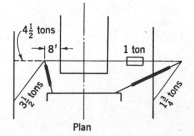

Plan

FIG. 7-6.

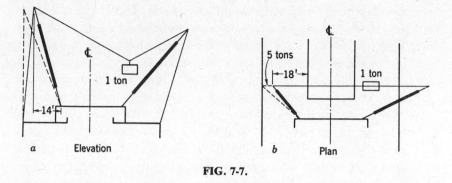

FIG. 7-7.

through its heel, the least stress on the guy will occur when it is at right angles to the boom or approximately in line with the fall.

A Dangerous Condition

A particularly dangerous condition is illustrated in Fig. 7-7. This happens when an effort is made to move the head of the boom aft (or forward, depending on which end of the hatch the gear is located) without topping it. This effort may result from setting the guys and heaving on the topping lift. Regardless of how it is done, the boom is swung far outboard with the guy secured in line with the fall. It will be noted in the plan view, *b,* that the guy makes a very wide angle with the boom and in the elevation, *a,* that the guy is very nearly vertical.

Under these conditions, tremendous stresses can be built up. If the heaving on either the guy or the topping lift is continued as far as possible, the boom will approach a position indicated by the dotted lines. The head of the boom will be almost vertically above the point where the guy is secured. In this position, the guy would be pulling directly against the topping lift. Theoretically, an infinite stress would be placed on the guy by the horizontal pull of the falls. *This is a situation that should be watched for and avoided.* It is particularly likely to occur where the beam is at a minimum. It may also occur, on some vessels, where the heels of the boom are closer to the rail than they are to the centerline.

Maximum Loads Function of Winch Strength

As an example, let us say that the stress on the guy is three times the weight of the load on the falls at a fall angle of 120°, as it proved to be by experimentation on the model with the guys secured at a particular point. With a 5-ton load, then, the stress on the guys will be 15 tons. Now if any given load is raised to a fall angle of 140°, the stress on the guys may go up to 3.3 times the weight of the load. If we were hoisting a 1-ton load, the guy stress would be 3.3 tons instead of only 3 tons. However, the important point here is to note that with 5-ton winches we *cannot lift more than* 3.4 tons to a position with a fall angle of 140°. Therefore, the guy stress is limited by the winch power to 3.4 times 3.3, or a maximum of 11.2 tons. As seen in Fig. 7-8, the horizontal pull at a fall angle of 120° is 4.3 tons, as compared with 4.7 tons with a fall angle of 140°. It should be pointed out here that if the booms were rigged with gun tackles in an effort to make the falls capable of handling heavier loads safely, the winches would then be able to hoist heavier loads to greater fall angles and the danger of exerting excessive stresses on the guys would exist. The officer of the ship must be careful of the set of his rigging under such conditions. Rigged with gun tackles, the stres on the guys could go up to a maximum of 22.4 tons, because theoretically a 6.8-ton lift could be hoisted to a fall angle of 140°.

From the above it is seen that although the stresses on the guys increase greatly in terms of weight of the load as the fall angle increases, the weight of the load that can be raised to these higher angles falls off; therefore, the stresses have a limit based on the capacity of the winches.

When hoisting long lengths to high fall angles, such as when loading structural steel or piling, and when the loads are 1 ton, the stress on the guys may be higher than if hoisting 5-ton loads to their maximum fall angle. This is because we can lift the lighter loads to higher angles and increase the horizontal pull between the heads of the booms. At any given fall angle the stresses on the guy are proportional to the weight of the load. Maximum stresses, however, are determined by the *strength of the*

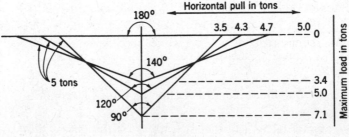

FIG. 7-8.

winches and not by the weight of the load. If gun tackles are used, the strength of the winch is increased.

The same situation exists for the stresses on the falls. A 1-ton load at a fall angle of 160° puts a 2.9-ton stress on each fall. The maximum stress that can be built up is the power of one winch.

The above discussion is correct theoretically, but ship's officers should be aware of the fact that the usual electric winch does not begin to trip out until a line pull of 8 to 12 tons in high gear is reached. During the appreciable time lag required to trip the breaker, the line pull continues to increase. Thus it can be seen that most electric winches may exceed the 5-ton limit mentioned in the examples and correspondingly exceed the guy stresses.

It must be emphasized that the maximum stresses mentioned above are for static loads and do not take into account the increases caused by jerking winches or temporary overloads that a winch can carry long enough for them to act on the parts of the rigging. Stresses from these causes are large but have not been considered in this study. Following this discussion of static stresses, the elementary principles involved in a study of the effect of dynamic forces will be set forth.

Maximum Stresses

It is well known that tight-lining of the falls causes extremely heavy stresses on the guys. Undoubtedly, a question may arise in the minds of many officers as to the magnitude of the stress when the falls are tight-lined with an empty hook. Because of the limitation placed on the power of the winch, the stress on the guys depends entirely on the fall angle when the winch is pulling its maximum.

Figure 7-8 indicates the maximum load that a pair of 5-ton winches can lift to various fall angles and the corresponding horizontal pulls between the heads of the booms. It is, of course, this *horizontal pull* that puts the stress on the guys. The horizontal pull will be at a maximum when the winches are pulling against each other with an empty hook, and with 5-ton winches would be equal to 5 tons. Thus, an empty hook is potentially more dangerous than a heavily loaded one.

When 5-ton winches are holding a 5-ton load at a fall angle of 120°, the horizontal pull is 4.3 tons. The stress on each fall under this condition would be 5 tons also; thus, it can be seen that this is as high an angle as these winches can possibly lift such a load.

Topping or Jackknifing of Booms

Topping or jackknifing of booms can have very serious and costly results. Although the extreme conditions under which a boom will or will not top are well known, the dividing line, or *danger point,* is not well known. Experiments with a model such as those carried out with respect to guy stresses have established the line of demarcation between jackknifing and not jackknifing and a means of locating this danger point on board ship.

When the two booms are at different heights, which is usually the case, their respective tendencies to top are different. The fall leading upward from the lower to the higher boom will tend to lift the lower boom. For this reason the working guy of the lower boom must be placed closer to the heel than is necessary for the higher boom. The working guy of the lower boom must be so placed that a line of sight (dotted line in Fig. 7-9) from the pad eye to the head of the higher boom will pass beneath the lower boom somewhere ahead of the gooseneck of the lower boom. If the line of sight is behind the heel of the lower boom, the boom will definitely top. If the line of sight passes above the boom but ahead of the heel, the boom will top until it reaches this line of sight and then it will stop.

Because the higher boom is being somewhat pulled down by the fall, its guy need not be as close to the heel as that of the lower boom and the line of sight from the pad eye to the head of the lower boom may pass behind the heel of the boom. However, the working guy at the height of the gooseneck must be ahead of the line of sight between the gooseneck of the higher boom and the head of the lower boom. If the high boom's working guy is behind this line of sight at the height of the gooseneck, the boom will top.

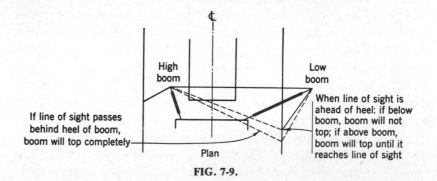

FIG. 7-9.

As a practical means of adopting this rule it is suggested that a man place himself on the line of sight and in line with the guy. Since the goosenecks on most ships are about 8 ft above the deck, there is danger of topping if the guy is over his head. If the guy is below the top of his head when he is on the line of sight, the boom is safe.

In cases in which the booms are at about the same height, the safest procedure is to secure the guys so that the line of sight crosses the respective booms. The line of sight referred to here is that from the pad eye where the guy is secured to the head of the opposite boom.

Any change in the position of a boom not only affects its own tendency to top but also affects the tendency of the opposite boom to top, since the line of sight moves with the head of the boom. It is for this reason that frequently when one boom tops the other will follow it.

The working guy of the yard boom is frequently located in a position that will make topping possible. The tendency to top can be observed by a little slackening of the topping lift. This may not be dangerous; however, it should be checked carefully. At the same time, the working guy of the hatch boom must be checked to make certain it is not located so close to the critical point that moderate topping of the dock boom will cause the hatch boom to jackknife.

Summary of Discussion of Guy Location

Several important facts that all officers should keep in mind when their ship is working cargo have been pointed out in the above discussion. These are summarized below.

a. The closer the heel of the boom is to the rail, the greater the stresses on the guys and booms for the same fall angle.

b. The increased drift obtained by spacing heels widely may not fully compensate for the increased stress on the guys.

c. When a boom is angled outboard from the fore and aft line through its heel, the stress on the guy decreases as it is moved back of the heel. The practical amount of this movement is limited by the critical point for jackknifing.

d. When a boom is on the fore and aft line through its heel or swung inboard from this position, the stresses on the guy and the boom are least when the guy is in line with the fall or at right angles to the boom when viewed from above. Under most conditions, there will be little difference in the amount of stress at these two positions of the guy.

e.   Depending on the spacing of the heels, the stress on the working guy of the hatch boom at a fall angle of 120° may be as much as $4\frac{1}{2}$ times the weight of the load. Improper leads of guys can increase these stresses considerably.

f.   At any given fall angle the stress on the guys will vary directly with the weight of the load, but the *maximum* stress that can be placed on a guy in any given case is determined by the capacity of the winches, not by the weight of the load. In fact, the lighter the load, the greater the maximum possible stress.

Additional Stresses

As mentioned before, the experiments on the model provided for measuring the stresses on the guys only. There was no provision made for actually measuring the stress on the topping lift or the compression on the boom. These values, as well as the guy stresses, have been calculated by a method described later (see p. 385) and are presented in Table 7-1. The stresses were calculated for the same conditions as were described for the experiment, that is, a 1-ton load with a fall angle of 120°. The figures referred to are those mentioned above. The boom compression shown in the table includes the compression caused by the stress on the single hauling part of the fall as well as the forces created by the guy, the topping lift, and the cargo fall from the head block to the hook.

Note that in all cases except Fig. 7-7, where the high stress on the model might have pulled the boom head far enough inboard to lower the guy stress, the calculated stresses on the guys are less than those found by experiment. It is believed that these differences can be attributed to the fact that in the model, as under actual conditions, the various parts are grouped around the boom head some distance from the centerline of the boom used as the reference point in the calculations. It is probable that the boom compression and topping lift stresses actually exceed the calculated values also. The actual stresses on these parts will be further increased by the weight of the boom and fittings at the head.

It is interesting to note that the stresses produced by the arrangement shown in Fig. 7-7 are cut almost in half by placing the guy at right angles to the boom in the plan view. They are cut in half again by leading the guy behind the heel. The values of the stresses involved for Fig. 7-7, with all values calculated considering the hatch boom, are as follows:

	Guy Stress (tons)	Boom Stress (tons)	Topping Lift Stress (tons)	
Guy in line with fall	10.0	16.2	6.9	(1)
Guy 90° to boom	5.6	9.7	3.2	(2)
Guy behind heel	2.4	4.2	0.3	

Note that moving the guy from position 1 to position 2 involves a shift of only 4 ft, but the change in the guy's angle with the boom is great and

TABLE 7-1

STRESSES UNDER CONDITIONS SHOWN IN FIG. 7-4[a]

	Guy Stress		Boom Com-pression	Topping Lift Stress
	Calculated	Experimental		
Fig. 7-4				
Hatch boom with guy:				
Behind heel	3.3	4.0	4.8	0.3
In line with fall	2.4	3.0	5.0	1.7
Yard boom guy	1.7	1.75	3.9	0.5
Fig. 7-5				
Hatch boom guy behind heel	1.9	2.25	3.4	0.2
Yard boom	1.0	1.0	3.3	0.7
Fig. 7-6				
Hatch boom with guy:				
Behind heel	2.8	3.5	4.3	0.2
In line with fall	3.6	4.5	6.6	2.1
Fig. 7-7				
Hatch boom guy in line with fall	10.0	8.0	16.2	6.9

[a] The values listed are given in tons.

the effect on the stresses is also great. Because the boom is angled out-board, far lower stresses are produced by leading the guy well back.

The stresses on the midship or hatch gear can be reduced drastically by swinging the head of the boom inboard until it plumbs the centerline of the hatch. In this position, the stresses are as follows:

	Guy Stress (tons)	Boom Stress (tons)	Topping Lift Stress (tons)
Guy in line with fall	1.7	3.9	1.3
Guy 90° to boom	1.0	3.3	1.8

Here again it will be noted that placing the guy at right angles to the boom in the plan view greatly reduces the stress on the gear. In this case the two locations of the guy along the ship's rail are 34 ft apart; this is because we are working so close to the heels of the booms with the hatch boom. Locating the guy behind the heel in this setup not only would produce a stress of 3.7 tons on the guy but also would cause the boom to top before the load reached an angle of 120° between the falls.

All of the above stress values were obtained with the gear rigged with fixed booms, as it is normally worked; therefore, they are quite significant when compared with the stresses that would result if the boom was used as a swinging boom. The guy stresses, of course, would then be negligible, but the boom compression and topping lift stresses would fall off too. With a swinging 55-ft boom on a 43-ft mast, the compression on the boom is about 2.3 times the weight of the load, including the pull of a single-part fall along the boom, in any position, whereas the *maximum* stress on the topping lift is no more than 1.7 times the weight of the load. This maxi-mum stress is incurred when the boom is horizontal.

Additional Conclusions

a.   Excessive stresses caused by faulty rigging of the gear cannot be safely compensated for merely by the use of preventers or by otherwise increasing the strength of the guys. To use this method of compensation may cause the boom to fail.

b.   The smallest stresses are produced (because of increased drift and good guy angles) on gear with widely spaced king posts when the head of the hatch boom is well toward the centerline of the ship and the working guy is at right angles to the boom in the plan view.

    c.   Since the pull of the fall along the boom makes up a large percentage of the total compression on the boom, maximum benefits are obtained by doubling the fall rather than putting the winch in low gear when heavy weights or lack of drift makes either necessary.

DYNAMIC LOADS

In connection with the consideration of stresses on the cargo gear, it is of great importance to take into account the effect of acceleration and deceleration on the total load. We will show in the following discussion that the weight of any given load is virtually increased when the speed of hoisting is increased or the speed of lowering is decreased. The magnitude of this increase can be dangerously high if the winches are not operated carefully.

Newton's Second Law of Motion

Newton's second law of motion states that *if an unbalanced force acts upon a body, the body will be accelerated; that the magnitude of the acceleration is proportional to the magnitude of the unbalanced force; and that the direction of the acceleration is in the direction of the unbalanced force.* From this fundamental concept we obtain the equation

$$F = Ma \tag{27}$$

where F = the force acting upon the body.
      M = the mass of the body.
      a = the acceleration of the body.

    If the weight (W) of a body is the only force acting upon it, it is a freely falling body and has an acceleration equal to 32 ft/sec/sec (or 32 ft/sec^2). This is called the acceleration due to the force of gravity, since it is caused by the pull of the Earth on the body. It is always denoted by the letter g. Since the force acting on a freely falling body is its weight, Eq. 27 becomes

$$W = Mg \tag{28}$$

from which we may obtain

$$M = \frac{W}{g} \qquad (29)$$

If the mass of the body does not enter explicitly into the problem, it is possible to eliminate it from Eq. 27 by substituting its value from Eq. 29, obtaining

$$F = \frac{W}{g} a \qquad (30)$$

In Eq. 30, F and W are expressed in the same units. In our consideration of loads on the cargo gear we will use pounds or tons force. Also, a and g are expressed in the same units; we will use feet per second per second.

Equations of Motion

Now we must also consider some equations of motion for constant acceleration, all of which we derive from the basic definitions of velocity and acceleration.

The speed or velocity of a body is defined as the distance transversed divided by the time elapsed. Hence we have

$$\text{Speed} = \frac{\text{Distance}}{\text{Time}}$$

or in symbols

$$\bar{v} = \frac{d}{t} \qquad (31)$$

Note that we have denoted speed or velocity by $\bar{v}$, read "vee-bar." This is to prevent confusion when using other symbols for velocity, such as plain v for final velocity of a body after being accelerated from an initial velocity u. Thus $\bar{v}$ is the *average* of v and u, or

$$\bar{v} = \frac{v + u}{2}$$

and

$$\frac{v + u}{2} = \frac{d}{t}$$

$$t = \frac{2d}{v + u}$$

Acceleration is defined as the change in velocity divided by the time during which the change takes place. If the velocity of the body initially is u and it changes so that after a time interval t it is v, then acceleration is, from the definition:

$$\text{Acceleration} = \frac{\text{Change in velocity}}{\text{Time}}$$

or in symbols

$$a = \frac{v - u}{t} \qquad (32)$$

Now, going back to Eq. 30, if the load on a cargo hook is the body and it is traveling up or down at a constant speed without any acceleration except that needed to *balance* the acceleration due to gravity, it should be evident that a in

$$F = \frac{W}{g} a$$

is equal to g. Therefore, *in this case*

$$F = W$$

However, if the *rate* of hoisting is increased or the *rate* of lowering is decreased, then a becomes greater than g and F becomes greater than W. Thus we see that the load on the hook becomes greater than the actual weight on the hook.

From the equations of motion 31 and 32 we obtain, when $u = 0$, the equation becomes

$$a = \frac{v^2}{2d} \qquad (33)$$

Now we have three equations: 30, 32, and 33.

$$F = \frac{W}{g}\, a \tag{30}$$

$$a = \frac{v - u}{t} \tag{32}$$

$$a = \frac{v^2}{2d} \tag{33}$$

With these equations we can investigate the change in the load on a cargo hook due to acceleration and deceleration. Using the word deceleration here is merely a matter of convenience. In the problems considered in this type of study, we would most often conceive of the load as being lowered at some set velocity and then brought to a velocity of zero by retarding the lowering rate. The increase in W would be the same if we were hoisting the load and started from zero and reached the given velocity in the same time or within the same distance. Thus, it becomes unnecessary to apply negative or positive signs to the values but important to keep the concepts clearly within focus. The load is increased above W when decelerating only when lowering the load.

The magnitude of the increase in a over the value of 32 ft/sec^2 is found by solving for a with either Eq. 32 or 33. After solving for the increase in g [call this value a' (a prime)] add the calculated value to g, substitute the new value for a in Eq. 30, and find the force on the cargo hook (F).

NUMERICAL EXAMPLES

EXAMPLE 1

Given: A load of weight W is being hoisted at the velocity of 1 ft/sec. Within the time interval of 0.3 sec its speed is increased to 4 ft/sec.

Required: What is the force acting on the cargo hook during the period of accelerated motion?

Solution: We use Eq. 32 to determine a'. Thus

$$a' = \frac{4 - 1}{0.3} = 10$$

also

$$a = 10 + 32 = 42$$

Now, using Eq. 30 with 42 as the value of a, we obtain

$$F = \frac{W}{32} \, 42$$

$$= 1.31 \, W$$

EXAMPLE 2

Given: A load of weight W is being lowered into a hold at the rate of 4 ft/sec. The winch driver stops the downward motion within a distance of 1 ft.

Required: What is the force F acting on the cargo hook during the period of deceleration?

Solution: We use Eq. 33 to determine a'.

$$a' = \frac{16}{2} = 8 \text{ ft/sec}^2$$

Note then that:

$$a = 32 + 8 = 40 \text{ ft/sec}^2$$

which is the value to substitute in Eq. 30 for a. Thus

$$F = \frac{W}{32} \, 40$$

$$= 1.25 \, W$$

CARGO WINCHES

Comments and data regarding three types of cargo winches will be presented in the following sections. The three types of winches found on ships are steam, electric, and hydraulic powered. The steam winch has been replaced by the electric winch on almost all modern vessels. The hydraulic winch has certain advantages that make it a competitor of the electric winch.

Electric Cargo Winches

Electric cargo winches are installed in two different ways. In the first method, the winch itself, its motor, and its motor brake are mounted on the deck or on top of a masthouse whereas the resistors and control panels are mounted inside the masthouse or below deck. In the second type of installation, which might be referred to as the *unit installation,* the winch is constructed as a complete unit, with the brake, resistors, and control panel in one housing. The master switch, which is the controller, is located on deck close to the hatch in both installations.

Operation

The operation of electric cargo winches is simple, and maintenance is less exacting than it is on the steam winches. Before operating electric winches, they should be lubricated, but the number of points requiring attention are few. All gears and jaw clutches are totally enclosed in oil-tight fabricated cases and are splash-lubricated. Alemite or Zirk pressure grease fittings are provided for the drum shaft bearings, intermediate gear, and pinion bearings. These points should be serviced at least once for each 4 hours of continuous operation.

To put an electric winch into operation is extremely simple. On the two-speed winch, the speed change lever should be shifted to the required load position. In shifting this lever, light pressure should be placed on it and the motor rotated slowly with the master switch on the first point until the clutch jaws are fully engaged. The locking pin should be placed in the hole in the end of the lever before the winch is operated. All hoisting or lowering is done by use of the master switch or controller.

Rating and Capacity

Table 7-2 presents data concerning the line pulls and speeds of typical electric winches along with some other interesting information.

Electric brakes as well as conventional band brakes are incorporated in most electric winches. These magnetic brakes automatically act to hold the load as soon as the current to the motor is shut off either by power failure or, when the master controller is placed in the off position (vertical position), by the winch operator.

The master switch or controller is usually located on deck close to the hatch so that the operator can properly observe the load. In normal opera-

TABLE 7-2
RATING AND CAPACITY OF TYPICAL ELECTRIC WINCHES[a]

Regular Duty, Two-Speed, Electric Cargo Winch

	High gear	High gear	Low gear
	219 fpm	279 fpm	112 fpm
	Line speed	Line speed	Line speed
Line pull	7450 lb	3720 lb	14,440 lb

Rope capacity: 700 ft of $\frac{7}{8}$-in. wire rope in five layers. Total weight of winch, motor, and brakes: 9050 lb.

Heay Duty, Two-Speed, Electric Cargo Winch

	High gear	High gear	Low gear
	218 fpm	305 fpm	75 fpm
	Line speed	Line speed	Line speed
Line pull	7450 lb	3720 lb	20,800 lb

Rope capacity: 800 ft of 1-in. wire rope in five layers. Total weight of winch, motor, and brake: 12,800 lb.

[a] Courtesy of American Hoist and Derrick Co.

tions, to lower a load into the hatch the operator pushes the controller toward the hatch or away from himself. With the controller in the vertical position the current is off and the load is held by the magnetic brake. To hoist a load, the operator pulls the controller toward himself or away from the hatch. Both the hoisting and the lowering of the load are done under power.

Hydraulic Winches

The hydraulic winch affords extreme smoothness of operation with precise control. Great acceleration is also possible, but this is a questionable advantage because of the possibility of inexperienced or simply careless winch drivers placing shock loads on the cargo gear. With light loads it would be an advantage.

An interesting comparison can be made between the equipment used on the Mariner type ships, with their electric winches, and the equipment that would have been required with hydraulic winches. The Mariner type ship was equipped with AC auxiliary power, but the problem of speed control on AC motors resulted in the installation of DC winches on the ships. This, of course, required the use of motor generators to convert the ship's AC supply. This appears to be a good example of how the cargo handling functions of a cargo ship are given secondary consideration by the planners of the ship in the early stages. The AC auxiliary units were judged desirable, evidently, but in providing advantages the AC installations multiplied the electrical equipment needed for cargo handling by almost three times. This, of course, made the cargo handling activities of the ship three times as vulnerable to breakdown.

General Description of the Hydraulic Installation

The hydraulic cargo winch consists of two fluid motors, each driving one end of the winch drum through a semirigid coupling. A control valve is included for handling the winch in operation. All operations of hoisting, lowering, braking, and stopping are accomplished by movements of the control lever in much the same way as the throttle valve is handled on the stream cargo winch.

The two fluid motors are capable of providing a wide range of torque–speed characteristics (see Table 7-3). For heavy loads, the oil flow is divided between the two fluid motors, obtaining full torque at one-half maximum speed. For lighter loads, the oil flow is directed into one motor only, and maximum speed at one-half full torque is available.

Hoisting Operation

Hoisting is accomplished by pulling the control lever back towards the operator and directing the oil partially or fully (depending on speed requirements) into both fluid motors simultaneously. A patented check valve feature of the control valve causes the inlet pressure to build up to overcome the load on the winch the *instant* the oil flow starts through the fluid motors. If increased speed is desired, the control level is pulled farther back to cause all the oil flow to travel through one fluid motor. However, if the load is too great to be handled by one fluid motor, both will remain cut in regardless of the *fast* position of the control lever. The oil pressure automatically adjusts itself to whatever the load on the winch requires. There is no time lag in the pickup of power, and the control valve can be operated *as slowly* or *as quickly* as desired, with prompt response at the winch. Rapid acceleration is combined with a smooth elastic start.

TABLE 7-3

TYPICAL PERFORMANCE RATINGS:
LINE PULL AT LAYER OF $\frac{5}{8}$-IN. DIAMETER WIRE

Hoisting	Inlet Pressure, lb/in.^{2}	Line Pull, lb	Maximum Line Speed, fpm
Using one	250	2580	230
fluid motor	300	3100	230
	400	4100	230
Using two	250	5160	115
fluid motors	300	6190	115
	400	8200	115

Lowering Operation

Lowering is accomplished by pushing the control lever away from the operator. The initial phase of lowering, providing there is a load on the winch, consists of braking by restricting oil circulation through the fluid motors. Further movement of the lever will cause the pressure to be built up and the winch to run in reverse. The complete speed range at one-half maximum torque is available when lowering.

Stopping

Stopping involves locking each motor hydraulically by preventing oil circulation through the fluid motors. In the stop position, oil from the pump by-passes the motors by going through the control valve back to the suction side of the pump. With heavy loads hanging on the winch, a slight creep can be expected. This is due to internal leakage past the vanes within the fluid motor. In case of accidental stoppage of the pump with the control valve in hoist position, the load will stay suspended, because the reverse action of the winch automatically locks the fluid motors hydraulically by closing the check valve feature in the control valve. This prevents reverse circulation. The load can then be safely lowered at no power under full control of the winch operator.

Piping Requirements

The pump size and speed are selected to obtain the flow required to attain the maximum desired line speed at the winch. The piping consists of a closed circuit with an expansion tank piped to the suction side of the pump. The expansion tank maintains a head of oil on the system and eliminates any possibility of air being drawn into the oil. The pump delivers a constant flow of oil through the circuit at a pressure proportional to the load on the winch. Only one pump is required for each pair of winches. The two winches are piped in series and the load at each winch is overcome by a proportion of the total pressure available. Pumps can be driven by a takeoff from a main or auxiliary engine or by a constant-speed electric motor.

Additional Features

Overload protection is obtained by a built-in relief valve at the pump. Excessive line strains can be prevented by proper adjustment of the relief valve.

As in any hydraulic transmission system, heat is generated through throttling, internal leakage, and other fluid friction losses. Deck equipment operating for short periods of time, such as the anchor windlass or warping winch, normally do not require cooling of the oil to maintain a favorable heat balance. However, the constant operation of cargo winches will require cooling of the oil at approximately 15% of the flow at average sea water temperature. However, the constant operation of cargo winches will require cooling of the oil at approximately 15% of the flow at average sea water temperature. If such a cooling device is employed in the return line, the oil temperature can be maintained between 100 and 120°F.

Some General Precautions

In a previous section some discussion was made of shock loads due to acceleration and deceleration of loads. Another danger, and one more likely to occur in actual practice, arises when operating a cargo winch with an empty hook while the runner is being slacked off. If the slack is run off too rapidly, the runner will not feed off but will simply form a number of loops. These loops may become tangled and caught on parts of the winch. When a load is applied to the winch again, the new turns will override the loops. Later, when slacking off with a new load, these loops will be released and the runner will slack suddenly. These shock loads can cause considerable stress on the runner and may be the direct cause of gear failure.

Electric winches have an overload safety device that causes a circuit breaker to cut off the current to the winch motor when an attempt is made to hoist a load that exceeds the capacity of the winch. Unfortunately, these circuit breakers do not act precisely when the line pull reaches the rated capacity of the winch. There is a gradual build-up of the torque until finally the current is automatically cut off. When operating a 5-ton electrical winch, for example, the line pull may reach 8 to 12 tons before the circuit breakers stop the winch action. During tight-lining of loads with the yard and stay rig, this can cause serious gear failures.

CALCULATION OF STRESSES

Calculation of Stresses Using Vectors

It is often convenient to be able to solve for the stress placed on a given part of the ship's rigging by graphic methods. We will discuss here vectorial methods for doing this.

The physical quantities of length, time, and mass can be described quite nicely by using numbers with appropriate units, such as 9 ft, 9 min, and 9 lb. These numbers are known as scaler units. However, when dealing with a physical quantity that has a number *plus a direction* related to it, it is more convenient to use vector quantities. Vector quantities are represented by properly *directed* lines the lengths of which represent the *magnitudes* of the vector quantities.

When a number of forces are acting upon a body in various directions, the total force and the direction of that force can be calculated by adding the various vector quantities. The single force thus calculated is known as the *resultant force*. If the resultant force is zero, then the body will remain at rest. For example, when a cargo hook is suspended between two booms with two falls and a loaded pallet attached to it, we know that the resultant of all these forces is equal to zero because the hook is not moving. It is at rest. Furthermore, it is obvious that the upward forces taken alone must produce a resultant force precisely equal to the downward forces. Therefore, if we have a known weight on the hook, we know that the falls are pulling so that they produce an upward resultant equal to the weight in suspension. With this information we can solve for the stress on each fall. Alternatively, if we are given the stress on each fall, we can calculate the resultant force itself.

One of the most common methods of calculating the resultant force is known as the parallelogram method. In the problems occurring in our study of the stresses on the parts of cargo rigging, we will ordinarily know one force (the load on the cargo hook) and be seeking the value of two forces acting to oppose the given force. These two forces will be the stresses on the two cargo falls. We will also deal with stresses on topping lifts, compression on booms, and approximate stresses on guys. Figure 7-10a shows a weight being suspended by two falls AC and CB with a fall angle of 120°. The weight is suspended at C. We know that the forces acting in the direction of A from C and in the direction of B from C are holding the weight in suspension.

The forces involved may be drawn as shown in Fig. 7-10b. F_1 and F_2 are opposed to the weight W; thus, F_1 and F_2 are creating the resultant R that opposes W. To solve for the stress on F_1 and F_2 by the parallelogram method, R is drawn opposite to W equal to W. In this example let W equal

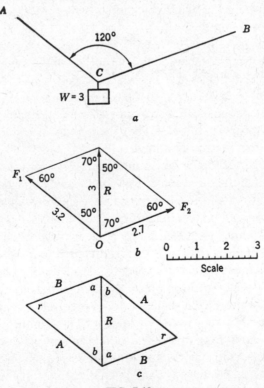

FIG. 7-10.

3 tons. Then R and W would each be drawn 3 units long. We start from any convenient point O. From the point O we draw F_1 parallel to CA and F_2 parallel to CB. Next, the end of the vector R is connected to the vectors F_1 and F_2 so as to form a parallelogram. Then the lengths of F_1 and F_2 from O in the chosen units give the stresses acting on the two parts. In this case, F_1 equals 3.2 tons, and F_2 equals 2.7 tons.

If the values of F_1 and F_2 and the direction in which they acted from the point O had been known first, the resultant force R could have been found by reversing the process.

Trigonometrical Solution

The values of the resultant or the forces can be found by using trigonometry when the angles a, b, and r are known (see Fig. 7-10c). These angles would be known because the directions of the forces and the resultant are known. In the above example, angle b is 50° and angle a is 70°. Thus, with R equal to 3 we can set down these elements in the diagram shown by Fig. 7-10c. By the law of sines we have

$$\frac{A}{\sin a} = \frac{R}{\sin r} \quad \text{and} \quad \frac{B}{\sin b} = \frac{R}{\sin r}$$

where $A = F_1$, $B = F_2$, $a = 70°$, $b = 50°$, $r = 60°$, and $R = 3$.
Since $\sin 70° = 0.94$, $\sin 60° = 0.87$, and $\sin 50° = 0.77$:

$$A = \frac{3 \times 0.94}{0.87} = 3.24 \qquad B = \frac{3 \times 0.77}{0.87} = 2.65$$

The slight differences in values arrived at by graphic and mathematical methods are due to the inability to read the units on the measuring instrument to the third significant figure in the graphic method.

NUMERICAL EXAMPLE

Given: Two falls are attached to a 1-ton lift and they both make an angle with the vertical of 60°.

Required: The stress on each fall using the parallelogram method.

Solution: Figure 7-11 illustrates the situation as a vector diagram. The answer is obtained graphically as 1 ton.

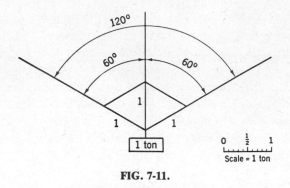

FIG. 7-11.

Effect of Fall Angle

It is important that the reader note that with both falls making an angle of 60° with the vertical, the stress on each fall is equal to the load being suspended. By making a diagram as shown in Fig. 7-12, it can be seen that as the angle between the falls increases above 120°, the stress on each fall goes above the weight being suspended, and that as the angle between the falls decreases below 120°, the stress on each fall goes below the weight being suspended. This is true when both falls make equal angles with the vertical. The lowest stress possible is equal to one-half the weight being suspended, and occurs when the angle between the falls is equal to zero. The stresses on both falls will be equal only when both falls make the same angle with the vertical.

In Fig. 7-13a it can be seen that although one fall makes an angle of only 60° with the vertical, it is supporting a stress greater than the weight being suspended; this is because the other fall is making an angle with the vertical that is greater than 60°. Note that the fall with the *least angle* from the vertical carries the *greatest* percentage of the total load. Figure 7-13b shows the result of decreasing the angle on the right side until it is 15° less than 60° rather than 15° greater. Now the load on the fall making an angle of 60° with the vertical is less than the weight being supported; note also that the stress on the other fall, which is making an angle with the vertical of less than 60°, is greater than the stress on the first fall. An analysis of these facts points up the great importance of limiting the drift between the hook and load while increasing the drift from the head of the boom to the deck of the ship.

Figure 7-13a and b shows the variation of stress on each fall as the

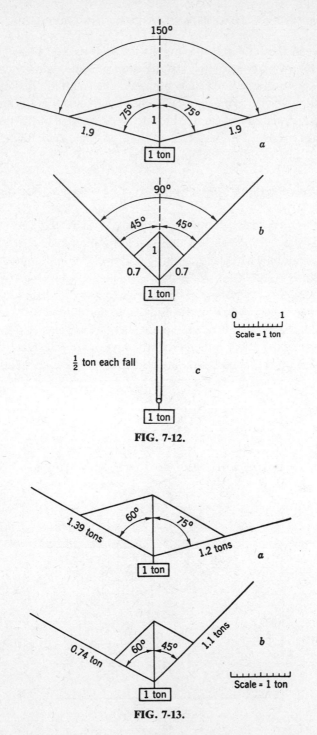

150°

75° 75°

1

1.9 1.9

1 ton

a

90°

45° 45°

1

0.7 0.7

1 ton

b

0 1
Scale = 1 ton

½ ton each fall

c

1 ton

FIG. 7-12.

60° 75°

1.39 tons

1.2 tons

1 ton

a

60° 45°

0.74 ton

1.1 tons

1 ton

b

Scale = 1 ton

FIG. 7-13.

377

angle between the falls and the angle that each makes with the vertical vary. The following conclusions can be drawn from this brief discussion: (1) The drifts should be such that the angle that each fall makes with the vertical will never be more than 60°. This is of special importance when the load being hoisted is equal to the safe working load of the fall. (2) The rig and slings used should be set up so that the angle each fall makes with the vertical is as small as possible.

Calculation of Stresses on Parts of a Swinging Boom

The parallelogram method can be used to calculate the stresses on the topping lift, the fall, and the shackles holding the head, heel block, and topping lift lead block in place; and the compression on the boom when a given weight is being lifted. An example is illustrated in Fig. 7-14 using a 50-ft boom topped at an angle of 45° whose topping lift lead block is 50 ft above the boom gooseneck. We will assume that a load of 1 ton is being supported by the cargo fall. With the angles between mast, boom, topping lift, and parts of the fall as indicated in the figure, we will start by calculating the stress on the head block shackle and its direction relative to the boom.

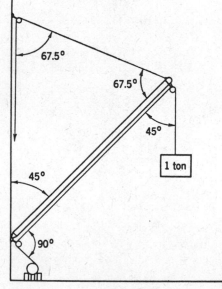

FIG. 7-14.

First we draw a separate diagram as shown by Fig. 7-15*a*. In this diagram, the following facts apply:

1. *BC* equals the load of 1 ton.
2. *BA* equals the stress on the hauling part of the cargo fall. Note that it has been given a value of 1.1 tons. The increase of one-tenth over the actual weight results from using an arbitrary constant to account for frictional resistance contributed by the sheaves revolving on their pins when the load is actually being hoisted.
3. *BD* equals the resultant of the two forces *BC* and *BA*. *BD* gives the stress on the head boom shackle and direction of this stress. The stress equals 1.95 tons.

The problem can also be solved by trigonometry. The trigonometric solution will be more accurate than the graphic method because of errors caused by the measuring instruments in the latter. Accuracy with the graphic method can be increased by using a large scale for the units involved.

We now solve for the compression on the boom and the stress on the topping lift bale. We draw another diagram as in Fig. 7-15*b*. *BD* is equal to the stress on the head block shackle, 1.95 tons, and is in a direction 21° from the boom *BE*. We know also that the topping lift is 67½° from the boom, and this is also drawn in. Drawing *DE* parallel to the topping lift direction, we obtain *BE* equal to the compression on the boom, 2.1 tons. Drawing *EF* parallel to *BD*, we obtain the stress on the topping lift bale, 0.75 tons.

In the same manner as before, we can solve for the stress on the shackle holding the topping lift lead block in place (see Fig. 7-15*c*). We know the stress on the topping lift is 0.75 tons, as solved for in Fig. 7-15*b;* we also know the directions of the topping lift parts *HG* and *HJ*. The stress on *HJ* is equal to that on *HG* because it is a static load. With *HG* and *HJ* laid down, we complete the parallelogram and draw *HK,* the resultant. On our scale, *HK* has a value of 1.25 tons.

The above examples should illustrate the methods of proceeding when solving for resultant forces with all parts in the same vertical plane, as they are when working with a swinging boom. The problem is considerably more difficult when working with a fixed boom and solving for stresses on guys as the loads are burtoned across the ship.

The height of the topping lift lead block above the deck relative to the

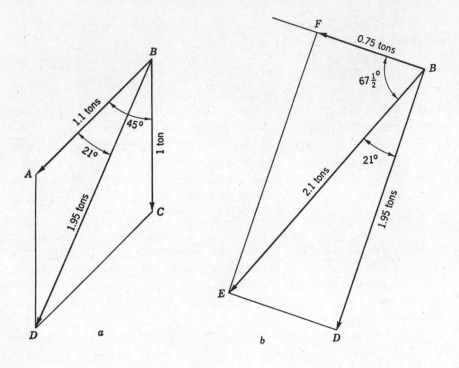

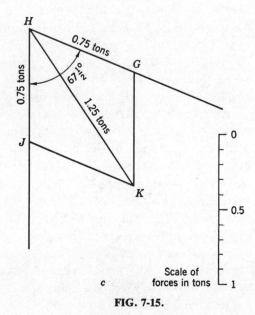

FIG. 7-15.

length of the boom has a great effect on the stresses placed on the topping lift, the boom, and the parts of the system supporting them. As an exercise, the reader should examine the stress on the topping lift and the thrust on the boom of Fig. 7-15 as the height of the topping lift lead block is reduced at 2-ft intervals and should then plot the values on a piece of graph paper.

Stresses on Fixed Gear

A method of solving for the stresses on the parts of a set of fixed cargo gear is described below. Four diagrams are needed to obtain all the data necessary to solve for the stresses. To set up the problem, certain important dimensions of the ship and her cargo gear must be known. These required facts are as follows:

1. The beam of the ship.
2. The width of the hatch.
3. The position of the boom goosenecks.
4. The position of the guy pad eyes.
5. The points over which the booms will be spotted.
6. Length of the booms.
7. Distance between heel of boom and topping lift lead block.

It is also necessary to set a desired fall angle or a desired drift over some point between the heads of the booms. If *both* of these values are set, then item 5 above must be picked from the construction of the diagrams and cannot be specified. If *only one* of these requirements is set, then item 5 must be specified and the diagram will give the other. In the following example we will want to know the stresses when the fall angle is 120° and the booms are spotted as depicted in Fig. 7-16. We will assume 60-ft booms and the other dimensions as shown on the diagram.

The first diagram to make is the plan view as shown in Fig. 7-16. Next we construct the diagram in Fig. 7-17*a*, which is an elevation of the gear on a plane vertical to the ship's deck and passing through points *AR* of Fig. 7-16. The first step is to draw the deck line and erect a perpendicular to represent the mast. Measure off *QH* equal to *RA* of Fig. 7-16. At *H* erect a perpendicular dashed line. Point *P* on the mast is where the heel of the boom is stepped, in this case 8 ft above the deck. Setting 60 ft on our dividers we measure an arc from *P* to intersect the dashed vertical line at

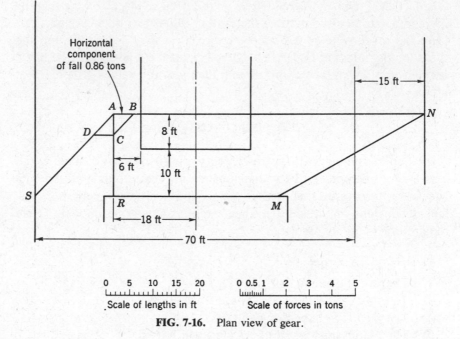

Horizontal
component
of fall 0.86 tons

8 ft

6 ft

10 ft

15 ft

18 ft

70 ft

0   5   10   15   20   0 0.5 1   2   3   4   5
Scale of lengths in ft   Scale of forces in tons

FIG. 7-16.   Plan view of gear.

H. We mark this intersection E. Now HE is the distance of the hatch boom's head above the ship's deck and PE is the true length of the boom. Next we make PO equal to 43 ft, the distance from the heel of the boom to the topping lift lead block, and draw OE. Now OE is the true length of the topping lift when the boom is topped to this angle. We now have the boom, topping lift, and mast in their actual lengths as they would appear on the vertical plane passing through AR of Fig. 7-16.

We now measure off HG on Fig. 7-17b, making this distance equal to AS of Fig. 7-16. Draw in GE to indicate the guy. It is of great importance, to avoid confusion at this point, to note that the guy GE is placed on Fig. 7-17 as it would appear on a vertical plane passing through AS of Fig. 7-16. GE is *not* a true projection on the plane through AR. This is done because the projected guy on the plane through AR will not be useful, but

---

FIG. 7-17.   (*Opposite*) Elevations of hatch boom. PE = boom = 60 ft; OP = heel to topping lift = 43 ft; QP = 8 ft; QH = 18 ft. (*a*) Difference between compression on boom with standing load and with running load. If EX = 1 ton, EY = 1.4 tons with a standing load secured to the boom head. But if the 1-ton load is supported by a fall through the head block, EY will increase by about 1-ton, to 2.4 tons. (*b*) Diagram for calculation of various forces on hatch boom.

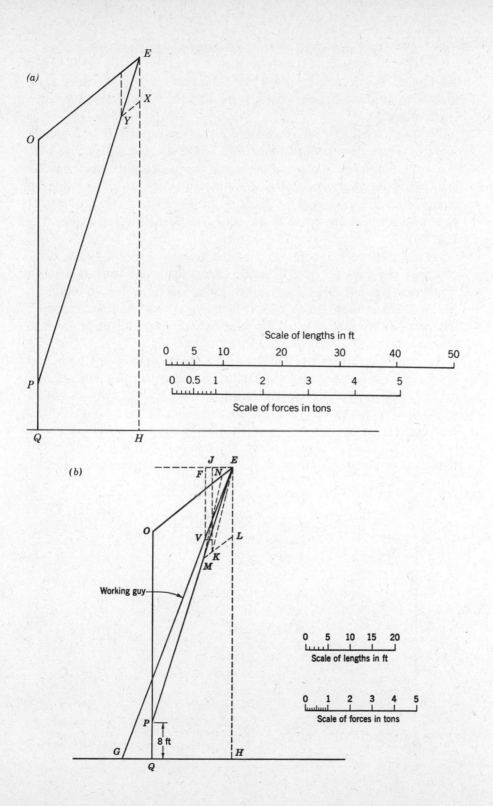

(a)

Scale of lengths in ft

0 5 10   20   30   40   50

0 0.5 1   2   3   4   5

Scale of forces in tons

(b)

Working guy

8 ft

Scale of lengths in ft

0 5 10 15 20

Scale of forces in tons

0 1 2 3 4 5

the true guy on the plane through *AS* is needed. Actually, we have combined two diagrams to make one and reduced the work required to solve the problem.

Next we construct Fig. 7-18. Figure 7-18 is similar to Fig. 7-17, except that it is for the dock boom. *MN* on Fig. 7-18 is equal to *MN* of Fig. 7-16. We will not use Fig. 7-18 except to obtain the height of the dock boom's head above the ship's deck. Hence, we measure *NU*, which we will need to construct an elevation of the gear on a plane vertical to the deck and at right angles to the centerline of the ship. This last elevation is shown in Fig. 7-19.

All the necessary data for drawing the diagram in Fig. 7-19 are available from the three previous diagrams. From the heads of the booms we measure angles 30° down from the horizontal and draw lines to represent the falls. Where these lines intersect is the point at which the cargo hook will be when the fall angle is 120°. The distance from this point down to the deck is the drift, and we note that it is 41 ft.

We now construct a parallelogram of forces acting on the hook and obtain the stress on each fall and the horizontal and vertical components acting on the head of the boom. Because of the 120° fall angle, we have a stress on each fall equal to the load on the hook. In this case it will be 1 ton, because we assume a load of 1 ton. Further analysis of the forces involved will show us that the horizontal component *AW* is equal to 0.86 tons and the vertical component *WX* is equal to 0.5 tons. These forces,

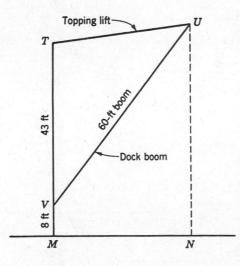

FIG. 7-18. Elevation of dock boom.

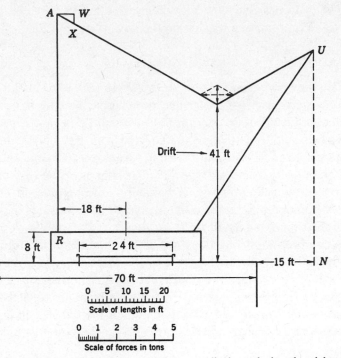

FIG. 7-19. Elevation of complete gear on plane perpendicular to deck and at right angles to ship's centerline.

AW and *WX,* are the forces acting on the boom due to the load on the hook (Fig. 7-19).

We now turn to Fig. 7-16 and analyze the forces acting on the head of the boom in a horizontal plane. First we lay down *AB* along the fall equal to the horizontal component of the force from the fall acting on the head of the boom. Thus, we make *AB* in Fig. 7-16 equal to *AW* in Fig. 7-19. Drawing *BC* parallel to the guy and *CD* parallel to the fall, we complete a parallelogram of forces from which we pick off *AD*. *AD* is equal to the horizontal component of the total tension on the guy as it resists the force *AB*. The resultant of the forces *AD* and *AB* is *AC*. *AC* represents the horizontal component of the forces acting on the boom.

Guy Stress

Using Fig. 7-17*b,* we proceed to solve for the guy stress, which is a tension stress. Draw *EF* equal to *AD* of Fig. 7-16 and parallel to the ship's deck. At F drop a line vertically downward to intersect the guy at V. Now, EV represents the tension on the guy, about 3.4 tons.

Boom Compression

To obtain the boom compression, draw EJ in Fig. 7-17b equal to AC from Fig. 7-16. At J drop a vertical line downward, making it equal to the vertical component of the guy tension, which in this case is represented by FV. Add additional length to the line from J equal to the vertical component of the fall tension, which in this case is 0.5 tons and is represented graphically by WX of Fig. 7-19. The sum of FV and WX is equal to JK. Now, draw a line through K parallel to the topping lift OE until it intersects the boom EP at the point M. Now EM represents the compression on the boom and amounts to 4.2 tons.

This compression figure does not take into account the force applied along the boom by the hauling part of the fall as it runs from the boom head block down to the heel block under a tension stress. This would increase the boom compression by an amount approximately equal to the force being applied to the hauling part of the fall, which in this case is equal to 1 ton. Hence, the compression on the boom would be about 5.2 tons.

Equivalent Load on a Swinging Boom

Confusion may easily arise when one discusses loads on swinging booms. In practice, such loads are suspended from the boom on a fall that runs through a head block and down to a heel block and finally to the winch drum. When considering forces on booms theoretically, it sometimes happens that the load is considered as hanging from a line that is secured directly to the boom head as if by a short strap, one end of which is shackled to the boom's spider band. The reader should satisfy himself that the compressions on the boom are different in these two cases, and that the forces acting on the connecting link at the bottom eye of the spider are also different in magnitude and in direction. See Fig. 7-14 and the discussion on p. 378.

To differentiate between these two types of loads, we will call the load placed on a boom by a load hanging from a fall running through a head block and to a heel block as a *running load*. The theoretical load placed on a boom by a load suspended from a line secured to the head we will call a *standing load*.

To obtain the equivalent *standing* load on the boom in Fig. 7-17, we extend the line *MK* parallel to the topping lift so that *MK* intersects *HE* at *L*. Now *EL* represents the equivalent standing load on a swinging boom. In other words, the compression on this boom as a swinging boom with a standing load equal to the value of *EL,* which is about 3 tons, would be equal to the compression that the boom is now experiencing. This compression, of course, would be without the additional compression caused by the hauling part of the fall.

To obtain the equivalent *running* load on the boom in Fig. 7-17, it is necessary first to calculate the compression on a swinging boom with the same dimensions and rigged in the same way when it supports a running load of 1 ton. The reader should do this to verify that the compression would be about 2.4 tons, including the 1-ton force applied along the boom by the hauling part of the fall. Next we divide the *total* compression we have already calculated to be on the boom (5.2 tons) by this value of 2.4 tons. The quotient is equal to the equivalent running load on this boom. Thus we obtain:

$$\frac{5.2}{2.4} = 2.2$$

Therefore, we see that the 1-ton load on this *fixed* boom is producing a compression on the boom equivalent to a load of 2.2 tons on a *swinging* boom. Since the safe working load of the boom is based on its being tested as a swinging boom and not as a fixed boom, this value of *equivalent running load on a swinging boom* is most significant.

Topping Lift Tension

Finally, to obtain the tension stress on the topping lift we draw *MN* parallel to *EK* in Fig. 7-17*b* to complete the parallelogram *EKMN*. Now *EN* measured on the topping lift represents the tension on the topping lift. In this case it is equal to 0.6 tons.

Note that if point *K* falls below the boom on a diagram such as Fig. 7-17*b*, the boom will not jackknife, but that as *K* rises, the topping lift tension diminishes. If *K* falls on the boom, *EN* becomes a point and the tension on the topping lift is zero. If *K* falls above the boom, the topping lift tension becomes negative and will actually lift the boom up, or in other words, cause the boom to jackknife. The hatch boom will usually go all the way once it starts to jackknife, due to other forces involved. The dock boom may rise part way and then stop with the topping lift hanging in a bight with the forces on the boom, guy, and fall in balance. However, there is no universal pattern; it all depends on the forces, their directions, and their magnitudes. If the winch driver notices the dock boom starting to rise, he can stop it by stopping the winch and lower the boom by slacking back. If he sees the hatch boom start to rise, as at the beginning of jackknifing, he should slack back quickly and be alert and ready to jump for safety. With booms rigged properly, of course, there is no danger.

WORKING WITH ABOVE-NORMAL LOADS

Limiting Factors

The limit on the weight of the load that is safe when working with any cargo handling rig is dependent on the size and type of wire being used for the fall. (The type of tackle is a factor also, but this is discussed separately under the discussion of doubling up.) In this section, reference is to the single whip as normally used with the yard and stay system. Although the fall is the limiting factor when initially hoisting the load on the up-and-down lift, it should be remembered that the guys may be the limiting factor when burtoning the load across the deck. The guys are thrust into the role of a limiting factor whenever they are not positioned properly or whenever the drift between the hook and the bottom of the load is excessive. We will consider only the fall as a limiting factor in this section. Table 7-4 on breaking strengths is presented for reference when considering the rigging of any given gear. Equations for finding breaking strengths should be used only when tabulated data are unavailable.

A single strand of $\frac{5}{8}$-in., 6 by 19, plow steel wire rope has a breaking stress of 12.9 tons and a $\frac{3}{4}$-in. wire has one of 18.5 tons. Many ships now use improved plow steel wire rope, for which the following breaking strengths are: $\frac{5}{8}$-in., 14.9 tons; $\frac{3}{4}$-in., 21.2 tons. From these figures, it is evident that improved plow steel wire rope is about 1.15 times stronger than plow steel. This is an increase of 15%. Ships such as the Mariner type ships are equipped with improved plow steel for all running as well as standing rigging. The fiber core is used almost universally for the running and standing rigging on merchant ships.

TABLE 7-4
WIRE ROPE BREAKING STRENGTHS[a]

Rope Diameter, in.	Breaking Strength, in long tons		
	Improved Plow Steel	Plow Steel	Mild Plow Steel
$\frac{1}{2}$	9.5	8.3	7.3
$\frac{5}{8}$	14.9	12.9	11.2
$\frac{3}{4}$	21.2	18.5	16.0
$\frac{7}{8}$	29.7	25.0	21.7
1	37.3	32.5	28.2
$1\frac{1}{8}$	47.0	40.8	35.5
$1\frac{1}{4}$	57.7	50.2	43.6
$1\frac{3}{8}$	69.3	60.3	52.5
$1\frac{1}{2}$	82.1	71.4	62.1
$1\frac{3}{4}$	110.7	96.4	83.6
2	142.8	124.1	108.0
$2\frac{1}{4}$	178.5	155.4	

[a] The data presented in this table apply to 6 by 19 classification ropes with fiber cores either preformed or nonpreformed. For the breaking strengths of galvanized ropes, deduct 10% from the strengths shown. For wire strand cores and independent wire rope cores, add $7\frac{1}{2}$% to the listed strengths.

Running Rigging and Safety Factors

When determining *safe working loads* for running rigging, a safety factor of at least 5 should be used. This gives safe working loads of 3 tons for the $\frac{5}{8}$-in. improved plow steel and 4 tons for the $\frac{3}{4}$-in. improved plow steel wire rope of 6 by 19 construction with fiber cores. (The ton referred to in this discussion is the ton of 2240 lb.) The safety factor of 5 may seem excessive, but it must be remembered that this is the factor only when the line is new. During normal use with shipboard rigs, the wire rope is subjected to treatment that will reduce its original breaking stress rapidly.

Boom Ratings

Tests of fully rigged cargo gear are made by the American Bureau of Shipping and equivalent authorities in other countries. These tests have become necessary since various nations began enforcing regulations relative to tests and periodic inspections of cargo handling gear as recommended by the International Labor Convention of 1932. Without proper gear certification, ships are subject to delays in many ports of the world in order to undergo tests and inspections locally.

The rules of the American Bureau of Shipping specify standards to be complied with for certification of the ship's cargo gear and maintenance of that certification. These standards also apply to the upkeep of chain and wire slings and certain other loose gear used during the longshore operations. Periodic annealing of wrought iron gear is required, and all blocks, wire, and loose gear should be furnished with certificates of testing by the manufacturer. The American Bureau of Shipping specifies a safety factor of 4 for all gear.[1]

After the gear is installed, the complete assembled rig is tested by swinging of live proof loads of amounts greater than the rated capacities in the presence of a surveyor. The live loads are required for the rating of newly built ships. On satisfactory completion of such tests, the heels of the booms are stamped with the safe working loads and the date. It should be remembered that this test is given to the gear used as a swinging boom, but that the boom is almost always used in a fixed position. This fact changes the stresses on all parts and should be kept in mind and investigated by the ship's officers using various diagrams of the gear and its arrangement on their particular vessel.

Gear Certification

The surveyor issues a certificate covering the proof loads and angle at which each boom is tested. This certificate, together with all manufacturer's certificates regarding loose gear, is inserted in the Ship's Register of Cargo Gear. This register has pages for entering certification data for the original proof loads, the annual surveys, and the surveys every 4 years, when the proof loads have to be applied again. Pages are also

---

[1] L. C. Host, "Certification of Ships' Cargo Gear in the United States," *The Log,* November 1953.

available for recording the annealing of all wrought iron gear. This simple bookkeeping process provides evidence of the condition of the cargo gear and the maintenance given it.

The American Bureau of Shipping rules require that new ships be furnished with a diagram showing the arrangement of the assembled gear and indicating the approved safe working load for each component part of the gear. With these diagrams, replacements for loose gear may be ordered to suit the safe working loads.

For the certification of cargo gear on existing ships, the use of spring or hydraulic balances to register the proof loads required to be applied to the assembled rig is permitted. Figures 7-20 and 7-21 show such test arrangements. In an actual 5-ton design, the thrust load in the boom and the tension loads in the topping lift and cargo falls are calculated under the application of the safe working load of 5 tons and of a test proof load of $6\frac{1}{4}$ tons at the cargo hook.

Doubling Up

Whenever the load to be hoisted by the yard and stay rig exceeds the safe working load of the rigging, some alternative rig should be employed that will increase the safe working load. The safety factor of 5 is recommended

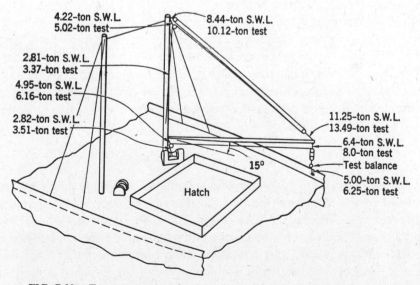

FIG. 7-20. Test arrangement for 5-ton boom. (S.W.L., safe working load.)

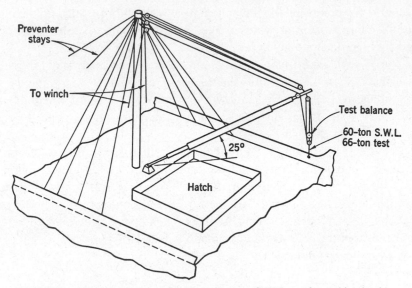

Preventer stays

To winch

Test balance

60-ton S.W.L.
66-ton test

25°

Hatch

FIG. 7-21. Test arrangement for 60-ton boom. (S.W.L., safe working load.)

when arriving at safe working loads with running rigging. Using the alternative rigs discussed in the following sections is known as *doubling up*. If the wire fall has just been renewed and the mate is confident of the skill of the winch drivers, he may not require doubling up for loads close to the limit. For example, if the mate considers a safety factor of 4 to be sufficient, he might permit hoisting a $3\frac{1}{4}$-ton load with a $\frac{5}{8}$-in. plow steel wire rope. The doubling up process is slow and costly, and freight rates generally reflect the cost of handling heavy items. It is often a great temptation to attempt the loading of items whose weights are just over the limit of the regular gear. Because of the risk involved, the practice of overloading the gear should be discouraged.

Yard and Stay with Gun Tackles

When one is forced to double up with a swinging boom, the time required to transfer the load from the dock to the ship or vice versa is greatly increased. This increase in time can be cut down by using systems that retain the advantages of fixed booms rather than changing to a live boom operation. This is especially important when a large number of loads just over the safety limit are to be handled. One of the easiest methods of augmenting the load limit of the gear is to rig both booms with gun tackles.

This requires two additional 14-in. cargo blocks, four shackles, and two cargo falls that are double the length of the regular falls (see Fig. 7-22).

The first step is to run the old fall off the winch drum. Next, the new fall is placed on the winch drum and led up to the head block via the heel block. The fall is then rove through the traveling block and the standing part is secured to the spider band at the head of the boom. This is done with both booms. The added traveling blocks should have beckets on them to permit the attachment of a common ring or separate swivels of a standard cargo hook assembly. The booms have to be lowered to the deck to make this conversion. When the booms are topped again, the rig is ready to handle cargo with fixed booms under the standard yard and stay system (see Fig. 7-23). If the booms are 5-ton booms with $\frac{5}{8}$-in. improved

FIG. 7-22. Doubling up of the yard and stay with gun tackles: close-up of the arrangement of the blocks at the cargo hook. Note the small chain between the tops of the blocks to prevent the slack block from toppling over and fouling itself. The cargo hook is of the Liverpool pattern. Courtesy of Clark Equipment Co.

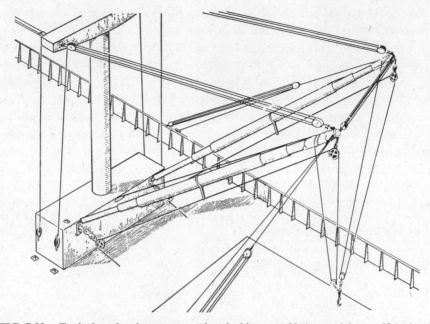

FIG. 7-23. Typical yard and stay system rigged with gun tackles on each boom. If the head block has no becket, the runner should be secured with two round turns to the boom itself about 4 ft back from the head. The eye splice at the end of the runner is shackled to the link band at the head of the boom. The turns around the boom must be started from the inside to prevent chafing against the runner leading from the head block to the heel block in operation. If 5 tons are burtoned with this rig, every possible precaution must be taken to insure minimum guy stresses and equal strain on all parts. Courtesy of U.S. Navy.

plow steel wire rope falls, the load limit should be kept at 5 tons. If the booms are 10-ton booms with a $\frac{3}{4}$-in. wire fall, the load limit should be kept to 8 tons. The greatest danger point with this rig are the guys. The guys should be checked carefully, and the positions chosen to give the least strain to the guys. Slings should be applied carefully so that a minimum of drift is allowed between hook and load. In this way the hook will not have to be hoisted an excessive distance above the deck and the angle between the falls can be kept to a minimum, thus relieving the strain on the guys. The angle between the falls should be kept below 120°.

Four Fixed Booms

If a hatch is double rigged, that is, if it has a set of gear at each end, it can be doubled up to work with fixed booms with the same load limits as for the yard and stay with gun tackles but in a much shorter time and with more safety. The time for rigging is cut down because the booms do not have to be lowered to the deck. Another advantage is that the stresses on the guys and booms are shared by the two sets of gear. The gear required is the same as for the yard and stay with gun tackles.

First, the two hatch booms are topped over the hatch and the two dock booms out over the dock in the conventional manner. The only difference would be that their heads should be in a fore and aft line and about 2 ft apart. The long cargo fall is then rove off on one set of the gear without being lowered by using the old cargo fall as a messenger. The lower end of the new, longer fall is rove through the additional cargo block, and the shackle is used to connect the end of this fall to the end of the regular fall on the opposing gear. This connection is hoisted to within a few feet of the head block on the other gear, and the winches of that gear are secured. The additional cargo blocks, now acting as traveling blocks, are secured to a cargo hook as in the rig of Fig. 7-23 (see Fig. 7-24). The load is hoisted and burtoned by the use of one set of cargo winches. The stresses on the

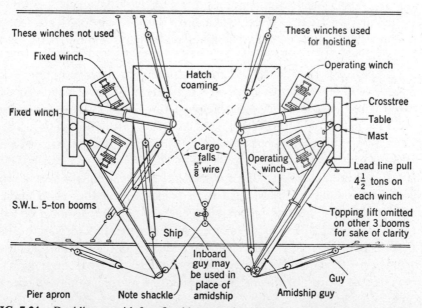

FIG. 7-24. Doubling up with four fixed booms: plan view of a hatch fitted with two sets of gear and doubled up for hoisting heavy lifts. The capacity of this rig with $\frac{5}{8}$-in. improved plow steel wire rope would be 6 tons with a safety factor of 5. If $\frac{3}{4}$-in. wire rope were used, the capacity would be 8 tons. Courtesy of U.S. Army.

guys will not be as great as on the first rig, but the falls will receive the same stresses; hence, the load limit is 6 tons with $\frac{5}{8}$-in. wire or 8 tons with $\frac{3}{4}$-in. wire.

The use of a traveling block insures an even strain on all parts. If the traveling blocks are eliminated, the four falls shackled directly to the cargo hook, and all four winches used, there is some possibility of an unequal strain being placed on one of the falls during the operation. However, if the winch drivers are all experienced and the operation is done slowly, it is much faster than rigging with the traveling block. If a number of lifts are to be handled by this means, it is recommended that the traveling block be used. If only one lift is to be handled and all four falls are attached directly, the operation should be done with care.

One Swinging Boom

If the load is over the above-mentioned limits or the circumstances of the operation call for more direct control of the load, it may be necessary to resort to the use of the swinging boom. One good reason for using the swinging boom instead of attempting to rig up with gun tackles on two fixed booms is that it is faster when only one or two loads are to be handled. Another reason is that the loads on the guys and boom are kept at a safe level.

To prepare this rig, it is necessary to drop the dock boom to the deck and reeve off a fall of double length as a gun tackle, with the standing part secured to the spider band at the head of the boom. This is the same procedure that is required for the yard and stay with gun tackles. The topping lift hauling part can be placed on the drum of the outboard winch drum and used as a working topping lift. The working guy can be left where it is located normally. If a spanner guy is rigged, it can be used to pull the boom inboard by leaving it in position with the hatch boom winged to the offshore side. The hauling parts of both guys should be led to the gypsy heads of the winches of the gear immediately forward or aft of the swinging boom by using snatch blocks as fair-leads where necessary. Figure 7-25 illustrates this rig. There is very little stress on the guys during this operation because they are used only to guide the swinging boom.

The limit of this system is 5 tons if $\frac{5}{8}$-in. improved plow steel wire rope or $\frac{3}{4}$-in. wire rope is used on a 5-ton boom. If $\frac{5}{8}$-in. wire rope is used on a 10-ton boom, the limit is 6 tons; however, if $\frac{3}{4}$-in. wire is used on a 10-ton

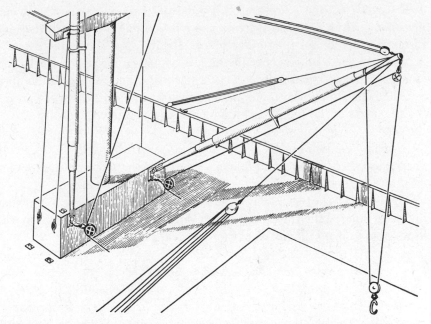
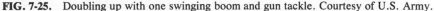

FIG. 7-25.   Doubling up with one swinging boom and gun tackle. Courtesy of U.S. Army.

boom, the limit is 8 tons. If new wire is used with the 10-ton boom, these limits can be raised slightly; the responsibility rests entirely on the ship's officer in charge of the operation.

Two Swinging Booms

By using both booms of a set as swinging booms for picking up a single heavy-lift item, the load limit can be increased insofar as the boom as a limiting factor is concerned. However, the limit as set by the fall size is still the same. In other words, if two 5-ton booms are swung together and rove off with ¾-in. falls, the limit is greater than 5 tons. The limit is double the *safe working load* of the wire fall; in this case, it would be 8 tons.

When one stops to consider the risk entailed in attempting to lift loads slightly over the safe working load, it does not make good sense to make a practice of doing so. A safety factor of 5 should be set and conscientiously adhered to. If a fall starts to strand or parts suddenly without even the warning of a parted strand, there is a great possibility of loss of life and a certainty of heavy damage to the ship and cargo.

Figure 7-26 shows two ways of rigging two booms to operate as swinging booms. Both of these systems make it possible to increase the load insofar as the boom is a limiting factor. Figure 7-26*a* shows the use of a traveling block. The heads of the booms should be about 3 ft apart horizontally; their vertical separation is not of major importance. If the horizontal distance between the boom heads is too great, the angle between the two parts of the traveling whip increases and this reduces the size of the load that can be carried.

Swinging the two booms in an arc is not easy. It must be done slowly and requires the topping and luffing of the booms; therefore, the topping lifts must be powered.

Figure 7-26*b* shows the use of an equalizing beam. This accomplishes the same thing as the traveling block. Using a beam in this manner makes it possible to keep the heads of the booms farther apart.

The Heavy-Lift Boom

Almost all ships are equipped with at least one *heavy-lift* or *jumbo boom*. As can be seen from the above discussion, when the weight of the unit to be taken aboard is above 3 tons, one of the doubling up rigs can be used. By using these one may avoid using the jumbo rig, which is very slow in operation and may require an hour or more to prepare for use. However, when the weight of the unit goes above the safe working load of any of the doubling up rigs, there is nothing left to do from the standpoint of safety but rig the jumbo boom. Now and then, for expediency, loads exceeding the safe working load will be taken on board with the conventional gear. Such practices may prove to be very costly instead of economical.

The heavy-lift boom is usually shipped in a pedestal mounted along the ship's centerline. There are usually two ratings for its capacity; the lowest is without a preventer stay being rigged. The maximum rating is with the preventer stay rigged. The preventer stay is a large wire rope that is normally carried unrigged on the mast supporting the jumbo rigging. Two stays are generally used. The Mariner type ship uses a preventer stay of $2\frac{1}{4}$-in.-diameter improved plow steel 6 by 19 wire rope. The jumbo boom has a capacity of 25 tons when not using preventer stays. When the stays are rigged, the capacity is raised to 60 tons. The purchase and topping lift are led to winches with specially built derrick drums capable of handling long lengths of wire. The purchase of this 60-ton boom is 1500 ft long and made of 1-in. 6 by 19 improved plow steel wire rove off on an upper

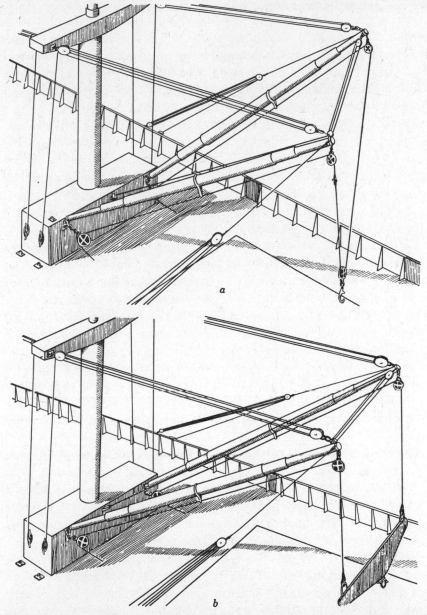

FIG. 7-26. (*a*) Doubling up with two swinging booms and traveling block. The load limit with this rig, insofar as the boom is concerned, is twice the safe working load of one boom. The limiting load as far as the fall is concerned is twice the safe working load of the fall. If gun tackles are used, the fall limit becomes four times the safe working load of one part. Thus, using a gun tackle and only 5-ton booms with this rig, the load that can be taken on board is 10 tons. With 10-ton booms and ¾-in. wire rope, the load limit is 16 tons. (*b*) Doubling up with two swinging booms and an equalizing beam. Courtesy of U.S. Navy.

quintuplet block and a lower sextuplet block with sheaves of 20-in. diameter. The topping lift is 1190 ft long and made of $1\frac{1}{8}$-in. improved plow steel 6 by 19 wire rope rove off on sextuplet blocks with sheaves of 20-in. diameter. The hauling parts of the purchase and topping lift pass through two lead blocks and then to the drums of the derrick winches.

The guys of this heavy-lift boom are rove off on threefold blocks and are made of $\frac{5}{8}$-in. 6 by 19 improved plow steel wire rope 650 ft long. No pendants are used on the lower ends of these guys because during certain stages of the use of the boom it becomes necessary to almost two-block the guy blocks with a short pendant at the upper end. Besides this fact, the hauling part of the guy is led, by the use of fair-leads, to the winch drum of the regular cargo winch of this gear.

With this rig, as shown in Fig. 7-27, four men operating the two guy winches, the purchase, and the topping lift can control the boom. The capacity of a given rig on a given ship is stated on the ship's capacity plan or on a blueprint showing the plan and elevation views of the rigging.

Handling a heavy lift requires careful planning. The method of slinging the load up must be considered carefully. In many cases a sling weighing several tons may be required to lift a load weighing 40 or 50 tons. The weight of the sling must always be considered when calculating the total load on the boom. If a load of 24 tons is to be hoisted and the sling weighs 3 tons, it is necessary to rig the preventer stays even though the load being lifted is less than the rated capacity of the boom without preventers rigged.

The chief officer should be advised of the method being used to hoist, stow, and secure a given heavy lift, and should make it his business to be on hand during the rigging and handling of all heavy lifts. If he finds it impossible to be on hand, he should designate another officer to observe the operations. The importance of officer supervision is even greater when the ship is not operating in her home port.

When using the heavy-lift boom, the load should be taken up only a few inches at first and held steady. All rigging should be checked while the load is held suspended in this position. Then the load can be hoisted very slowly. At no time should there be any attempt to rush the process. If any particular part of the procedure can be said to be more dangerous than any other, it is when the load is being lowered and it becomes necessary to stop the lowering. If this is done with a jerky motion or too suddenly, the stresses on the gear may be raised to a dangerous level and contribute to the failure of some part of the rigging. The stresses can be raised also

FIG. 7-27.   The heavy lift boom on the Mariner type ship. Note the fair-leads rigged for guying the boom during the slewing operation. The double guy on the far side of the ship would be either on the dock side or lighter side depending on where the heavy lift was being taken from or discharged. Using the double guy enables the slewing of the boom out over the dock. Courtesy of U.S. Maritime Administration.

by hoisting the load too rapidly. Hoisting and lowering of heavy lifts should always be carried out slowly and carefully.

At single-rigged hatches, where it is necessary to lead the guy hauling parts to winches at the next hatch, care must be taken to assure that drivers of the guy winches can see the signal man clearly. There have been several cases in which the guys parted because one guy winchman heaved while the other was either heaving or holding fast.

When slewing the jumbo boom, it should be at the lowest position possible for easy operation. The boom should not be slewed when topped so as to reach the hatch area immediately in front of the pedestal. The forces applied to the guys place great strains on the gooseneck, pedestal, and pacific iron when the boom is so high. At lower positions it rotates more easily.

All of the parts of the heavy-lift rigging should be carefully stowed away or secured in the rigged condition when not in use. Because of the tremendous weight of the blocks and hooks used, the gear is not generally stripped and secured. The boom is topped as high as it can go, and the purchase and topping lift are secured by lashings after slushing them down with grease. Generally a collar is provided aloft in which the boom is secured. The large blocks are secured by using short wire pendants and turnbuckles. It is also important that the gooseneck and pin of the jumbo boom (as of all booms) be kept well lubricated.

Stuelcken Boom

The wishbone type Stuelcken heavy-lift boom (see Fig. 7-28a) is becoming a common sight on modern merchant vessels. These booms have

(a)

FIG. 7-28. (a) 70-ton Stuelcken boom aboard *S.S. Del Monte*. Courtesy of M/N E. K. Harrell, Class of 1984, USMMA. *Opposite*: (b) Parts of the Stuelckenmast®. Courtesy of Blohm + Voss AG.

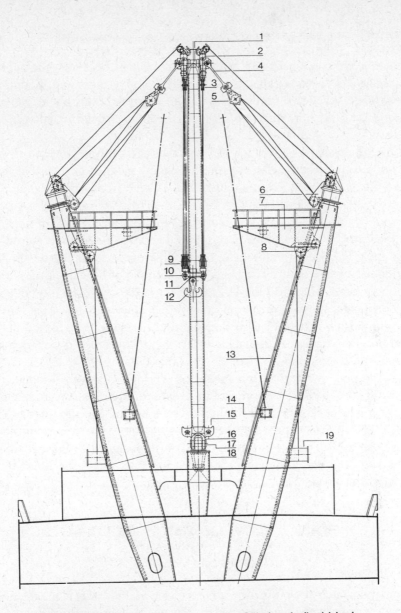

1 — Derrick head fitting
2 — Pendulum block fitting with guide rollers
3 — Upper cargo blocks
4 — Connecting flats
5 — Lower span block
6 — Span swivel
7 — Cross tree
8 — Inlet for the hauling part
9 — Lower cargo blocks
10 — Connecting traverse
11 — Swivel eye for flemish hook
12 — Flemish hook
13 — Ladder
14 — Gooseneck pin socket
15 — Fastening device for lower cargo block
16 — Heel fitting
17 — Derrick pin
18 — Gooseneck and gooseneck pin socket
19 — Winches

(b)

FIG. 7-28. (*Continued*)

403

capacities ranging up to 525 tons. The Stuelcken boom on the typical merchant vessel has a capacity of about 80 tons. A heavy-lift boom has yet to be designed that improves upon the features of this boom. Several variations exist in the means of moving the boom from one hatch to the other.

The Stuelcken rig (see Fig. 7-28*b*) consists of a centrally located swinging boom supported by two king posts (usually inclined outboard at their tops). Each king post is fitted with a swivel head from which a multiple-purchase topping-slewing tackle is led to pivots at the boom head to prevent twisting of the tackle. Four winches are utilized: two for hoisting the load and two for each of the topping tackles. The boom is raised by both topping winches and swung by hauling on one topping winch and paying out on the other. The chief characteristic of Stuelcken rig is that the boom head, when fully raised, can be flopped forward (or aft) between the king posts, allowing the boom to work the adjacent hatch. This shift of working positions takes 3 to 5 min, without rerigging. The forked boom head prevents fouling of the tackle when the boom is moved between the king posts to the opposite side.

In addition to its main advantage of working adjacent hatches, the advantages of the Stuelcken rig are its greater lifting capacity, its requirement for less deck gear, and the increased speed of the cargo hook. The typical 60-ton Stuelcken boom has a hoisting speed of 20 fpm and light hook speed of 45 fpm. The Stuelcken rig is designed for fast loading, unloading, and spotting of heavy cargo and thus far is the ultimate in shipboard heavy-lifts.

MODIFICATIONS OF THE MARRIED FALL

Wing and Wing

Rigging the hatch boom over the offshore side of the vessel while the dock boom remains over its conventional spot on the pier apron is known as *wing and wing* rigging. It is a simple variation of the regular yard and stay rig used when loading or discharging deck loads and its use is necessary when handling loads on both sides of the ship.

The House Fall

At some ports, a number of piers are equipped with a cargo block made secure on a steel structure on the face of the dock shed in such a way that

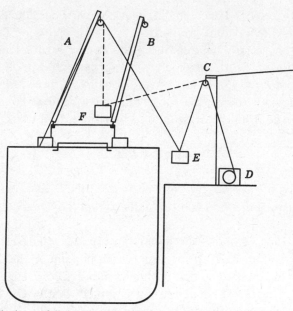

FIG. 7-29. The house fall rig. The house fall may be led from block *C* to a dock winch *D* as shown or to a ship's winch at the foot of boom *B*. When the load is in position *F*, the operator of the winch of boom *A* is in control of the load and has a good view of it. Boom *A* should be used for the ship's fall, because its working guy is rigged to handle the stress of burtoning.

a fall rove through it plumbs the dock apron. The structure may be a steel outrigger or a short boom stepped on the side of the building. Such an arrangement is used mostly on piers with very narrow aprons, which are in some cases less than 3 ft wide. They are also very convenient on piers with two decks. The fall used with this rig is furnished by the pier, because the regular fall on the ship is too short. The house fall may lead to the ship's winch via proper fair-leads or, in some cases, to a winch on the pier. The hatch boom with its fall is used as it is normally. Figure 7-29 is a diagram of this type of rig.

The advantages of the house fall are as follows:

a. On piers with very narrow aprons, it eliminates the possibility of fouling the head of the dock boom against the face of the dock shed.

b. It provides a steady spotting area under the permanently installed block on the dock.

c. On piers with second decks, it may provide the only means of working the second deck platform, because of the limited drift of the ship's booms. When the ship is deeply loaded and the booms are short, the drift over the second deck landing platform may be reduced to only a few feet if the ship's booms are used.

d.   If the house fall block can be made fast at a sufficient height, it becomes possible to work extremely wide aprons with a spotting point at the dock shed doorway or at any one of as many as three railroad spurs. When working with booms, the dock boom must be lowered to a dangerously small angle from the horizontal to work the third track.

Split Fall Rig

Two variations of the split fall rig will be described. Essentially, it consists of unmarrying the yard and stay falls so that they can work independently.

In one rig, the two booms are guyed loosely; we will describe this one first. The dock boom is guyed so that it plumbs a point on the dock, and is allowed to swing inboard so that it plumbs another point slightly inboard on the stringer plate. The hatch boom is guyed tightly. Obviously, this type of rig is easier to set up if inboard guys are used on both booms instead of the midship guy.

The rig operates as follows, assuming a discharging operation: The single fall of the hatch boom brings up a load and a man stationed on deck pulls the load outboard just enough to allow it to be landed on the deck of the ship. The hook is removed from the load and it goes below for another load. In the meantime, the dock fall is hooked onto the load on deck. A strain is taken and the load slides outboard. The dock boom, being guyed loosely, swings inboard as far as its working guy will allow it. This brings the dock boom over the stringer strake of deck plating. The winch driver hauls away on the fall, dragging the load under the boom, where he stops it momentarily before hoisting it clear of the deck. The man on deck pushes the load outboard and the load is landed on the dock. When loading, the steps are reversed.

In another arrangement of almost the same type, the two booms are guyed tightly in a fixed pattern. The point plumbed by these booms in both split fall arrangements must be only a few feet clear of the ship's side and the hatch coaming.

In this system, again assuming a discharging operation, the hatch boom brings a load up and stops. The winchman holds the load with the hook about level with the top of the hatch coaming. The man on deck takes the hook of the dock fall and puts it on the sling. Now both falls work together and take the load on deck or out over the dock, at which point the hatch fall is released. The load is landed with the dock fall while the hatch fall goes below for another load.

As can be seen, the only difference between the two methods of using the split fall is that in the first system the load is moved across the ship by being pushed or pulled manually or being dragged with one fall. In the second system the load is burtoned by temporarily married falls.

With longshoremen trained in the use of the system and with certain types of cargoes, the split fall rig has proved faster than the married fall system. When handling baggage stowed in deep lower holds on passenger ships, it is quite likely to be more productive than the yard and stay system. Full cooperation of the longshoremen is mandatory, of course.

Yard and Stay Jury Rig

This rig utilizes two booms with a single winch. The hatch fall is the regular wire rope, but the dock fall is a four-stranded 4-in. manila line. The wire fall is made fast to the winch drum as it would be ordinarily. The manila line, however, is worked on the winch gypsy head by a man known as the *burton man* (see Fig. 7-30).

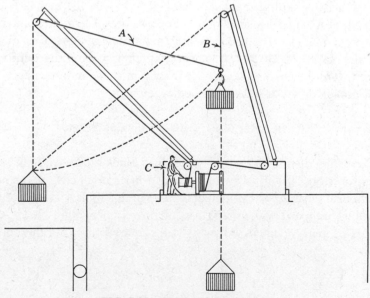

FIG. 7-30. Yard and stay jury rig.

To describe its use, let us assume that we are discharging. First the load is hoisted from the hold as with the regular rig. The manila line *A*, of course, is married to the cargo hook along with the hatch fall *B*. To retrieve the slack on this line, the burton man *C* takes one round turn on the gypsy head and hauls away. Thus, when the load is above the hatch coaming, the manila line is quickly removed from the gypsy head and three turns are taken in the opposite direction to the turns on the winch drum. Now, as the winch driver slacks away on the wire fall, the burton man hauls away on his line. This action burtons the load across the deck, and if the load was taken up to the right height the manila line will have all the weight of the load while the wire fall will be completely slack. At this point the wire is slacked off a little more by surging the manila line on the gypsy head. When sufficient slack has been obtained, the winch is stopped and the burton man slacks off on his line until the load is landed gently on the dock.

When a load is taken aboard, the burton man lifts the load and then surges his line and slacks off while the wire burtons the load across the deck until the hatch is plumbed. Once the wire has all the weight, the burton can be removed from the gypsy head and the load landed in the hold.

A description of this system makes it seem a good deal more difficult to use than it is. A skillful seaman is needed to handle the manila line, but this is not difficult with a small amount of practice. The authors have used this system on tankers to discharge and load drums of lubricating oil at the small dry cargo hatch forward and remember it as being surprisingly efficient. It is a system that may be placed into use on any ship when a winch breaks down or in a similar emergency.

Off-Center Block with a Swinging Boom

An interesting rig utilizing an off-center lead block on the cargo fall can be set up for working with a swinging boom in emergencies or for normal operations if desired. The principal feature of this rig is that the boom can be swung outboard without the use of power, providing the off-center block is properly located. Power is still required to swing the boom inboard.

Operation

In Fig. 7-31 the cargo fall (1) is rove through the off-center block (2), through the heel block (3), and to the winch. The distance between the

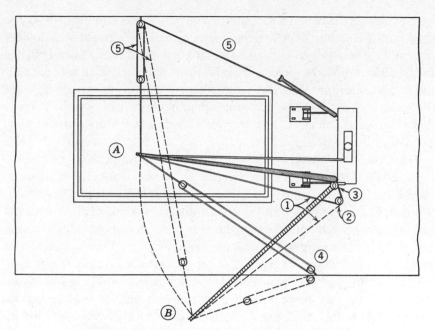

FIG. 7-31. Off-center block with a swinging boom.

off-center block (2) and the heel block (3) should be approximately 4 ft. Its exact location may be determined by experimentation; it can be shifted easily provided pad eyes are available. The inboard guy (5) is led at right angles to the boom when plumbed over the hatch. This guy must be long enough to allow the boom to swing out over the dock from position A to position B. The hauling part of the inboard guy is led to the offshore winch. Now, assuming a discharging operation, when the load has been hoisted high enough to clear the hatch coaming and bulwark, the stress on the fall leading to the off-center block will tend to pull the boom outboard. Hence, as the inboard guy is slacked, the boom will slew outboard. The slack of the outboard guy should be taken in by hand. When the load has been landed on the pier, the boom is swung inboard by power on the inboard guy.

The Farrell Rig

The first real improvement in many years in the yard and stay system of rigging the ship's gear for burtoning cargo was advanced by Captain V. C. Farrell about 1947. This improvement over the conventional gear consists of placing the heels of the outboard guys or vangs and the heels of the booms on a common axis (making them coaxial). The greatest usefulness from this rig is obtained by also installing topping lift winches and placing the hauling part of a twofold topping lift through a lead block secured near the ship's centerline. With this setup, all that must be done to top or lower the boom is to press the button controlling the topping lift winch. There is no need to tend the guys. The boom head will move along a straight line parallel to the centerline of the ship.

Once the guys have been secured to the short vang posts and pulled tight and the fall has been made clear in the normal manner, one man can hoist all the gear on a ship in less than an hour. If this is done while at sea, the falls must be run up on the winch drums and pulled tight with the cargo hook secured amidships or the falls crossed and shackled into pad eyes to keep the gear from breaking loose as the ship rolls. The latter procedure is necessary regardless of the system used to guy the gear.

With willing and able longshoremen and winch drivers, the full advantage of this rig may be realized. One set of gear may be used to service two spotting points on the ship. With fingertip control of the topping lift winch, the winch driver can alternately top up and lower the boom so as to spot a point first at one end of the hatchway and then one at the opposite end. Hence, without having to wait for the spot to be cleared each time, the hook can continue to work.

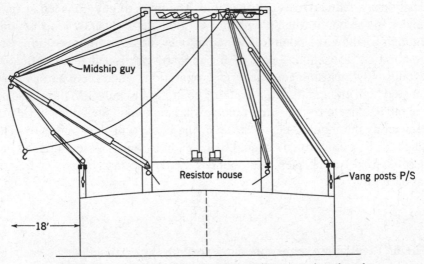

FIG. 7-32.   Elevation of the Farrell rig. P/S, port and starboard.

Opening or closing hatches with beams or pontoons takes much less time because of the ease and speed of positioning the hatch boom over the exact spot necessary to pick up the beam or pontoon. This may be done without having to pull the hook into line with a guide line.

Figures 7-32 and 7-33 show the slight change that is necessary to rig the conventional gear with the Farrell rig. The erection of the vang posts is all that is necessary if the ship already has topping lift winches.

Data gathered during a carefully observed voyage showed that the hatch served by the Farrell improved burtoning gear handled 39% more cargo per hour than one served by a pair of booms rigged in the conventional manner. The anomalous situation of more cargo per hour passing through the single-rigged hatch than through the double-rigged one came about via a combination of the diminishing returns experienced with double ganging and the fact that cargo handling requirements were sometimes such that it would not have paid to activate both sets of booms on the double-rigged hatch.

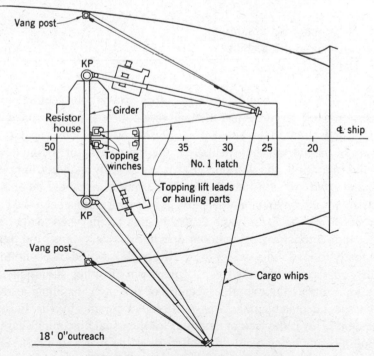

FIG. 7-33.   Plan view of the Farrell rig.

The advantages experienced with the Farrell gear included a 70% reduction in the time required to open and close the hatches and an 80% reduction in the time required to place the booms in operation or secure them for sea. The increased spotting ability, it is believed, reduces cargo damage, and the gear in general is *safer* than gear that does not employ topping lift winches.

The Ebel Mechanical Guy Rig

This rig was devised by F. G. Ebel, Senior Naval Architect, Maritime Administration. It was installed on the *S.S. Schuyler Otis Bland,* a prototype ship designed prior to the Mariner type ships. The cargo handling gear on this ship represents an outstanding improvement over the conventional rig.

Although the married fall system is an efficient system for transferring a great variety of cargo types from shore to ship and back again, there are three outstanding disadvantages to the conventional method of setting this rig up. These are:

1. Maximum load limitations.
2. Lack of spotting ability.
3. Dangers involved.

For reasons that become quite clear after studying the section covering stresses on fixed cargo booms, the load that can be safely burtoned on the regular rig is limited to about 50% of the rated boom capacity. The strength of the working guys usually is the limiting factor. Being designed for hand operation, the working guys are generally made of light equipment and rove off with manila line. If designed to handle larger loads, their size would make them unwieldy. As a result, it has been the custom to expect to rig the fixed boom as a swinging boom when loads become greater than 3 tons. Swinging boom operation is very slow, and thus it is very costly to load units over 3 tons. Although it would be economical to handle some cargo in units over 3 tons from all other standpoints, this limitation reduces the overall efficiency of the cargo handling gear.

The second disadvantage was partially overcome by the Farrell arrangement. This is the lack of flexibility of the usual burtoning gear due to the boom heads being fixed in position over the pier and hatch. This, of course, makes it necessary to bring all cargo directly under the boom head that is going to hoist it and to haul it away from a fixed point of deposit under the boom that lowers it. Changing the positions of the boom heads with the unimproved rig entails the use of several men and considerable time, and is dangerous.

The third disadvantage of the old system is the probability of gear failures. Many longshoremen have not the slightest idea of the safest point at which to secure the guy when setting the gear for work. Consequently, if a competent officer is not present to check on the location of guys, excessive stresses may be incurred, with disastrous results. The guys are usually made as light as possible, and after a little use they become weak and fail when loads near the limit are hoisted. Because of the time and trouble involved in converting the fixed boom to a swinging boom, there is a temptation on the part of officers to attempt hoisting loads just a little over the safe limit. Careless or inexperienced winch drivers tight-lining a load, spreading the boom heads in the wing-and-wing position when topped very high, and using excessively long slings are additional possible causes of excessive stresses on the guys and booms.

Further dangerous elements with the older rig are the amount of slack line lying about the decks and the necessity to shift the topping lift from cleat to gypsy head and back again when positioning the booms.

The Ebel rig adequately overcomes all three of the above mentioned disadvantages. The gear on the *Bland* was designed:

1. To handle all loads up to the full capacity of the booms (5 and 10 tons) by the burton system.
2. To provide for complete *power positioning* of the unloaded boom.
3. To eliminate all manual handling of lines.
4. To increase safety.

Figure 7-34 shows schematically the arrangement of the topping lift on the 10- and 5-ton booms. Figure 7-35 shows the arrangement of the mechanical guys on the 10- and 5-ton booms. The topping lift is offset inboard at a point near the centerline of the ship to control the swinging of the boom in the outboard direction. The hauling part is led down the inboard side of the king post through a lead block to the drum of one of the small king-post-mounted topping lift winches.

The standing part of the guy is secured to the extreme outboard end of the crosstree; run over a sheave at the boom head, down to a sheave at the bulwark, back around the second sheave at the boom head, and back

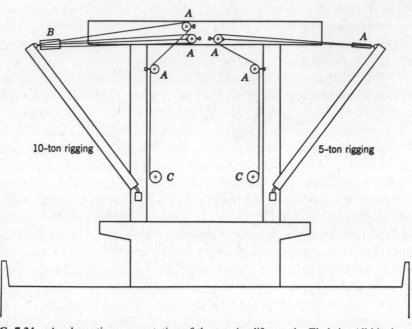

FIG. 7-34. A schematic representation of the topping lifts on the Ebel rig. All blocks are single (*A*) except for the boom topping lift block on the 10-ton rig (*B*). The hauling part is led to a topping lift winch (*C*), which is controlled from the position where the winches are operated. All blocks have 14-in. sheaves and are rove off with $\frac{3}{4}$-in. 6 by 19 improved plow steel wire rope.

to the crosstree through a lead block; and then led down the outboard side of the king post to the drum of the electric guy winch, which is also mounted on the king post.

With this setup, there is complete mechanical control of the boom. The boom can be lifted from its cradle and swung to any desired position over the hatch or pier with mechanical power. All hand operations are eliminated, and the entire operation is done by one man.

Tests have shown that it takes 1 min 13 sec to raise a pair of 5-ton booms from their cradles to the working position. It takes 56 sec to shift a pair from the working position on one side to the working position on the other side. The 10-ton booms require slightly longer times, 2 min 36 sec and 1 min 21 sec, respectively. The winch driver can perform all these shifts without moving from his regular operating station. To do the same operation with the conventional gear takes several men 10 to 15 times as

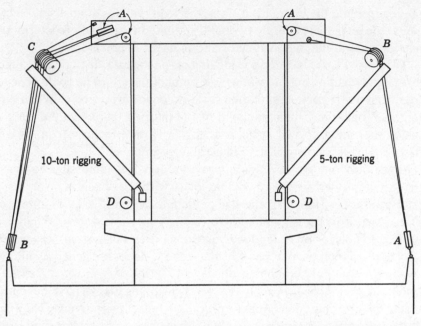

FIG. 7-35. Mechanical guys: a schematic representation of the guys on the Ebel rig. All blocks marked with an *A* are single; *B* indicates a double block, and *C* a quadruple block. *D* marks a guy winch that is controlled from the position where the winches are operated. Note that the standing parts are secured to the crosstrees. All blocks have 14-in. sheaves and are rove off with ⅜-in. 6 by 19 improved plow steel wire rope.

long, and the danger of a boom dropping or a man being injured in handling the lines is ever present.

As mentioned before, with the older method of rigging, the married fall system, there is always the danger of accidental hoisting of the load to greater heights than the gear was designed for. If this happens, the guy stresses become very great, especially when a poor guy location is selected by the longshoremen. The consequence may be a parted guy, a collapsed boom, or a jackknifed boom. All of these failures of the gear will result in damaged cargo and damaged gear, and may result in injury or death to personnel.

The Ebel rig is designed so that the effective guy resultant force keeps the stresses moderate even when handling 5- and 10-ton loads. If the load is hoisted to excessive heights, the dock boom head will rise by *riding up* in the bight of the guy tackle until it reaches a position of equilibrium with the load. While the boom head rises, the load remains almost stationary. Thus, the angle between the two falls is limited to a fixed predetermined maximum, and the overloading of any element is precluded.

When the load is lowered by slacking off on the falls, the boom resumes its original position by *riding down* the bight of the guy. The boom is never free from the tensioned guy; therefore, it cannot drop freely. When burtoning 5 tons, the minimum height of the married fall above the deck at which the riding up will occur is about 30 ft, and for lighter loads it is higher. This is ample drift to handle almost any load.

Since only one guy location is provided for, it is impossible to locate the guy improperly. When handling loads with the single fall, the maximum load is determined by the size of the fall: With ⅝-in. 6 by 19 improved plow steel, it is 3 tons, and with ¾-in. wire rope, it is 4 tons. When it is desired to hoist 5- or 10-ton loads by the married fall system, it is necessary to use a multiple purchase, because the winches are generally limited to a single line pull of about 3 tons.

On the *Bland,* two specially designed burtoning blocks have been provided for burtoning loads up to 5 and 10 tons, respectively. They are illustrated in Fig. 7-36. The one with single sheaves is for handling loads up to 5 tons with two-part lines. The other, with double sheaves, is for handling 10-ton loads with four-part lines. These blocks are self-overhaul-

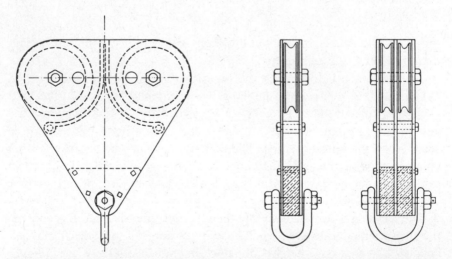

FIG. 7-36. Nontoppling blocks. These blocks are used for doubling up the yard and stay rig to be used with fixed booms. Instead of having to use two separate blocks at the hook, this single, specially built block suffices.

ing and nontoppling and are equipped with roller bearing sheaves. These blocks prevent the twisting and toppling that sometimes take place when two separate blocks are married for burtoning with a multiple purchase.

ADDITIONAL METHODS OF HANDLING CARGO

The Ship's Crane

The question of which is the most efficient, the yard and stay rig or ship's cranes, would appear to be answered in favor of the crane if one were to judge from the number in use at this time.

The factors involved in deciding the merits of one system as compared with another are: (1) initial cost, (2) cost of maintenance, (3) the productivity of the system, (4) the flexibility of the system, and (5) safety. It is evident that in order for anyone to be in a position to say anything conclusive about either gear, they would have to have a modern cost accounting system and industrial engineering organization to obtain the facts on which to base a comparison.

Types of Installations

Early installations of cranes on Swedish and German ships were of about 3-ton capacity with a 30-ft radius of reach. In 1950 the *M.V. C-Trader,* owned and operated by the W. R. Chamberlin Company in the packaged lumber trade between Oregon and California, was equipped with two 10-ton-capacity cranes with reaches 38 ft in radius. In 1953 two cranes with capacities of 5 tons and a 38-ft maximum radii were installed on the FS-790, an Army Transportation Corps self-propelled barge (see Fig. 7-37). A second barge for refrigerated cargoes was constructed and equipped with these cranes in 1954. Until the original installation on the *C-Trader,* no U.S. ship had been equipped with such heavy-duty shipboard cranes. Cranes on general cargo ships today range from 5 to 100 tons in capacity and are most hydraulicly powered (see Fig. 7-38).

Cranes have been installed on either side of and on the centerlines of ships. When they are installed on the centerline, it is with the intention of one crane being able to work both sides of the ship and one end of two adjacent hatches. This is more economical and weighs less. However, if the crane is on the centerline, it must be much larger to provide ample reach over the side.

FIG. 7-37.   Hydraulically powered ship's crane. This is a view of the type of crane placed on the Army Transportation Corps' cargo barge FS-790. These are similar to but smaller than the cranes on the lumber schooner *C-Trader*. Courtesy of Colby Steel & Manufacturing, Inc.

Cranes are becoming more and more commonplace aboard U.S. general cargo ships (see Fig. 7-61). It seems that there are many things in their favor. Figure 7-39 shows two possible installations on large ships.

Operating Experience Data on C-Trader

On the *C-Trader,* the average rate of loading per hour per crane was 80,000 board ft, as compared with 20,000 board ft/hour/set of gear with the yard and stay gear. The average rate of discharging was 110,000 board ft/hour/crane as compared with 25,000 board ft/hour/gear. The *C-Trader* carried only two cranes to service three hatches; with standard gear at least four sets and possibly five would be installed. The operating company reported that the *C-Trader's* port time was reduced to one-half of what it would be with the regular gear.

After 3 years of operating experience, during which the *C-Trader's* owners used able seamen shipped by the Sailor's Union of the Pacific as

FIG. 7-38. (*a*) Five-ton crane between hatches 3 and 4 of the *S.S. Del Monte*. Courtesy of M/N E. K. Harrell, Class of 1984, USMMA. (*b*) Typical modern multipurpose cargo vessel with twin hatches and twin slewing cranes. Courtesy of Schiller International Corp.

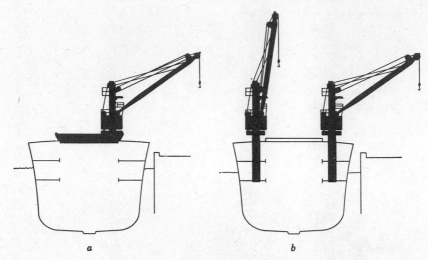

FIG. 7-39. Because of the limited radius of the boom on cranes it is necessary to install them so that an ample reach over the side of the ship is obtained. These sketches illustrate two means of achieving this end. (*a*) The application of a movable crane mount. (*b*) The mounting of two cranes on each outboard of the ship's centerline instead of one on the centerline. Courtesy of Colby Steel & Manufacturing, Inc.

crane operators, they reported no appreciable loss of time, no excessive repair costs, and no problem in breaking in the sailors as crane operators. Most of the sailors had never operated a crane of any kind before working these shipboard cranes. Because of the simplicity of operation, there was no problem in training new men. This could never be said of the yard and stay system.

Advantages of the Ship Crane

Spotting area.   The crane is able to pick up and drop loads over a greater area than can the yard and stay, which should, under certain conditions, reduce delay time of the hook. Figure 7-40 indicates the working zone of the crane as compared with the working zone of the conventional burtoning gear.

Safety.   Cranes are installed without guys and their supports do not include shrouds and stays. All supports are built into the mounts for the cranes. This leaves the decks of the ship clear of numerous lines of running and standing rigging and thereby adds to the safety of operating the ship.

Simplicity of operation. The operator needs only a few minutes of instruction, and one man can prepare or secure the gear in a few minutes.

Disadvantages of the Ship Crane

Not only do the above advantages leave little room for argument, but the disadvantages often listed may not truly weigh against the crane. One disadvantage that seems to be at least partially valid is that the crane is not as flexible as the yard and stay. In other words, the crane cannot handle *all* the cargo that the yard and stay can.

Two features of the first cranes placed on ships that proved to be detrimental to speedy operations were (1) the pendulous swinging of the load when the crane was slewed between ship and shore, and (2) the need to adjust the height of the load as a separate operation when luffing the boom. (Luffing the boom means topping or lowering it.)

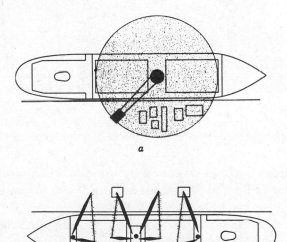

FIG. 7-40. The spotting ability of the crane is graphically compared with that of conventional gear in this sketch. (*a*) The spotting area of the crane is that covered by the shaded section. (*b*) The shaded areas show the spotting areas of the conventional gear. Note that this figure indicates the possibility of one crane doing the work of four sets of gear provided the hook could keep up with the men in the hold. Although this would not ordinarily be possible, it is conceivable that two cranes could take the place of four sets of gear. Courtesy of Colby Steel & Manufacturing, Inc.

FIG. 7-41.   The two falls leading to the hook dampen all pendulous swinging created by tangential forces when slewing the crane boom. Courtesy of Colby Steel & Manufacturing, Inc.

The first objection was overcome by using two falls leading out over sheaves at the end of the jib instead of a single fall to support the load. This arrangement can be seen in Fig. 7-41.

The second problem was overcome by designing into the controlling machinery a device that would allow the boom to be luffed while the load remained at a constant height. This is known as *level luffing*. It prevents the load from swinging when the boom level is adjusted and requires much less power when topping the boom with a load on the hook.

The Overhead Crane

As an example of an experimental rig that was tried and found to be inefficient for today's general cargo carrier, we will look briefly at the

Harrison overhead crane. This gear was placed on the *S.S. Sea Hawk* about 1946 at hatches no. 1, 2, and 3 (see Figs. 7-42 and 7-43). Conventional gear was installed at no. 4 and 5. The *Sea Hawk* was a C-3 type ship.

This crane was a type of modified shop crane placed on runways that were mounted on deck athwartships at both ends of the hatch. The crane consisted of a bridge girder supported by the athwartship runway girders. Two trolleys mounted on the bridge girder carried the winches on which the cargo falls were rove off. The operator's station was placed under the bridge girder clear of the hook. All machinery for lowering and hoisting the hook, moving the bridge girder athwartships, and moving the trolleys longitudinally on the bridge girder was contained in the trolley. Cantilever trusses were hinged to the outboard ends of the runway girders, and when the ship docked, these trusses were swung outboard so that the bridge girder could reach out over the dock when working cargo. When the ship was ready for sea, these trusses had to be swung inboard and secured in a fore and aft position.

FIG. 7-42. Harrison overhead crane as installed on *S.S. Sea Hawk*.

FIG. 7-43. Overhead crane, showing the hinged cantilever jobs rigged for working cargo. The need for wide apron piers with this gear is obvious.

On the *Sea Hawk,* one 5-ton trolley was installed over hatch no. 1, two 5-ton trolleys over hatch no. 2, and two trolleys capable of being adjusted to lift either 3 tons or up to 15 tons over no. 3. Hook selection was made by throwing a single lever on the trolley. This arrangement allowed fast lifts of loads up to 3 tons or slower lifts of loads up to 15 tons. By using an equalizing strong back, a 30-ton lift could be accomplished by using two hooks without any change in the rigging.

Some tests made involving the cycle times of this gear as compared with those of the standard gear showed that when handling loads of 3 tons or less, the burtoning system was about 10% faster. Figure 7-44 is a chart that illustrates the results of all the tests taken on the *Sea Hawk.* From the curves on this chart it is evident that the overhead crane gear as installed was much faster when handling loads over 4 tons. Perhaps one reason why this gear or some variation of it has never been used is that on almost all ships, under normal operating conditions as practiced today, the average load is well under 3 tons. The majority of the loads are about 1 ton.

Besides being very fast with heavy lifts, the overhead crane gear gave much better spotting ability over the dock and in the hatch.

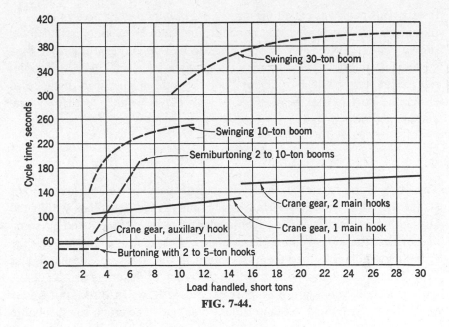

FIG. 7-44.

The gear's disadvantages were as follows:

a.   The overhang of the cantilever jibs was fixed in such a way that very wide camels were required to work the ship at docks with narrow aprons.

b.   The drift was limited. This made it very difficult or impossible to handle certain types of cargo.

c.   On the large hatches where double gear had to work to keep the hatch hours equalized between the large and small holds, both trolleys were carried on the same bridge girder; thus, the top speed of both gangs working in the hold was governed by the slower of the two.

d.   The transverse movement was relatively slow due to the large mass of the load, bridge, and trolleys.

e.   The deck run was almost completely lost for the stowage of big deck loads, which in recent years has grown in importance.

Extensible Boom Overhead Cargo Gear

Another type of cargo handling gear that was designed to overcome some of the disadvantages of the overhead crane was proposed but never tested. This gear was called the *extensible boom overhead cargo gear*. In this proposal, the bridge was carried on rollers by a longitudinal fixed runway girder. Extensible booms were to be hung under the bridge by a system of rollers. These booms, independent of each other at double-rigged hatches, could be put out over the side any desired distance to either the port or starboard. Figure 7-45 shows a hatch with only one extensible boom installed.

The Dock Crane

The dock crane is used extensively in European ports and many other sections of the world, but it is found at only a few terminals within the United States. Of the many advantages of the dock crane listed by some authorities, the most valid is that the area of deposit is much larger. The crane may pick up a load from any point in the square of the hatch and deposit it on the pier anywhere within the reach of the crane jib. On docks with especially wide aprons that are served with two or three spur tracks of a belt railroad, this is an important advantage. The crane can discharge

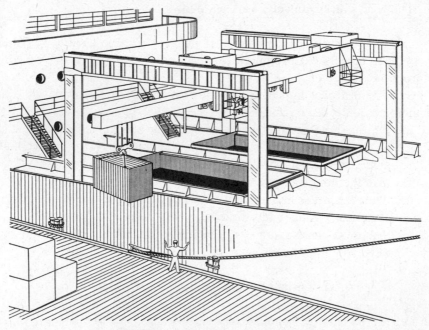

FIG. 7-45.   Extensible boom cargo gear.

into or load from railroad cars on any track or from the doors of the warehouse shed. The ship's booms cannot do this because of their limited reach.

Other arguments in favor of the dock crane are not always valid. Situations dependent on the operating conditions determine whether these arguments truly describe advantages. For instance, it is said that the crane is better because only one man is needed to operate it; therefore, it requires less labor. This argument is not valid if the employers must hire a standard-sized gang regardless of the operation. It is true that perhaps the extra man could be used to advantage at some other operation, but if there *must* be two winch drivers then each will work only half of the time.

With dock cranes, there is always a means for unloading lighters that come alongside the dock when a ship is not present. This is definitely an advantage in European ports where a large amount of cargo is brought to the dock by lighters while the ship is not there. Lighters may be unloaded with crane trucks and forklift trucks also, but the dock crane is more flexible.

The strongest objection to the dock crane is that it requires a very large initial investment, and this investment is to duplicate equipment that will always be provided by the ships being serviced. The ship must carry gear for handling cargo because she will go to many places in the world where no cranes are available; indeed, there may not even be any dock in some of her ports of call.

On docks that have smooth road beds and where forklift trucks, tractors, trailers, and similar materials handling equipment is available and used to its fullest efficiency, cargo can be handled just as fast with a small spotting area as with a large one. The only thing that slows up the conventional gear is failure to clear the position under the hook, which results in hook delay. Proper use of equipment on a modern pier will result in no hook delay from this cause.

Why then, in the face of what has been said above, are so many dock cranes found throughout the world? The authors believe the answer lies in the lack of modern materials handling equipment of an auxiliary nature plus the fact that much lightering work is done at these ports and the dock crane finds more frequent use than it would in most ports of the United States. In efficiency of discharging ships, the yard and stay system and dock crane are not far apart. Ships must be equipped with the yard and stay or ship cranes; therefore, it appears that whether dock cranes should be installed depends on local conditions.

Three general types of dock cranes are used for aiding ships in discharging and loading at piers handling general cargoes. One of these is a cantilever crane mounted on the roof of the dock shed with a level luffing jib. This makes it possible to pick up cargo from the ship and deposit it on the dock with just two motions: (1) the vertical motion of the hook; and (2) the luffing of the boom. Two or more cranes can be used at a hatch without interfering with each other or with the ship's rigging.

The other types are gantry cranes with jibs that may or may not be level luffing but are always capable of slewing. Slewing is the term used to describe the action of revolving the crane assembly. The word *gantry* simply indicates that the crane is mounted on a structure that spans something. The semigantry crane has one set of legs supported by a rail along the face of the pier shed wall or on the top edge of the wall. The legs rest on wheels, of course, that run on tracks. The other type of gantry crane is

FIG. 7-46.   Level-luffing dock cranes. These are dock cranes of the full portal or gantry type with capacities of 3 tons. They are level luffing, but because of the narrow ledge between the outer edge of the crane support and the dock edge, part of the advantage of level luffing is lost. Courtesy of Wellman Engineering Co.

FIG. 7-47. An example of the *transporter* crane generally used for discharging at bulk handling docks. This particular crane is rigged with a 9-ton bucket and is located at Western Maryland Port Covington Pier, Baltimore, Maryland. Note the operator's cage below the boom and to the left of the bucket. Courtesy of Wellman Engineering Co.

the full arch gantry with all of its legs supported on wheels that run on tracks laid on the dock apron (see Fig. 7-46). These last two types must slew their loads to get them from the ship to the dock or vice versa, unless they are level luffing and set back from the dock's stringer piece.

A special type of crane used to discharge bulk products such as iron ore and coal, known as the *transporter,* consists of a built-up boom that projects out over the ship horizontally. A trolley runs along the length of the boom carrying a winch with a grab bucket on the falls. The operator's cabin may or may not move with the trolley. The capacities of these booms may vary considerably, but they average somewhere between 5 and 20 tons. The transport crane is used at many terminals on the Great Lakes (see Fig. 7-47).

The Siporter

On passenger ships there are holds for the stowage of cargo many decks below the passenger quarters, and long narrow trunkways must be cut through these decks to reach the holds with the conventional yard and stay system or with ship cranes mounted on deck. The *siporter* was developed to help solve this problem. It consists of an extensible boom on which a trolley is mounted for carrying the cargo fall and hook. The operator rides along in a carriage that contains all the hoisting machinery. The entire assembly is fitted within a side port opening into the ship's hull (see Figs. 7-48 and 7-49). The side port, of course, is well below the decks used for passenger accommodations and working rooms. Loads are picked up and taken inside through the side port and lowered down into the hold a vertical distance of only one or two deck levels.

The Hulett Unloader

The Hulett unloader is a special type of combination crane and grab bucket, as shown in Figs. 6-18 and 6-19. The grab buckets are capable of picking up 20 tons of iron ore in a single bite. With these rigs a load of 10,000 tons of iron ore can be discharged in 4 hours. When the iron ore is somewhat depleted within the hold and not enough depth remains for the bucket to get a full bite, the Hulett unloader is used to pick up a large bulldozer and place it in the hold to push the remaining iron ore into a pile under the hatch. In this way the Hulett unloader can continue working at a high rate of productivity for a longer time.

Floating Cranes

For handling heavy lifts some high-capacity cranes are mounted on the decks of barges so that they can be brought alongside a ship on the offshore side while the ship is working cargo from the dock. The barge carrying the crane may or may not have a large deck space for transporting the heavy lift. The decision as to whether to use the ship's heavy-lift boom or to go to the expense of chartering a heavy-lift crane is entirely a matter of economics. The cost of chartering the crane and its crew must be compared with the cost of using the longshoremen to break out the ship's heavy-lift boom and the accompanying delays on the ship. Of course, if the heavy lift exceeds the capacity of the ship's gear, then a

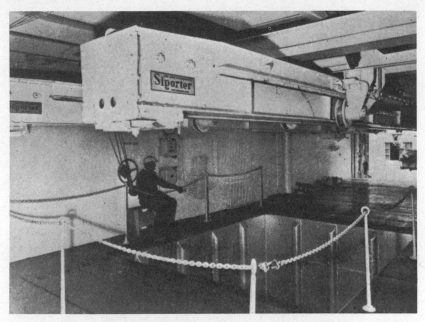

FIG. 7-48. View of the siporter from within the ship. The man on the left controls the load when it is being lowered into or hoisted out of the hold. Courtesy of Lakeshore Engineering Co.

FIG. 7-49. View of the siporter from out on the dock. The trolleys and hoisting blocks with the short sling pendants and hooks are clearly visible. In other models of the Siporter, there is one operator who rides a carriage back and forth with the load. Here, however, a second operator can be seen to the left. He controls the load when it is extended out over the dock. Courtesy of Lakeshore Engineering Co.

431

FIG. 7-50. A typical floating crane being used on the offshore side of a ship while the ship's gear and a house fall work on the dock side. Courtesy of Dravo Corp.

suitably large floating crane must be used. The capacities of these cranes range from 50 to 100 tons, with a few in the major ports of the world being capable of lifting as much as 400 tons (see Figs. 7-50 and 7-51).

Conveyors

Many types of conveyors are used for assisting in the loading and discharging of cargo. Two types are used to accomplish the loading and discharging completely. These are the endless canvas pocket type and the air conveyor. Full loads of bananas are discharged and loaded by the use of canvas belt conveyors. Figure 8-21 shows the type of conveyor used with bananas. Copra and grain in bulk are discharged by the use of air conveyors. Auxiliary conveyors for bringing cargo from or taking cargo to the wings of the hold and for similar uses on the pier are discussed in Chapter 8.

Aerial Ropeways

Aerial ropeways have been used in many places for loading and discharging ships. Figure 7-52 is a diagram of a ropeway that has been used

FIG. 7-51. The floating crane *Monarch* has a 250-ton lifting capacity. Here it is shown lifting 130 tons. The lift is another crane's cabin, engine room, and turret as a built-up unit. As would be expected, the *Monarch* is listing about 10°. Courtesy of Todd Shipyards Corp.

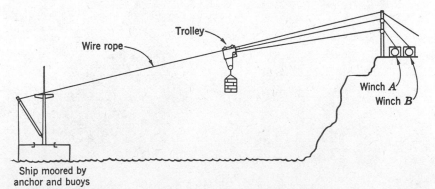

FIG. 7-52. Aerial ropeway. The trolley position is controlled by the operation of winch *B*. When the trolley is directly over the landing position on the beach or over the ship's hatch, winch *A* is operated to lower or hoist the load.

433

successfully to load and discharge ships at an open roadstead. This idea was used at the port of Kukuihaele, Hawaii, for loading bagged sugar and discharging general cargo from a ship moored about 200 yd off the beach onto a rocky cliff rising 200 or 300 ft up from the beach edge.

During World War II the British Ropeway Engineering Company developed a very simple system for unloading from a distance of 200 ft from a harbor quay after the quay itself had been destroyed by enemy action. A wire rope was attached to the mast of the ship at the crosstree level. The other end was anchored ashore after it had been pulled taut. Patented equipment was used to hoist the load out of the ship's hold and deposit it on an independent trolley (see Fig. 7-53). Owing to the inclination toward

FIG. 7-53. View of the patented trolley to which the load is transferred from the ship's cargo fall. After the transfer is made, the trolley runs down the inclined wire rope to the beach. Courtesy of British Ropeway Engineering Co.

the beach, when the trolley was released with its load it rolled to the beach end, where it was deposited. On the beach, the trolley was removed from the wire rope and returned to the ship on a separate line. On the actual installation, loads of 1500 lb were handled, but greater loads could easily be accommodated.

An installation for loading bauxite at Takoradi Harbour, Gold Coast, West Africa, consists of a ropeway carrying buckets that feed a hopper that in turn feeds a traveling belt transporter. The belt on the transporter deposits the bauxite into the ship's hold. The ropeway buckets discharge into two transporters alternately so that fore and aft holds are loaded at the same time. This system averages about 330 tons/hour. Figures 7-54 through 7-57 show the elements of this bauxite loading station.

An interesting bicable plant for loading ships with pyrites was erected on the south coast of Cyprus in the middle 1930s. This ropeway had a capacity of 200 tons/hour. Its length was 1800 ft. The full buckets traveled on wire rope 2 in. in diameter; the empty buckets returned on $1\frac{1}{4}$-in. wire. A 30-hp motor drove the Lang lay hauling rope at a speed of 274 fpm. Figure 7-58 illustrates this installation.

At Avonmouth, England, an aerial ropeway is used in conjunction with cranes and hoppers to discharge bulk coal, ore, zinc concentrate, phosphate, and superphosphate. This ropeway is 5070 ft long and designed for

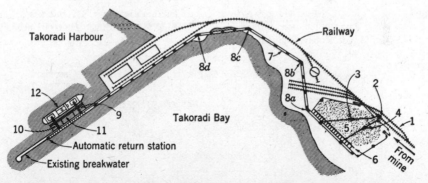

FIG. 7-54. Bauxite loading station at Takoradi Harbour, Gold Coast, West Africa, using an aerial ropeway with an endless belt transporter. (1) Railway supply line. (2) Wagon tippler, 200 tons/hour. (3) No. 1 drag scraper, 200 tons/hour. (4) No. 2 drag scraper, 200 tons/hour. (5) Storage pile, 40,000 tons. (6) Bunker front and ropeway loading chutes. (7) Bicable ropeway, 300 tons/hour. (8a,b,c,d) Automatic angle stations. (9) No. 1 traveling belt transporter, 200 tons/hour. (10) No. 2 traveling belt transporter, 200 tons/hour. (11) Steel wharf structure. (12) 10,000-ton vessel. Courtesy of British Ropeway Engineering Co.

FIG. 7-55. View of the aerial ropeway and dock at Takoradi Harbour. The buckets on the left are going down full; those on the right are returning to the storage pile to be refilled. Courtesy of British Ropeway Engineering Co.

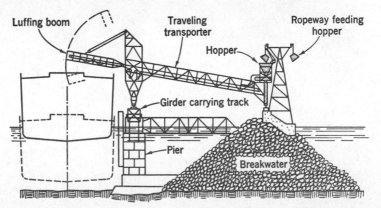

FIG. 7-56. A diagram of the transporter with its endless belt at the dock end of the Takoradi Harbour bauxite loading station. Courtesy of British Ropeway Engineering Co.

FIG. 7-57. View of a ship alongside the dock at Takoradi Harbour being loaded with bauxite. Courtesy of British Ropeway Engineering Co.

a maximum capacity of 250 tons/hour when handling zinc concentrates. The capacity is less when handling the other products.

Figure 7-59 shows the ropeway-transporter built at Al-Fatha, Iraq. This transporter spans the Tigris River and can accommodate vehicles up to 42 ft in length and handle a gross weight of 53,000 lb. The total weight of the carriage and load car is 38,000 lb, so the maximum total rolling load on the main cables is 91,000 lb. The total length of the cable is 2278 ft from the engine and winch house on one side to the anchorage on the other.

FIG. 7-58. Aerial ropeway at Cyprus erected in 1930. Courtesy of British Ropeway Engineering Co.

FIG. 7-59.   The Tigris ropeway-transporter. Courtesy of British Ropeway Engineering Co.

PROPOSALS FOR THE FUTURE

Just what the future will bring to the field of loading and discharging equipment is, of course, unknown. There has been much conjecture, but little positive change. Going up and over the gunwale of the ship and then down into her depths as a way of getting the cargo from the dock into the ship is deplored by the materials handling theorist. This system violates the fundamental principles of efficiency in handling. As loading is done today, the cargo is brought to the pier apron to a point only a few feet from where it will be stowed on the ship. Then we send it to its resting place not in a straight line but along a circular route. The movement should be *straight,* direct from the pier to the ship. This idea definitely favors use of the side port, which has been neglected in past years rather than used more extensively.

Some men who have thought seriously about what should be done to improve productivity in the field have proposed radical changes, such as changing not only the cargo handling gear but also the piers to accommodate a specially designed ship. The greatest obstacle to a radical change is the tremendous initial cost. When and if a great change will come about and the probable nature of that change are matters of endless discussion among interested groups.

The Ellis Ship-to-Shore System

One proposal that has been made entails a redesigned terminal as well as a redesigned ship. This is the Ellis ship-to-shore system. The idea was proposed by the Charles Ellis Engineering Company and although as yet it has not been attempted, it is well worth serious consideration. Perhaps some modified version of the system might someday be used to put a ship on a paying basis in some particular trade.

The system consists of a series of small rail carriages each having one-third to one-half the capacity of a normal railroad car. Each carriage weighs 600 lb and will support 15 tons of cargo. These carriages are loaded with pallets or containers on the dock before the ship arrives. Each carriage can handle three pallets about 6 ft by $7\frac{1}{2}$ ft by $6\frac{1}{2}$ ft high. Each carriage has about 900 ft^3 of storage space, or a floor space of 18 ft by $7\frac{1}{2}$ ft. Thus, one carriage can handle passenger automobiles, small trucks, and large crates.

The carriages are moved about the marine terminal or ship's hold and 'tweendecks by standard tractors. Travel from terminal to ship, via a brow or elevator, is effected by chain conveyor. The average freighter or combination ship would have one set of side ports and one elevator forward and one set aft.

The pier would have a system of carriage track installed for its full length. Across the pier, at two loading points, chain conveyors with brow or elevators leading into the ship are installed. The terminal to ship transfer gear is slow and rugged; it would take a carriage about a minute to pass through the side port. Thus, the total cargo moving between the ship and terminal is from 1500 to 1800 tons/hour.

The carriages, on arrival on the ship at the upper deck, stop on a platform that has four small turntables for the purpose of changing the direction of the wheels of the carriage. After stopping and having its wheel direction changed electrically, the carriage moves fore or aft to an elevator that takes it down to its stowage deck. At the correct level it moves fore or aft again, and finally into its stowage position either along the centerline of the ship or in the wings. Changes of direction are accomplished everywhere by the use of the small turntables.

One of the first objections that will be raised against the idea is that there are some cargoes that could not be accommodated, such as heavy lifts over 15 tons and pieces that are too large to fit through the side ports and within the limited heights of the decks. This objection might be met with the proposal that the ship be fitted with a heavy-lift boom forward and aft. Large and heavy pieces could be stowed on the deck of the ship without slowing up the regular discharging or loading procedure.

Another, and more serious, difficulty is that the pier where the ship is loaded, as well as the discharging pier, must be redesigned to accommodate the system. Thus, a company wanting to use the system would have to remodel every terminal at which the ship was to call. The ship could not be diverted to any other port or dock except those especially fitted out to work with the system. This might preclude the use of the system in normal foreign trade, where a ship must have extreme flexibility or sacrifice many cargo offerings. However, it does not completely remove the possibility of the system being worked into a specialized coastwise route with ships of a limited size.

An objection to the idea that will also be raised is that there is much lost space; the broken stowage would be excessive. While it is true that there would be much lost cubic in such a system perhaps it is worthwhile to compare the cost of trying to pack cargo into every nook and cranny of the ship with the savings due to a fast turnaround. To the average ship operator, the idea that space should not be utilized at all costs is somewhat difficult to accept. However, there are many examples in which space has been completely ignored in favor of a reduction in handling and a rapid turnaround. One of the most successful of these is the RO/RO type of operation.

PRIMARY CAUSE OF DELAY

As cargo is worked into and out of ships today, there is one very important cause of delay that raises the cost of the operation tremendously.
This is the *lack of spotting ability* in the hatch. The lack of spotting ability
means the inability to pick up or land a load from any point in the hold of
the ship. When discharging, the load has to be brought from the wings by
one method or another before it is picked up and taken from the hold.
Some of the methods used to get the cargo from the wings to the hatch
square are rough on the cargo and the cargo handling equipment and
result in much damage to both. Other methods are slow, and the lost time
costs much money during the life of the ship. Still other methods combine
both of these disadvantages. When loading, the problem is simply reversed in direction. The loads must be landed in the hatch square and
taken to the wings.

Statistical Study of Delay Values

Observing the loading and discharging activities of any ship using the yard
and stay system, ship crane, or dock crane makes it quite clear that the
hook is usually much faster than the men on the dock or those working on
the ship. The hook will nearly always be delayed at both ends of its travel.
This clearly points up the fact that the present gear we have on the ships is
fast enough as far as the actual travel time for the hook, in or out, loaded
or light, is concerned. The loading and discharging systems now employed on ships can best be improved by changing the methods used to get
the cargo up to the hook and away from the hook and for actually hooking
on or unhooking the load.

   An example of the type of observations that may be made by any
officer in any operation is given below. The data so obtained may be used
to discover the exact extent of the delay on the ship and/or dock. Besides
illustrating the truth of the statement that the hook is delayed by activities
of the longshoremen, such observations may lead to a greater understanding of the fundamental faults of any particular system. As a result of
greater understanding gained by the officer afloat who works with the gear
in many ports and under varied conditions, he may make valuable suggestions to the operations department regarding changes designed to improve
the productivity of the longshoremen.

The cycle referred to on the forms below is the period of time it takes the hook to travel in or out and back again between being attached to a load. The delay time should be counted as all the time that passes after the hook has arrived in the opposite position after the athwartship travel period. The delay time includes the time required for swinging the load into the wings, preparation for unhooking and hooking on, waiting for the landing spot to be cleared, dragging the cargo out of the wings, and any similar activity. In other words, we are assuming that in an ideal system the load would be landed, unhooked, the empty hook would return, and the new load would be hooked on with no delay except for the actual hooking on or unhooking of the sling. The value of the delay can be stated in terms of time, weight, or cost. The form as set up would result in obtaining the value of the total delay. Knowing the number of cycles per hour and the weight of each sling load, this delay can be stated in terms of tons per man-hour or gang-hour of lost productivity. Also, knowing the number of men in each gang and the rate of pay for each man, the cost of

TABLE 7-5
COST OF DELAY FACTORS

Description of Factor	Factor Symbols	Use
Total cycle time in seconds	C_d	Average of all cycles observed
Delay time in seconds	d	Average of all delays
Cycle time without delay	C	$C_d - d = C$
Pounds per sling load	L	By observation
Number of men per gang	N	By observation
Hourly rate of pay per man	S	By inquiry
Proportion of delay time in total cycle time	$P,\%$	$d/C_d \times 100 = P,\%$
Tons per gang-hour with delay	R_d	$1.6L/C_d = R_d{}^a$
Tons per gang-hour without delay	R	$1.6L/C = R$
Tons per man-hour with delay	r_d	$R_d/N = r_d = 1.6L/C_dN$
Tons per man-hour without delay	r	$R/N = r = 1.6L/CN$
Cost per ton with delay	K_d	$S/r_d = K_d = C_dNS/1.6L$
Cost per ton without delay	K	$S/r = K = CNS/1.6L$

[a] The constant 1.6 is simply the value of the ratio 3600/2400 that appears in the tons per gang-hour formula.

the delay time can be calculated. The equations necessary for making such calculations are given in Table 7-5.

Cost of Delay per Ton Equation

Using the symbols in Table 7-5, we can derive an expression for *the cost of delay for every ton handled*:

$$K_d - K = C_d NS/1.6L - CNS/1.6L = dNS/1.6L$$

Cost of delay per weight ton $= dNS/1.6L$

The officer interested in making such observations may prepare forms to suit his own operations, but the data he obtains should be similar to those shown in the example presented in Table 7-6. After obtaining such information as the weight of the load, number of men in each gang, and the rate of pay, he merely stations himself by the hatch with a stopwatch and takes a number of observations. The greater the number of observations and the more explicit the remarks explaining the causes of delay and other facts about the operation, the more valid and reliable will be the data obtained.

Measurement Ton Costs

The cost of delay per measurement ton may be obtained from the cost per weight ton by multiplying the latter value by the ratio of 40 to the stowage factor of the commodity:

$$\text{Cost of delay per measurement ton} = \text{Cost of delay per weight ton} \times \frac{40}{f}$$

where f = stowage factor of the commodity.

This value may be calculated directly by the equation:

$$\text{Cost of delay per measurement ton} = \frac{dNS}{90v}$$

where v = the volume of each load.

TABLE 7-6
OPERATIONAL DATA ON CARGO GEAR

Vessel: *S.S. Pioneer Bay*   Commodity: Cotton waste

Location: #3 T.D. Fwd. across   Packaging: Bales

Loading   Weight of each load: 1500 lb

Observation No.	In Lift, sec	Delay on Ship, sec	Out No Lift, sec	Delay on Pier, sec	Remarks and Summary
1	25	15	20	20	Sling was wire snotter with sliding
2	20	15	25	12	hook.
3	37	55	15	20	Distance from weather deck to pier
4	30	107	23	11	averaged 22 ft.
5	48	20	25	7	Electric winches, longshoremen were
6	38	14	18	12	average.
7	35	31	21	15	Most delay was caused by inability of
8	23	77	20	19	longshoremen in ship to stow bales
9	40	115	13	25	quickly.
10	31	155	24	13	Average cycle with delay: 114 sec
11	49	79	20	16	Average delay (total): 69 sec
12	29	16	19	39	Average cycle less delay: 45 sec
13	18	83	14	22	
14	39	33	17	19	20 men per gang. Total cost in salary
15	45	22	14	27	and benefits:
16	19	21	12	53	$4.50 hourly per man.
17	15	6	11	20	Cost of delay per weight ton =
18	22	30	12	19	
19	19	21	13	20	$= \dfrac{69 \times 20 \times 4.5}{1.6 \times 1500}$
20	24	43	16	18	
21	29	52	10	39	$= \$2.70$
22	23	112	15	25	
23	17	8	13	12	
24	21	95	15	10	
25	18	6	12	11	

444

Keeping records of this type will help the ship's officer build a store of knowledge about the cargo operations on his ship.

Improving the Spotting Ability

Some ways and means of improving the spotting ability of conventional gear are presented below. Some of these are merely proposals; others are actually in use on today's ships.

Farrell's Rig

When the yard and stay is equipped with vang posts and topping lift winches that can handle the boom with a load on the hook, the load can be landed or picked up from any place in the square of the hatch.

Ebel's Mechanical Guy Rig

With the guy and topping lift winches readily operated from the winch driver's operating station, the load may be landed or picked up from any point within the hatch square.

Farrell's Coordinated Rolling Wing Deck

This is a device that consists essentially of a movable deck arranged so that it rolls from the wings into the square and back out again with a full load of cargo. The rolling wing decks are large enough to cover the wings on both sides of the hatch square. They are mounted on 4-in.-diameter wheels that roll athwartship on flat bar tracks welded on top of the deck and hatch beams. The height of the rolling deck, including the tracks, is $5\frac{7}{8}$ in. A system of wire rigging coordinates the movement of the rolling decks so that when the deck on one side is rolling outboard or inboard the one on the opposite side also rolls outboard or inboard automatically. This makes it impossible for both decks to roll to the low side if the ship takes a sudden list.

When the ship is ready to load, the rolling wing decks are moved out of the recesses by heaving on a wire messenger with the cargo hook. The decks will meet at the centerline directly under the hatch opening. In this position, the cargo is loaded. When the rolling wing decks have been loaded, they are rolled back into the wing recesses using the cargo hook and runner again. It takes about 18 sec to move the loaded decks into the wings. When this has been done, the square is left clear and it can be loaded up; thus, all the work is done under the cargo fall and movement to the wings is eliminated. In discharging, the process is reversed.

These decks definitely increase the spotting ability because in effect they bring the wings of the hatch under the hatch square and enlarge the hatch opening.

Enlarging the Hatch Opening

From the standpoint of improving spotting ability and increasing the efficiency of work in the hold of the ship, it would be ideal to have a hatch with covers almost as large as the entire hold. It would be desirable, of course, to have a 4- or 5-ft border around the hold to assist in working the ship when the hatches are all open. Evidently the hatch openings have become about as large as they can because of the limitations placed on the size by longitudinal strength requirements. Although the lengths of hatch openings have come to represent a very large percentage of the total length of the hold, the widths have been limited in order to retain continuous longitudinal strength members.

Twin Hatches

One alternative to an increase in hatch opening area is the use of *twin hatches*. Instead of having one hatch opening, two smaller openings are

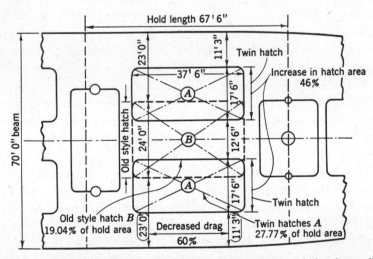

FIG. 7-60.   Plan of twin hatches for C3-S-DB3 cargo vessels. *A*, twin hatches, solid line. *B*, old-style hatch, dotted line.

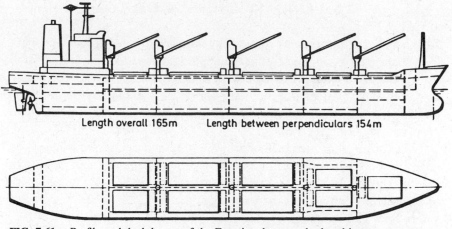

Length overall 165m   Length between perpendiculars 154m

FIG. 7-61. Profile and deck layout of the Frontier class standard multipurpose cargo vessel. Note the twin hatch arrangement at all hatchways except no. 1. Also note preference for cranes. Reprinted from L. Buxton, R. P. Daggitt, and J. King, *Cargo Access Equipment for Merchant Ships,* E & FN Spon Ltd., London, 1978. Courtesy of MacGregor-Navire Publications, Ltd.

provided, but with a total area that is greater than that of the single opening. Spreading these athwartships reduces the distance from the hatch opening to the wings. The twin hatches shown in Fig. 7-60 were proposed for use on the C-3 type ship built by the Maritime Commission. Using these double hatches, the drag distance to the wings would be decreased 60%, and the twin hatch area would amount to 27.77% of the total hold area. The single large hatch area amounts to 19.04% of the total hold area; thus, the hold hatch area in the double hatch system would be increased 46% over the old arrangement.

From the above information, it certainly appears that the twin hatch system would improve spotting ability. However, there would be a reduction of the *size* of the opening into the hold from the single large opening to just one of the twin hatches. This means that the upper limit on the size of a piece that could be lowered into the hold would be decreased. This seems like an insignificant disadvantage.

Although modern vessels may be found with hatch widths up to 80% of their overall breadth, general cargo vessels must take consideration of strength and 'tweendeck stowage. As a result, the modern general cargo vessel hatchway will be composed of two, or possibly three, openings side by side, each equipped with its own hatch cover (see Fig. 7-61).

THE SHIP'S LOADING AND DISCHARGING EQUIPMENT

SECONDARY CAUSE OF DELAY

Another cause for reduced productivity of longshoremen is the time spent in preparation of and in securing the apparatus used to load or discharge the cargo. Although this cause of delay in cargo operations has been rated as a secondary factor here, there are ports where the aggregate of all the hook delay times will be equal to or even less than the time used to prepare and set up the gear.

Some of the activities included in the preparation and securing of the gear are the positioning of the booms or cranes, opening and closing of hatches, laying of dunnage or metal plate runways, hoisting of materials handling trucks on and off the ship, rigging of save-alls, and setting up of safety lines and lights (see Table 7-7).

Reducing the Secondary Cause of Delay

There have been two important improvements made in reducing the secondary cause of delay, with many more proposed.

TABLE 7-7
BREAKDOWN OF DELAYS DURING CARGO WORKING HOURS [a,b]

Activity	Delay as a Percentage of Working Hours	Average Minutes per Shift
Opening and closing all hatches	4.0	21
Refreshment breaks	3.7	19
Weather delays	3.0	15
Setting up or shifting cargo gear, lighters, etc.	2.1	11
Waiting for cargo, stevedores, or lighters	2.0	10
Other delays (dunnage, restowing cargo, etc.)	4.7	24
Total	19.5	100

[a] Figures relate to average of loading and discharging general cargoes during a normal 8½-hour shift, which excludes a ¾-hour meal break.

[b] Information from L. Buxton, R. P. Daggitt, and J. King, *Cargo Access Equipment For Merchant Ships,* E & FN Spon, Ltd., London, 1978, p. 64. MacGregor-Navire Publications, Ltd.

The Topping Lift Winch

The topping lift winch is one of the most important improvements, because it not only reduces the time required to position booms but also makes the process much safer. One of the most dangerous activities connected with cargo operations was the topping and lowering of booms with a rig that required taking a hauling part of the topping lift to the winch gypsy head. Dropped booms and the accompanying rapidly overhauling lines on deck have taken their toll of seamen and longshoremen for years. With the topping lift winch this has been eliminated.

The topping lift winch is a small, 20-hp electric winch generally secured to the mast or king post so that it is out of the way. The topping lift hauling part leads directly to the winch drum and the winch may be operated by pressing a button or moving a small lever arm. Such controls are constructed so that the winch operates only while the switch is held in or down and will stop when pressure is taken off the control. This is an important safety feature.

Some of these winches require that the boom be unloaded when topping or lowering; other winches may be operated with a loaded boom. Some are prevented from backing off by rachet mechanisms, others by brakes that go into effect after the boom has been positioned. Some topping lift winches are drums on an extended shaft of the regular winch. To operate them, a lever is used to engage the topping lift winch drum with the powered shaft and disengage the cargo fall winch. After the boom has been positioned, the topping lift winch is disengaged and the cargo fall winch is engaged.

Hatch Covers

For years the traditional hatch beams and hatch boards were used to cover hatchways. It would take a gang of 10 men up to an hour to open or close a hatch. By comparison, an automatic mechanical hatch cover can be opened by one man in about 2 min, or two men can open a hatch operated by wires and winches in about 10 to 15 min (less if the wires are already rigged). By 1965 a large proportion of ships had mechanical weather deck covers; delays due to opening and closing such hatches had fallen to only 2.7% of cargo working hours, according to a major survey of general cargo operations carried out in Dutch ports. Thus during a typical 8½-hour shift, time lost through this cause amounted to only about 14 min. The corresponding figure for 'tweendeck hatch covers was 1.3%, showing

TABLE 7-8

CHARACTERISTICS OF PRINCIPAL HATCH COVER TYPES[a,b]

Cover Type	Usual Ship Types	Decks Applicable	Operating Mode	Guide to Minimum Coaming Height	Drive System	Cleating System	Depicted in:
Single pull	(D)	Weather	Rolling & tipping	Depends on section length, but more than (A)	Electric, hydraulic, or ship's cargo gear	(E)	Fig. 7-62a
Hydraulic folding	(D)	Weather/'tween	Folding	(A), (B), or (C)	Hydraulic	(E)	Fig. 7-62b
Wire-operated folding	(D)	Weather/'tween	Folding	(A), (B), or (C)	Winch	Screw or quick-acting	Fig. 7-62c
Direct pull	All ships with cargo gear	Weather	Folding	(A) or (B)	Ship crane or derrick	Automatic	Fig. 7-62d
Roll stowing	(D)	Weather	Roll stowing	Depends on drum diameter, but usually more than (A) or (B)	Electric or hydraulic	Automatic	Fig. 7-62e

Side and end rolling	All, but mainly large bulkers and oil/bulk/ore	Weather	Rolling	(A) or (B)	Electric or hydraulic	(E)	Fig. 7-62f
Lift and roll	All, but mainly bulkers	Weather	Rolling	(A) or (B)	Electric or hydraulic	(E)	Fig. 7-62g
'Tweendeck sliding	Multideck cargo ships	'Tween and car decks	Sliding/ nesting	(C)	Electric	"Token"	Fig. 7-62h
Pontoon (F)	Containerships/ Multideck cargo ships	Weather/'tween	Lifting	(A), (B), or (C)	Ship or shore crane	(E)	

a *Notes:* (A) Minimum coaming height allowed by 1966 Load Line Convention. (B) (A) Increased as necessary for fencing. (C) Flush with deck provided drainage is satisfactory. (D) All except combination carriers. (E) Screw, quick-acting, or automatic cleats can be fitted. (F) Without tarpaulins on weather deck.

b Information from L. Buxton, R. P. Daggitt, and J. King, *Cargo Access Equipment for Merchant Ships*, E &FN Spon Ltd., London, 1978. Courtesy of MacGregor-Navire Publications, Ltd.

that operating time for hatch covers had ceased to be a major problem following the widespread adoption of improved access equipment.[2] With present-day designs, the time lost has fallen even further. Table 7-7 shows other results of delays during cargo working hours from the same survey.

The principle types of hatch covers are listed in Table 7-8 and illustrated in Fig. 7-62.

The major types of hatch covers all show variations of operation within each type. The ship's officer should be knowledgeable of all aspects of the operation and maintenance of the hatch covers aboard his ship. In this regard he should refer to the manufacturers' instruction books and technical data sheets. The shipowner is faced with a wide variety of hatch covers and must take into account the type, size, and service of his vessel when selecting the equipment that will best suit his particular operation. In addition, the shipowner should pay specific attention to the minimum coaming heights derived from the 1966 Load Line Convention, which came into force in 1968.

End of the shelter deck classification.   Prior to 1966 there was an advantage to having a ship designated as an "open shelter deck" ship in that all the shelter deck space was exempted space and thereby reduced the vessel's gross and net tonnage. As a result, many of the fixed costs of operating a vessel that are based on the vessel's gross and net tonnage were reduced.

An open shelter deck ship had a tonnage opening in the weather deck (shelter deck). Thus, the space between the weather deck and main deck was excluded, provided there was an approved means for closing tonnage openings and that any bulkhead openings were of an approved pattern.

It became obvious to many maritime nations that tonnage openings were potentially unsafe. In 1966, the IMO, then known as the IMCO, dispensed with tonnage openings and made provisions for ships to be assigned two load lines corresponding to the vessel's GRT and NRT (gross and net registered tons). One load line is determined using the weather deck as the freeboard deck, with the 'tweendeck space included in the tonnage. A second load line is determined using the second deck as the freeboard and tonnage deck; the 'tweendeck space is regarded as a superstructure, but excluded from the GRT and NRT. A "tonnage mark" is then assigned, which is cut into the ship's side at a height corresponding

---

[2] L. Buxton, R.P. Daggitt, and J. King, *Cargo Access Equipment for Merchant Ships,* E & FN Spon, Ltd., and MacGregor-Navire Publications, Ltd., London, 1978, p. 64.

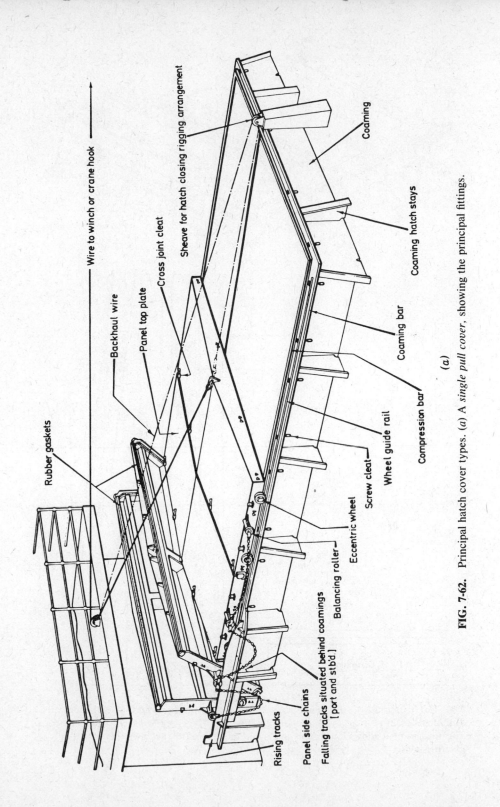

Wire to winch or crane hook ——→

Backhaul wire

Panel top plate

Cross joint cleat

Sheave for hatch closing rigging arrangement

Coaming

Coaming hatch stays

Coaming bar

Compression bar

Wheel guide rail

Screw cleat

Eccentric wheel

Balancing roller

Rubber gaskets

Rising tracks

Panel side chains

Falling tracks situated behind coamings [port and stb'd.]

(a)

FIG. 7-62. Principal hatch cover types. (a) A *single pull cover*, showing the principal fittings.

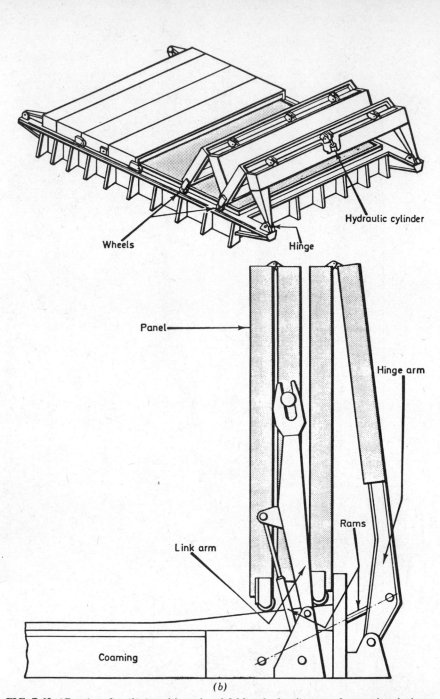

FIG. 7-62. (*Continued*) (*b*) A multipanel *end-folding hydraulic* cover for weather deck use. Multipanel external ram, *hydraulic folding cover*, of Kvaerner Multifold Crocodile type. This arrangement allows four panels to stow at one end of the hatchway, operated only by external rams.

454

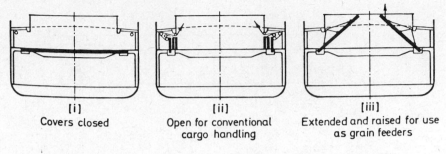

[i]
Covers closed

[ii]
Open for conventional
cargo handling

[iii]
Extended and raised for use
as grain feeders

(c)

FIG. 7-62. (*Continued*) (c) *Wire-operated side-folding covers*, which can be used as grain feeders when inclined as shown.

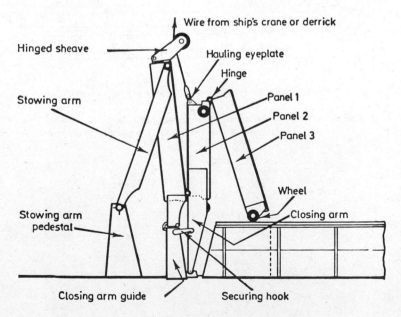

(d)

FIG. 7-62. (*Continued*) (d) *Direct pull cover* in the open position.

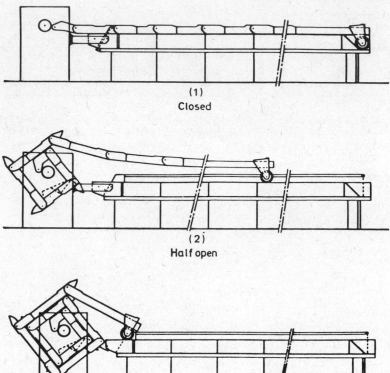

(1)
Closed

(2)
Half open

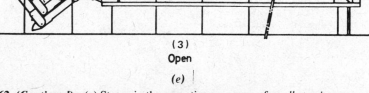

(3)
Open

(e)

FIG. 7-62. (*Continued*) (*e*) Stages in the operating sequence of a *roll stowing cover*. From the fully closed position (1) the cover winds on to the stowage drum, one panel at a time, until it is fully open (3).

456

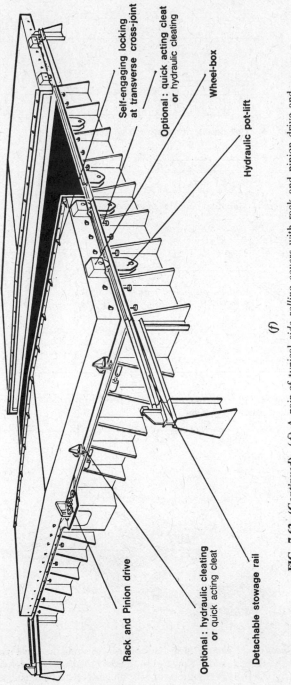

Self-engaging locking
at transverse cross-joint

Optional: quick acting cleat
or hydraulic cleating

Wheel-box

Hydraulic pot-lift

Rack and Pinion drive

Optional: hydraulic cleating
or quick acting cleat

Detachable stowage rail

(f)

FIG. 7-62. (*Continued*) (*f*) A pair of typical *side rolling covers* with rack and pinion drive and hydraulic lifting and cleating.

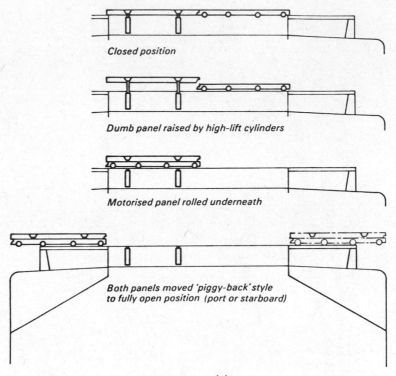

Closed position

Dumb panel raised by high-lift cylinders

Motorised panel rolled underneath

*Both panels moved 'piggy-back' style
to fully open position (port or starboard)*

(*g*)

FIG. 7-62. (Continued) (*g*) Operation of a *lift and roll cover* installed transversely. Note that only the wheeled panel is powered.

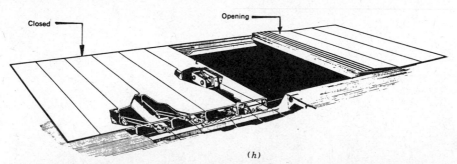

(*h*)

FIG. 7-62. (Concluded) (*h*) *Sliding 'tweendeck covers,* showing the drive unit. Reprinted from L. Buxton, R. P. Daggitt, and J. King *Cargo Access Equipment for Merchant Ships,* E & FN Spon, Ltd., London, 1978. Courtesy of MacGregor-Navire Publications, Ltd.

to this second, shallower draft, and if the mark is not submerged, the smaller values of both GRT and NRT are assumed to apply in assessing port dues and other costs. If the mark is submerged, the larger values apply. The higher tonnage applies in determining compliance with statutory regulations.[3]

As a result of the 1969 IMO Convention, a Universal Measurement System (UMS) for tonnage was drawn up. Gross tonnage is calculated directly as the volume of all enclosed spaces in a ship, with no exemptions or deductions, particularly of shelter 'tweendecks. Net tonnage is largely a function of cargo space volume and number of passengers. For many ships the new gross and net tonnages are expected to be fairly close to their existing tonnages. The principal exceptions are vessels whose 'tweendecks are currently exempt. Thus the adoption of UMS is unlikely to have any effect on cargo access equipment in most vessels, although "paragraph" ships with 'tweendecks may be affected. Paragraph ships are vessels whose characteristics fall right at the limit between those at which different regulations apply. For example, ships of over 1600 GRT are required to carry a radio operator and to have higher-quality lifesaving equipment. Consequently, many ships are designed with a GRT of 1599. Another border between regulations lies at 500 tons. Unless special arrangements are made, paragraph ships are likely to have considerably smaller overall dimensions as measured by the UMS than those measured by the previous tonnage rules. Thus, if the present "paragraph" tonnages remain, owners will have to decide in the future whether to build physically smaller ships with fewer decks or to disregard tonnage considerations and build whatever size of ship is required for a particular trade.[4]

[3] L. Buxton, R. P. Daggitt, and J. King, *Cargo Access Equipment for Merchant Ships,* E & FN Spon, Ltd., and MacGregor-Navire Publications, Ltd., London, 1978, p. 76.
[4] L. Buxton, R. P. Daggitt, and J. King, *Cargo Access Equipment for Merchant Ships,* E & FN Spon, Ltd., and MacGregor-Navire Publications, Ltd., London, 1978, p. 76.

CHAPTER 8

MATERIALS HANDLING PRINCIPLES AND EQUIPMENT

MATERIALS HANDLING AS A SUBJECT OF STUDY

When we consider the materials handling problems of the marine terminal and the ship, we are considering a special handling operation. Nevertheless, basic materials handling principles apply, and a brief survey of some of these should be helpful to the ship's officer, terminal manager, operations manager, and the stevedore. Two noteworthy benefits are obtained from a knowledge of these principles: (1) The most elementary yet worthwhile accomplishment is to make all personnel conscious of the fact that they are dealing with a problem for which basic principles do exist. Hence, the word is spread that there is a source of factual data concerning such problems. This is good, because it tends to reduce reliance on pure opinion and to emphasize facts. (2) Knowledge of the principles helps to explain *why* one way of doing a job is better than another and *why* one piece of equipment is better for a given job than another. This knowledge of *why* gives each man, from operations manager down to longshoreman, a feeling of self-confidence that encourages thinking and will, *in the long run,* result in better utilization of machines and manpower.

There are several excellent books on the subject of materials handling. They discuss guiding principles for the materials handling engineer in great detail and give numerous examples of their application. The reader desiring a more detailed coverage of the field of materials handling is advised to read one of these books. In the following pages, we will survey the most important basic principles and discuss some of their applications with reference to machines and techniques used in the maritime field.

Professor Immer has classified the various principles under four categories: *planning, operating, equipment,* and *costing* principles.[1] The authors have listed them here in what they consider the order of their relative importance. After the first few, it becomes difficult to judge their relative importance; in fact, the last eight or nine discussed may be considered equal in importance.

After reading the comments under each principle listed, the reader may reflect that the ideas set forth seem quite obvious and fundamental. This is true. But it is also true that adherence to the principles is the only effective way to produce an efficient materials handling operation. For example, the first principle is simplicity itself, yet compliance with the idea is one of the most promising ways of obtaining effective cargo operations.

The Operating Principle of Unitization

Materials handling becomes more efficient as the size of the unit handled increases.

Some of the most radical changes in ship operation are based on this principle. The SEABEE operation has carried this idea further than any other segment of the shipping industry. The *unit* here is a barge. Compliance with the unitization principle has produced a profitable operation despite the fact that other principles of ship stowage and materials handling may have been disregarded. An optimum application of this princi-

---

[1] J. Immer, *Materials Handling,* McGraw-Hill, New York, 1953, p. 23.

ple is the SPLASH (Self Propelled Lighter Aboard SHip) vessel, wherein cargo can float aboard as one unit (see Fig. 8-1).

The pallet load, the container, lighters, barges, and integrated tug barges are developments that are slowly being exploited more and more by foresighted operators the world over. They are all based on the principle of unitization.

FIG. 8-1. The SPLASH vessel *Mammoth Oak,* which is one of two similar Sumitomo-built ships in service with Central Gulf Lines, departs completely from the traditional LASH vessel design. It has adopted the float-on/float-off method of barge movement, avoiding use of the relatively expensive gantry crane for a fairly small ship. Operation is similar to that of a floating dock having a well deck, open stern, and double skin sides. By ballasting to a sufficient depth, heavily loaded barges can be floated in and out three abreast simultaneously. The odd barge, making up the 19 that it can carry, is floated under the wheelhouse. By covering the dock area with a deck structure aft of the superstructure, containers can also be carried.

Rigged as they are for handling the loads of yesteryear, there is a limit on the size of the units that can be handled economically on present-day general cargo ships. The unitization principle may produce radical changes in the future ship as the cost of labor and specialized routes make themselves more evident. Use of a containership with all containers loaded and waiting when the ship docks should and does lower the materials handling costs tremendously. If the overall operation was coordinated to such a degree that these containers were packed by the shipper and delivered to the terminal ready for loading, the efficiency will be even greater.

At any rate, the small single carton or single bag is being eliminated and combined into larger units. Those operators who become conscious of this fact and bend their efforts toward this end first will be the first to realize the great potential savings.

The Equipment Principle of Terminal Time

The shorter the terminal time, the greater the efficiency of the materials handling equipment.

The only time that a piece of materials handling equipment is making money for its owner is when it is actually moving cargo. This principle can be used by the operating manager as a guide to deciding on the merits of one type of materials handling equipment as compared with another. Other factors being somewhat equal, the system that reduces the terminal time to the smallest value is obviously the best. This idea applies to all materials handling equipment, large or small. It is the principle that makes the forklift truck and pallet combination superior to other systems.

Before the advent of the elevating platform truck and skid, the mechanically powered load carrying truck was used on marine terminals. The latter was simply a four-wheeled truck with a motor power attachment. This piece of equipment is very inefficient from the standpoint of terminal time. It carries the load between the pickup point and the discharge point with ease, but the truck and driver are tied up, accomplishing nothing, while the load is laboriously placed on board or taken off the truck. The simple, yet highly effective, improvement of an elevating platform used in conjunction with skids immediately rendered the fixed-platform powered truck obsolete for all except specialized uses. The elevating platform and skid, however, are slower than the tapered forks used with a pallet, and so the former quickly gave way to the latter. This change is primarily due to the validity of the terminal time principle, although there are other advantages. The pallet load may be tiered without using dunnage between tiers, because the pallet provides its own base. The pallet's overall height is less than that of the skid; this becomes important when considering the maximum height of a tier of unitized cargo. The pallet is lighter than a skid of equal strength.

Let us consider the world's largest piece of materials handling equipment, the *ship*. With proper cost accounting methods, the ship cost per unit of time while a forklift truck is standing idle should be obtainable; however, the authors have been unable to obtain an average figure for use as an example. There are some figures available for the cost per *day* of a ship, which is really just an oversized materials handling machine through which the cargo must pass in its journey from shipper to consignee. A ship standing at a dock costs between $15,000 and $30,000/day. The meaning, in dollars and cents, of terminal time now becomes quite clear.

It appears that there will be large increase in profits awaiting the first group that finds a successful means of reducing the terminal time of the general cargo carrier. The ideal arrangement would be to have a single container that is loaded and waiting for the ship when it arrives in port, as with a SPLASH vessel. Then, by pumping out ballast, a powered unit with a buoyant body could rise up and engage the cargo container. Once sufficiently strong connections had been made between the cargo containers and the powered buoyant part, the load could be taken to sea.

The Operating Principle of Gravity Utilization

Move materials by the use of gravity where possible.

At the marine terminals of the world, this principle has long been exploited when the cargo can be carried on the ship in bulk form. The best example of the use of gravity for loading ships is the ore carrier of the Great Lakes. Grain docks and some oil docks utilize gravity to a large degree also. Here we have a force provided by nature that will move cargo with a minimum cost if a suitable pipe or chute to direct the flow is provided. In some operations gravity has been used successfully to load bagged and cased products also. Case oil, which is the term applied to refined oil loaded in 5-gal containers and packed into uniform boxes, two containers per box, was at one time loaded by means of specially built spiral conveyors. This conveyor was actually a chute and the boxes slid down into the hold. The base of the spiral conveyor was equipped with a

FIG. 8-2. Great Lakes iron ore loading terminal: view of the chutes leading into the ship. Iron ore is running into the ship down the foremost chute. This operation is an excellent example of the full utilization of gravity. Courtesy of Pittsburgh Steamship Co.

ring of gravity rollers, and upon leaving the spiral conveyor, the individual cases were diverted along several short lines of gravity roller conveyors scattered through the hold of the ship. The length of the spiral conveyor could be adjusted by adding or removing sections. This system was used at refineries in Port Arthur, Texas, and New Orleans, Louisiana. Productivity of the longshoremen was almost doubled by using this system rather than the conventional yard and stay rig.[2]

Figure 8-2 shows the chutes at a Great Lakes terminal where iron ore is stored in large bins and then released in runs of 60 to 80 tons. When released, the ore simply pours into the ship's hold. In one test at one of these terminals, 12,500 tons of iron ore was loaded in 16 min. On the average, these terminals load the ore at the rate of 5000 tons/hour. The

---

[2] B. Stern, *Cargo Handling and Longshore Labor Conditions,* Bureau of Labor Statistics, U.S. Government Printing Office, Washington, D.C., 1932.

number of men required for the operation is less than that in a single gang of longshoremen handling general cargo.

The Planning Principle of Improved Flow or the Straight Line

Efficiency in materials handling is increased by the elimination of switch-backs and vertical movement.

This principle recognizes the simple fact that the shortest distance between two points is a straight line, and inasmuch as *motion is money,* the most economical flow for materials is in a straight line. This principle is badly violated on almost every terminal in the world. The cargo may flow almost in a straight line up to the ship's side on the pier apron due to a well-organized and carefully planned receiving operation, but what happens at the pier apron? The cargo's ultimate position in the ship is only a few feet away in a horizontal or almost horizontal direction, but as ships load today, the cargo travels in a circle to get there. To eliminate some of this inefficiency, it seems that a greater utilization of side ports should be made. There is no arguing the physical fact that it requires only a comparatively small amount of energy to move a ton of material along a smooth horizontal surface, whereas it takes a tremendous amount of energy to raise that ton only a few feet.

It is unlikely that perfect satisfaction of this principle will ever be approached with existing docks and ships, but all future development should be directed toward that end. RO/RO vessels and vessels such as the Delta Line ships utilizing side ports to load bananas come as close as possible to compliance with this principle (see Fig. 8-3).

The Planning Principle of Air Rights

Dock area is increased by utilizing the third dimension.

This principle may be considered too high on the list, but with the limited dock space on many existing terminals it is quite obvious and fortunate that the idea is being exploited fairly well on today's piers. Satisfaction of this principle is now possible with the use of high-stacking forklift trucks and pallets, whereas it was impossible before such equipment was available. Even with the cargo stacked as shown in Fig. 8-4, some piers are heavily congested. Imagine the effect of having to spread all the cargo out; all inefficiencies would be compounded and many present operations might even be rendered completely uneconomical.

FIG. 8-3. Loading bananas through a side port.

FIG. 8-4. Utilization of air rights increases effective dock area.

467

Hence, although it is almost taken for granted on today's terminals, the principle's importance, as well as its simplicity, cannot be denied.

The Costing Principle of Determining Handling Costs

For an intelligent analysis of any handling system, the costs of the handling operation must be known.

This is a difficult problem, but management should make every effort to determine the costs of every step in the terminal's operation. How else can one system be compared with another to determine which is truly the best from an economic standpoint? One system may be more productive than another, yet be more costly. This cost analysis must be done by experts in industrial engineering and cost accounting. The facts obtained may reveal astonishing weaknesses in any given system and allow better planning in the future or perhaps point the way to immediate desirable changes.

The Operating Principle of Safety

High productivity with economy is impossible without safety.

Everybody agrees that safety is a necessity, but too many people fail to remain safety conscious when planning and executing plans, including materials handling plans. Cargo operations on the marine terminal and especially on the ship have a high potential for developing unsafe conditions. Management and labor must be constantly reminded to enforce safety regulations.

One of the worst violations of a common safety rule at the marine terminal is the lack of proper lighting, especially in the hold of a ship. The authors cannot remember ever having seen a really well-designed lighting system for cargo hold interiors, with one possible exception. The large 100-ft holds on C-4 type ships converted to heavy-lift ships have about 30 lights scattered about the overhead of the hold. This arrangement is a large step in the right direction. A well-designed system would be one that provides light in all parts of the hold interior even when cargo is stowed in various bays. The light provided should be *bright*, not a yellow glimmer. This is not all; the well-designed lighting system must be reliable and rugged so that it cannot be made inoperative by working cargo into and out of the spaces. Finally, it must be accessible to the ship's officer and inaccessible to unauthorized persons. The use of a number of cluster

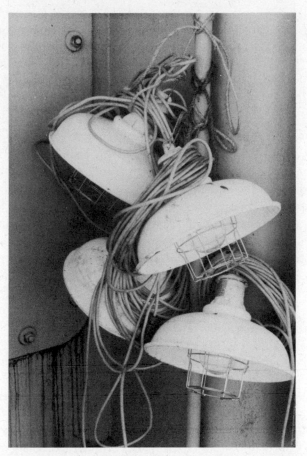

FIG. 8-5. Portable cargo lights. Notice protective wire guards around bulbs. Courtesy of M/N J. Robben, Class of 1986, USMMA.

lights (Fig. 8-5) in a hatch is inefficient; also, they are vulnerable to sudden breakage and constitute a fire hazard. Besides all these defects, their use normally results in leaving some spots that are not being actually worked in darkness, and these, of course, make ideal places for the sneak thief to go to work on the cargo. There is nothing so frustrating, and sometimes so futile, as the efforts expended by a ship's officer to keep all holds rigged with the minimum of three cluster lights on a busy cargo ship working all hatches on a rainy or foul-weather night. The cords break, the lights are smashed, and eventually the supply of lights becomes inadequate. Can it be that with all the modern materials and design talent in the lighting engineering industry, ships must continue to go their way shrouded in darkness? Any additional cost for a ship with a well-designed lighting system is warranted.

The Costing Principle of Equipment Amortization

Purchasing new handling equipment is warranted if the expense of the new equipment is exceeded by savings sufficient to amortize the cost in a reasonable time.

The economies realized as a result of abandoning old methods and equipment have in certain cases paid off the costs of new equipment in 3 months.[3] This, of course, is an extreme case but it serves to point up how costly the old methods must have been. Just what the amortization period should be when making a decision regarding new equipment is something that must be decided after considering many things, including such items as legal allowances for depreciation rates and period of expected use. Some managements have adopted the policy of approving the equipment that gives the lowest handling costs, regardless of the time required to pay for itself; this policy guarantees that long-term savings will not be overlooked. The life expectancy of most equipment is about 10 years, although for some it may be as high as 20 years. It seems sensible to set the amortization period somewhere near but below the life expectancy. Approximately 5 to 6 years is an average reasonable time.

Note, however, that it is difficult, if not impossible, to determine the potential savings to be gained by changing methods of operation unless the handling costs are known. This places emphasis on the importance of the principle of determining handling costs. A lack of knowledge of handling costs is probably one reason why mechanization was slow to come to the marine terminal and why, in general, new methods and equipment are not readily adopted.

The Equipment Principle of Flexibility

Materials handling equipment becomes more useful, and therefore more economical, as its flexibility increases.

This particular principle cannot be followed blindly, but should be used with caution in the face of all other facts about a given operation. Obvi-

[3] J. Immer, *Materials Handling,* McGraw-Hill, New York, 1953, p. 30.

ously, the forklift truck and pallet system is far more flexible than the elevating platform and skid. Forklifts may be used to handle skids as well as pallets if necessary, but elevating platform trucks can handle skids only. Besides this, the forklift truck can be fitted with attachments that make it possible to handle bars, pipe, rolls of paper, drums, bales, and many other items without pallets or skids. Thus, it is quite clear which is the better handling equipment.

The general cargo carrier that can handle truly *general* cargo of all types and sizes without delay for refitting is the best such carrier. For instance, for flexibility she should be able to carry general packaged merchandise, a limited quantity of bulk liquid cargoes, 50,000 to 100,000 ft^3 of refrigerated cargo, heavy lifts up to 50 or 75 tons, and bulk grain with a minimum of expense for fittings, long lengths, and dangerous cargoes and containers. Trailer ships represent the antithesis of flexibility; they can call at only one type of specially built dock and can carry only one type of cargo. In spite of their merits, they would be condemned if judged solely by the principle of flexibility.

The Operating Principle of Mechanical Equipment

Heavy units are moved most economically through the use of mechanized equipment when gravity cannot be used.

Practically all units on the marine terminal may be classed as heavy; therefore, mechanization should have great advantages. Mechanical equipment reduces worker fatigue, which is a safety consideration and also may affect productivity. Mechanization speeds the handling process and, if union regulations allow, reduces the number of workers. The reduction in the number of men has a twofold advantage: Both the possibility of injury and the cost of labor are reduced. It is important to point out that the men remaining on the job will not be required to do as much heavy labor, and that mechanization will produce jobs in the service area. Without mechanized equipment, it becomes impossible to utilize overhead clearances economically by stacking cargoes high on the dock.

One of the biggest obstacles to more and better mechanization of materials handling methods on the terminals and on the ships, especially on the ships, is the attitude of labor. This problem may be solved by education and training over a long period, but it will not be easy. What labor seems unable to accept is that more mechanization and better use of present mechanization makes industry more stable, gives it a healthy financial

foundation, and eventually will use more men. At the same time, the men
will be relieved of back-breaking manual labor. With labor and management cooperating fully on a completely mechanized coastwise operation,
this branch of water transportation would probably be able to flourish
once again. Such a flourishing would involve the full use of the unitized
load, side port operation, the latest industrial trucks, and an abandonment
of the standard-sized longshoreman gang. Under these conditions, the
longshoremen and the seamen would have more jobs, and for the longshoremen the work would be easier and safer.

Equipment Principle of Standardization

*After thorough experimentation, it is economical to standardize the
equipment and methods of materials handling.*

The advantages of following this principle are a reduction in the number and makes of parts for making repairs and an increase in the efficiency
of the personnel making the repairs. It is even desirable to standardize
such simple pieces of materials handling equipment as pallets; the objective here would be to economize not on the cost of the pallets, but on the
maintenance, handling, and storage costs.

It should be obvious that the maintenance and repair of a heteroge-
neous mixture of handling equipment is not as economical as that of
standard equipment. The important thing is to make certain that the
standard equipment chosen is the best, based on all the other principles
mentioned in this section.

Operating Principle of Maintenance and Repair

*It is economical to avoid breakdown through correct maintenance and to
anticipate repairs and replacements.*

It should be obvious that it is not good business to run equipment until
it breaks down and then perform corrective maintenance. It should be
equally clear that some regular inspection and check-off system, as with a
preventive maintenance program for lubrication and replacement, is necessary to prevent unexpected breakdowns. Even the simple pallet needs
attention. If one has a standardized set of pallets, a supply of boards cut to
fit the pallets can be maintained. When an end board or any other part of
the pallet is found to be cracked or damaged, this part should be replaced
immediately. In this way, the total number of actively used pallets is kept
high and greater damage is prevented.

Planning Principle of Coordination of Handling and Movement

Greatest economy is realized when the handling and movement of materials is coordinated.

This is merely another way of stating the fact that the greatest materials handling economies will be obtained through the complete overall planning so that all functions dovetail together in a smooth flowing organization. Here the importance of cooperation, communication, and coordination cannot be overemphasized. What the receiving clerk does with the cargo when it is delivered to the pier should not be done without consideration of the stevedore's activities when he picks it up for loading on board the ship. Full coordination is impossible without a well-prepared tentative cargo plan and without maximal scheduling of delivery to the pier. Companies that neglect to organize the flow of cargo to their dock and work without or with an incomplete tentative plan are violating this basic principle, and their activities are certain to be inefficient.

Planning Principle of Making the Organization Handling Conscious

For full cooperation, create full understanding of the philosophy that motion is money, extra handlings are evil, and reduced handlings are good.

This principle suggests the establishment of an educational program including management as well as labor. It may take the form of voluntary participation in a time and motion study, attendance at the showing of a materials handling film with discussions, suggestions relative to materials handling, or a series of lectures on the need for increased productivity through mechanization as a means of maintaining a high living standard.

MATERIALS HANDLING EQUIPMENT

In the following sections we will describe the various types of materials handling equipment and comment briefly on those items used most frequently. The changeover from hand-powered equipment to mechanically powered equipment took place from about 1932 to 1941. These years were the years for the first purchases of dock tractors and lift trucks, respectively. The first purchases of powered conveyors and crane trucks were also between these years. The year of the earliest recorded purchase of dock tractors is 1918, and that of lift trucks is 1928. These facts give some idea of the speed with which mechanization takes place along the water-front. It can hardly be called rapid.

Hauling and Handling Materials

To reduce the uneconomical use of manpower and thereby bring some degree of efficiency to the marine terminal, there must be a continued effort to reduce the number of handlings of all products carried by the ship and to develop machines to do a given job with the least inefficiency. Efforts along these lines have produced a variety of materials handling equipment. Each has several uses for which it is better fitted than is any other type of equipment. The two broadest classifications into which equipment can be divided are equipment which is best for *hauling* and that which is best for *handling*.

We will define hauling as the horizontal movement of cargo for a distance over 300 ft. Handling is defined as the vertical movement or relatively short (i.e., under 300 ft) horizontal movement of cargo.

Tractor–Trailer Combination as Compared with Forklift Truck and Pallet

The best contrast between equipment best suited for hauling and that best suited for handling is between the *tractor–trailer combination* and the *forklift truck and pallet*. The former is essentially a hauling combination, whereas the latter is a handling combination. In many simple operations, the costs can be made much higher than they should be merely by an improper decision regarding the type of equipment that should be used. Below are some figures on the use of fork trucks for hauls of between 300 and 1000 ft for similar commodities. Note the productivities in tons per man-hour obtained on the different haul lengths.

USING FORKLIFT TRUCK

Commodity	Length of Haul, ft	Tons per Man-Hour
Glass bottles in cartons	900	0.47
Flour in bags	300	3.58
Firebricks on skids	1000	1.05

The productivities when working with shorter hauls would be expected to be greater, and, as seen from the following table, the productivities *are* greater.

USING FORKLIFT TRUCKS

Commodity	Length of Haul, ft	Tons per Man-Hour
Glass bottles in cartons	65	1.27
Flour in bags	60	8.06
Firebricks on skids	160	6.79

It should be obvious that it is not economical for a fork truck to make a run of 1000 ft with one pallet load if a tractor could be hauling four to six such loads while a man and fork truck were loading other tractors.

The Two-Wheeled Hand Truck

The common hand truck will remain a useful tool in the handling of cargo no matter how mechanized the total operation may be made. When labor is very cheap, the purchase of mechanized units may not be warranted according to the principle of amortization. The hand truck can be used for the movement of packages too heavy to be moved by hand or to increase the unit load of small packages on occasional short trips. The hand truck can be used also as a pry or lever in much the same way as a crow bar. With two or more men working with hand trucks, a large crate or case may be raised several inches and transported a few feet very slowly. The nose is wedged under the case, the wheels are chocked, and the operator bears downward on the handles. This action raises the case on the nose of the truck, and with two or more trucks at opposite ends the entire case may be raised off the deck and moved slowly in a straight line (see Fig. 8-6).

Two or three men make the best combination for working with the hand truck. With three men, two remain at the loading terminal to load the truck while the third man acts as the mover. With two men, the operator of the truck also helps to load.

The load must be placed on the truck in just the right way to make it easy to transport. If the load is too low on the bed of the truck, the handles are difficult to keep down while pushing the truck. On the other

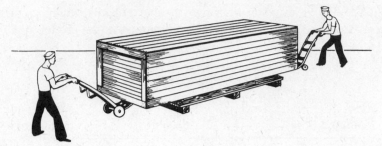

FIG. 8-6. An example of how two two-wheeled trucks can be used to move large crates and cases when necessary. Courtesy of the U.S. Navy.

hand, if the load is too high on the bed of the truck, the operator has a difficult time supporting the load while maneuvering the truck. If the load is placed on the bed so that it is well balanced when the truck is tipped backward about 60° from the vertical, surprisingly heavy loads can be handled easily by an experienced man.

The working loads may run anywhere from 200 to 600 lb depending on the size and type of truck. The trucks are able to support more than it is normally possible to transport on them. Models weighing about 45 lb have a capacity of 900 lb, while heavy-duty models weighing 155 lb have a capacity of 2000 lb.

The two-wheeled hand truck is economical only for very short occasional movement of materials, and then only if the cost of labor is extremely low or some special circumstances exist, such as a low volume of movement or limited working room. Two types of hand truck are shown in Fig. 8-7. Figure 8-8 shows a method of loading light bulky cases. The operator must experiment a few times to discover the best way in which to load any given container. Care must be taken not to place a fragile case or bag over the nose in such a fashion that the upper containers will damage it.

The Four-Wheeled Platform Hand Truck

This type of truck is also useful today where labor is very cheap or the volume of movement is too low to warrant the purchase of powered equipment. It is much faster and easier to use this truck than it is to use the two-wheeled type when working cargo to and from the pier apron on marine terminals. The loads can be made up on the platform of the truck within the pier shed and transported directly to the hook on the apron

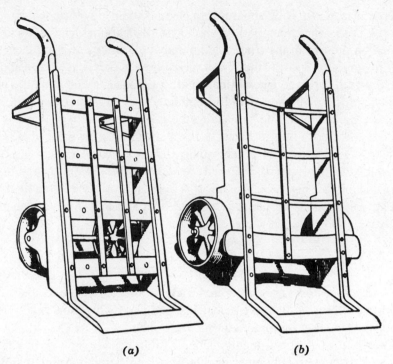

(a)   (b)

FIG. 8-7. Two types of hand truck. (*a*) Western style. (*b*) Eastern style. Courtesy of the U.S. Navy.

FIG. 8-8. With experience an operator of a two-wheeled truck can load surprisingly large loads on the bed. Here we see a load of 15 lightweight cases loaded on the bed of an Eastern style two-wheeled truck. Without experimentation to discover the best way to load such containers, it would have been possible to load only six cases. Courtesy of the U.S. Navy.

where the load is hoisted aboard. If the load is too heavy for one man, two may be used. The same method applies to discharging. The load can be landed on the bed of the truck and immediately moved away to clear the spot for the next load. If the two-wheeled hand truck is used, the spot is blocked until the hand truck is loaded after the load is landed on the pier. The use of the platform truck thus eliminates an additional handling and reduces the costs and the damage to the cargo.

Four-wheeled platform hand trucks also find use in ship's holds for moving loads between the hatch square and the wings. The load is landed on the bed of the truck and pushed to the wing or built on the bed in the wing and pushed out to the square. It is necessary to lay down a roadway of light steel plates or dunnage boards over which the truck can pass easily.

The Tractor–Trailer Combination

For hauling purposes the tractor and trailer combination is excellent. One important requisite for the efficient use of this combination is a smooth surface on the pier apron and in the dock shed.

The tractor is a short, highly maneuverable vehicle powered by gasoline with a sheet steel bumper face in front and a trailer-coupling device on the rear. Its primary purpose is to pull a number of trailers on a train. It may also be used for pushing railroad cars along spur tracks and similar work. Most tractors used on marine terminals are equipped with four wheels, and although they may be capable of greater speeds, they should not be driven over 5 mph, especially when hauling a trailer train.

The trailer is actually a special type of four-wheeled platform truck that is able to take heavier loads than the regular platform truck. Trailers are constructed in many ways: Some are fitted with four wheels that turn, others with only the two front wheels turnable, and still others with caster wheels. The bed of the trailer averages about 3 by 7 ft and is about 14 in. high. The bed is usually made of hard wood and bound with steel. Sometimes the bed is covered completely with a light steel plate for heavy-duty work.

When pulled as a train by the tractor, the trailers follow in the path of the tractor. One tractor can be used to keep the hook supplied with trailers, either loaded or empty depending on the operation. For example, in loading, the tractor will arrive at the hook on the pier apron with four to six trailers in tow. Dropping these trailers, the tractor removes the empties and travels to the point on the dock where the cargo originates. Here it drops the empties and picks up a loaded set. It then takes the loaded set to the pier apron again, arriving before the last of the previously loaded set has been emptied. Depending on the length of the haul and the coordination of the entire job, one tractor may be able to keep two or three hooks supplied. A forklift truck should be employed with the tractor-trailer combination to load the trailers or empty them on the dock. No forklift is needed under the hook, because the loads are hoisted from or landed directly on the trailer bed.

Assuming an ideal operation, then, the following would be needed: three trailer trains, one tractor, and one forklift truck. Only four men would be required, assuming all pallets were loaded and stacked on the dock, awaiting this operation. As the distance from the loaded pallets to the pickup point on the pier apron decreased, eventually it would become more economical to do away with the tractor-trailer combination and bring in another forklift. Any given operation must be analyzed carefully on its own merits.

The Forklift Truck

The forklift truck is the most widely used piece of handling equipment. The load is supported on a pallet that is carried on a pair of parallel bars that protrude in front of the truck body itself. These bars or forks are thin and slightly beveled and are attached to the truck by a sturdy frame. Almost all of these trucks used on marine terminals are equipped with hydraulic lifting devices capable of raising loads for high-level tiering.

The fork truck may be driven by electricity, gasoline, gas-electric power, diesel engine, or engines using liquified petroleum gas as a fuel. It is extremely flexible, handling loads on either skids or pallets. It handles some commodities without the use of a pallet, such as rolls of paper, bales, and lumber. Some models are able to stack loads with a 48-in. depth as high as 15 ft. There are many attachments for the fork truck that enable it to handle special commodities without pallets. Figure 8-9 shows some of the different types of attachments.

The capacity of the forklift truck. The capacity that a forklift truck can carry may be stated as a given weight with the center of gravity a given distance from the face of the forks. For example, it may be given as 4000 lb at 15 in. This means that the truck will carry a maximum load of 4000 lb with the center of gravity no more than 15 in. forward from the

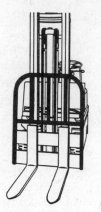
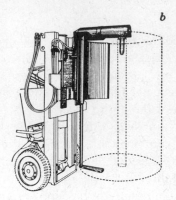

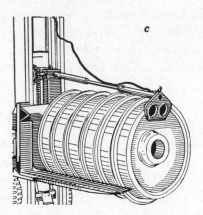
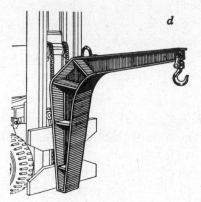

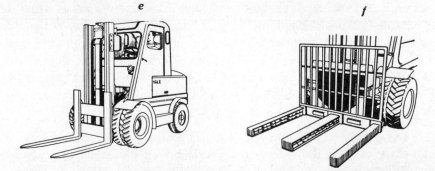

FIG. 8-9. Attachments for the forklift truck. (*a*) Load backrest. (*b*) Bartel device for rolls of newsprint. (*c*) Car wheel handling device. (*d*) Gooseneck boom. (*e*) Detachable cab. (*f*) Gripping forks for bricks and cinder blocks.

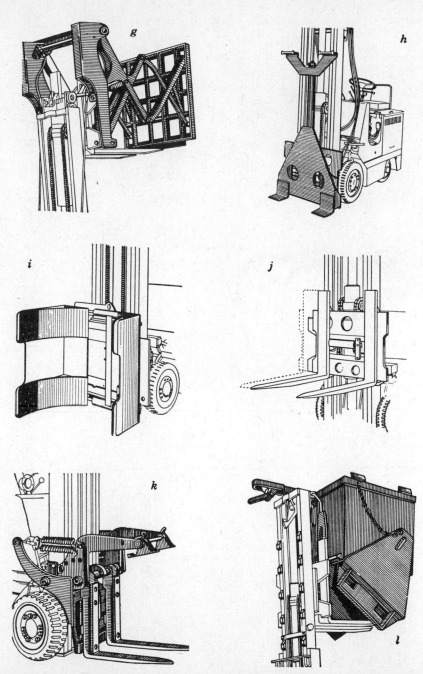

FIG. 8-9. (*Continued*) (*g*) Pallet unloader used with pallet plate eliminates need for pallets. (*h*) Hydraulic crate clamp device. (*i*) Revolving roll clamping device. (*j*) Side shifter attachment for positioning loads to left or right. (*k*) Clamp and fork attachment for handling tin plate safely. (*l*) Bottom dumping hopper.

481

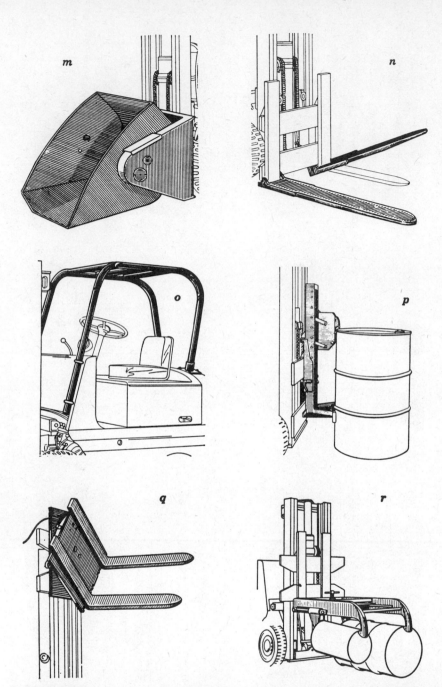

FIG. 8-9. (*Continued*) (*m*) Shovel scoop for handling loose free-flowing materials. (*n*) Fork extensions for handling materials of greater depth. (*o*) Canopy guard. (*p*) Vertical drum handling attachment. (*q*) Revolving forks for dumping skid bins. (*r*) Horizontal drum handling device.

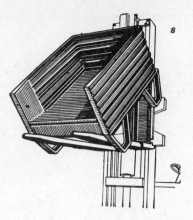

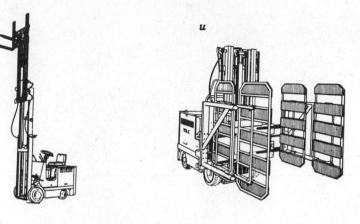

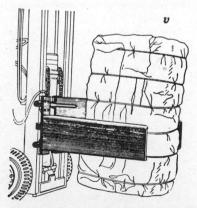

FIG. 8-9. (*Continued*) (*s*) End dump hopper for safe handling of steel scrap, forgings, and so forth. (*t*) Triple lift attachment increases the telescopic range of forklift to $16\frac{1}{2}$ ft. (*u*) Carton clamping device for handling large fragile cartons. (*v*) Hydraulic clamping device for bales, boxes, drums, and fragile containers.

483

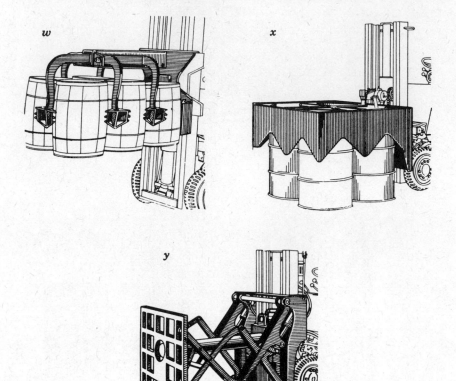

FIG. 8-9. (Concluded) (w) Multiple barrel handling attachment. (x) Multiple drum handling attachment. (y) Hydraulically operated pusher attachment equipped with ram for pushing off coils of steel, wire, and similar materials. Courtesy of Yale and Towne Manufacturing Co.

heel of the forks. If the center of gravity is closer than 15 in. the maximum load is still 4000 lb. If the distance is greater than 15 in., the maximum load is less than 4000 lb. This is explained below.

Inasmuch as the load is always carried ahead of the front driving wheels, the centerline of the driving axle is the pivotal fulcrum point for the load. The truck is usually designed so that there is a safety factor of 25% additional weight carried behind the fulcrum point when the maximum load is carried with its center of gravity 15 in. forward of the heel of the fork. If the load has a depth that is greater than 30 in., the center of gravity will be farther than 15 in. in front of the heel of the fork. The

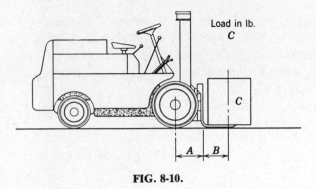

FIG. 8-10.

maximum load under these conditions will be equal to the quotient obtained when the sum of the distance from the axis of the drive wheels to the fork heel and the distance from the fork heel to the load's center of gravity is divided into the inch-pound capacity of the truck. The inch-pound capacity of the truck is found by multiplying the maximum load by the sum of the distance from the axis of the drive wheels to the fork heel and the distance stated in the manufacturer's rated capacity for the truck. This can be made clearer by referring to the diagram in Fig. 8-10. Let

A = inches from the center of the drive axle to the heel of the fork.

B = inches from the center of the maximum specified load to heel of fork.

C = maximum specified load in pounds.

D = the inch-pound capacity.

Then
$$(A + B)C = D.$$

Now, if the depth of the load changes B to a greater value, E, the maximum possible safe load is reduced and may be determined by the following equation:

$$\frac{D}{(A + E)} = X$$

Where X is equal to the new maximum safe load in pounds.

If the depth of the load changes B to a smaller value, the maximum load is still equal to C.

Aisle widths.   When space is at a premium, it is desirable to know the minimum aisle width allowable for working the fork truck when handling loaded pallets at right angles to the side. If this value is known, it may be approached when planning the layout of incoming cargo on the dock for maximum use of the dock area.

The following aisle widths are normally required for the efficient operation of fork trucks of the capacities listed:

Truck Capacity, lb	Aisle Width, ft
2000	10
4000	12
6000	14

These figures are based on a 48-in. load depth. For shorter loads, the requirements are slightly smaller. These are not the minimum possible aisle widths within which the fork truck can turn and place or remove a load; they allow clearance for smooth and rapid operation and for two-way traffic.

To determine the minimum aisle width with a clearance set by the terminal superintendent or other responsible officer, equations such as those given in Fig. 8-11 may be used.

Floor loads imposed by fork trucks.   The following table gives estimated average loads, in pounds per square foot, exerted on a floor by fully loaded fork trucks of the various capacities indicated:

Truck Capacity, lb	Floor Load Exerted, lb/ft^2
2000	260
3000	280
4000	320
5000	340
6000	380
8000	440

It must be remembered that the weight exerted by a four-wheeled vehicle and its load is not subject to exact calculation because, although the weight is exerted only at the four small points of contact between the

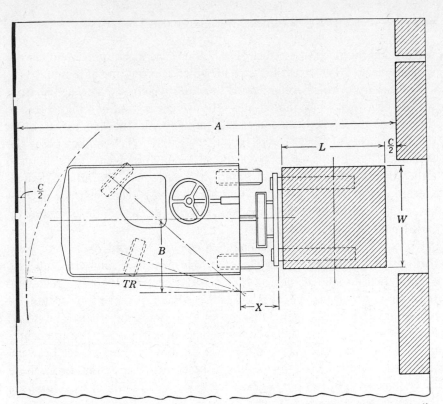

FIG. 8-11. Layout for aisle width equation. A = aisle width. B = distance from centerline of truck to parallel line through the point about which truck turns when steering wheels are in extreme cramped position. TR = turning radius. L = length of load. W = width of load. C = clearance, set at 6 in. X = distance from centerline of drive axle to face of fork. The following equations are sufficiently accurate for obtaining required aisle width (A) for finger fork truck operation when stacking loaded pallets of various sizes at right angles.

Examples: For loads having width (W) less than $2B$:

$$A = TR + X + L + C$$

Let B = 25 in., X = 14 in., L = 48 in., C = 6 in., and TR = 61 in. Then

$$A = 61 + 14 + 48 + 6 = 129 \text{ in.}$$

For loads having width (W) more than $2B$:

$$A = TR + \sqrt{(X + L)^2 + [(W/2) - B]^2} + C$$

Let B = 25 in., X = 14 in., L = 48 in., W = 60 in., C = 6 in., and TR = 61 in. Then

$$A = 61 + \sqrt{(14 + 48)^2 + [(60/2) - 25]^2} + 6 = 129.3 \text{ in.}$$

487

wheels and the floor, the effect on the floor is as if the same weight were spread over a larger area, the size of which can only be assumed.

Furthermore, the percentage of the total weight exerted at the front wheels of a fork truck increases as the weight of the load increases, so that, when the truck is loaded to the tipping point, there is no weight on the rear wheels at all. The accuracy of the estimate is complicated by the fact that even if two floors have the same rated capacity, they may be affected differently by concentrated moving loads. The above figures, therefore, can be considered only as approximate averages to be used as guides in determining the floor capacities required for the operation of fork trucks of various capacities.

It may be necessary to calculate the load imposed on a floor by a forklift truck carrying a load less than the maximum limit. In doing so, each of the four wheels may be considered to be supported by a rectangular area of which the wheel is the center. Each of these four rectangles would have an area equal to the wheel base times the tread. The wheel base is the longitudinal distance from the center of one axle to the center of the other. The tread is the transverse or athwartship distance between the centers of the contacting points of the treads on the tires. From these facts comes the following equation for calculating the floor load imposed (X) in pounds per square foot:

$$X = \frac{36 \times (\text{Weight of truck} + \text{Weight of load})}{(\text{Tread} \times \text{Wheel base})}$$

where the weights of truck and load are in pounds and the tread and wheel base are in inches.

Load-Carrying Powered Truck

For specialized work a four-wheeled nonelevating platform truck may be used. This type of truck was one of the first mechanized trucks used on marine terminals. Its greatest disadvantage for general cargo work on the marine terminal is quite apparent: The operator must rely on a second device for loading and discharging, and it transports only one unit at a time. Exactly the same function can be accomplished by one trailer with a tractor, and adding two additional trailers removes the necessity of the tractor being tied up at either end of its run; this is a great increase in efficiency. However, use may be found for this type of equipment on certain types of jobs: Those that are occasional in nature, require careful control of the load, and include very heavy loads. Some models of pow-ered trucks are capable of carrying 10 tons. The platforms of these trucks may be at variable heights off the deck. Some models have platform heights of between 20 and 24 in. Lower-platform models have heights of only 11 in.

The greater the weights that are carried, the more concern there should be about the height of the platform. A worker should lift packages weigh-ing between 150 and 200 lb no higher than the knee, and then only when he is trained to lift properly. Packages of 75 to 100 lb should be lifted only waist high, although they can be transported at shoulder level if loaded onto the man from shoulder height by helpers. Packages of 25 to 35 lb may be lifted shoulder high.

High- and Low-Lift Elevating Platform Trucks

These trucks are designed to pick up loads on skids, transport them, and deposit them without the aid of an assisting device. In other words, they are self-loading and -discharging. The skid is a platform built over side runners or stringers so that there is about 6 to 8 in. clearance under the skid bed. The skid also may be constructed of angle iron with stilts at the four corners.

This type of truck was first introduced in industrial plants about 1915, four years before the first forklift truck was built. The load on this type of truck is carried over the wheels and does not extend out in front of the body. This fact makes it possible to construct the trucks to handle very heavy loads without having to put equally heavy counterbalancing weights at the rear. Some models are capable of handling loads up to 30 tons, but these are obviously not the type of equipment to be found working on the average marine terminal.

The high- and low-lift elevating platform truck represents the first im-provement beyond the nonelevating platform truck. It is many times more efficient than the simple load-carrying truck because it is a self-loader and -unloader when used in conjunction with the skid. The terminal time is reduced to a mimimum.

Clamp Trucks

The clamp truck is a type of forklift truck. The forks are replaced by an attachment capable of gripping a container between two arms and raising it off the deck. The truck then transports the item and sets it down where desired. In other words, this device accomplishes all that the forklift truck does, but without the use of a pallet. Figure 8-12 shows a clamp truck picking up a roll of paper. The load may be a single item or in units. If in units, these should be strapped together or carefully stacked and handled with a broad-faced gripping device (see Fig. 8-9u). The containers must be capable of withstanding the pressure of the gripping plates without being damaged.

The development of flexible gripping arms made of spring steel has made it possible to lift palletless loads of compressed gas cylinders nine at a time, a stack of automobile tires, and other such units. The gripping arms can be placed on any lift truck clamp. Each pressure plate is loosely mounted on three spring-steel fingers. This allows the pressure plate to tilt in the direction or directions dictated by the load shape and provides for proper load distribution among the fingers. Further accommodation is permitted by the individual flexing of the fingers. The design of the arm structure is such that as seen from the front, each pressure plate toes in at the bottom. As seen from the top, the pressure plates toe in at the front. These features make it possible to accommodate variations in the size of objects in the load and also to apply a pressure tending to hold the load together. The pressure can be set to suit the load by means of a dial at the operator's position. A high-friction plastic coating bonded to the pressure plate reduces the necessary grab pressure. Capacities run between 1500 and 2700 lb.

Crane Trucks

The crane truck has found considerable use on narrow apron piers in ports. These trucks are able to pick up net slings or pallet sling loads and transfer them with ease. They are popular where the pier structure makes it impossible or inconvenient to utilize the much faster tractor-trailer combinations and/or forklift trucks with pallets. Their terminal time is lengthened because they are not self-loaders or -unloaders, and for the same reason they require extra labor at the terminal within the pier shed.

These trucks find their greatest usefulness when handling certain extremely bulky and awkwardly shaped materials that are best accommodated in a net sling. Figure 8-13 shows a typical crane truck handling a load. The boom on the truck shown in this illustration cannot be slewed; however, some models incorporate slewing ability into their designs. Because of the limited drift under the boom head and the length of the sling,

FIG. 8-12. (*a*) Rolls of liner board in hold, roll clamps, and forklift with special attachment. Courtesy of M/N J. L. Belle, Class of 1985, USMMA. (*b*) Clamp truck. Hydraulically operated clamps pick up vertical rolls of newsprint, rotate them 90°, and lay them down horizontally for easy discharge. Note use of crisscross two-layer dunnage floor to provide a firm and level surface for stowage of rolls. Courtesy of Clark Equipment Co.

FIG. 8-13.   Crane truck. Courtesy of Port of New York Authority.

the boom is fitted with a wide flanged roller on its tip so that the sling with all its parts can be hove up short for transporting the load.

Overhead Monorails

Overhead monorails are a means of moving materials about a dock shed of a marine terminal that keeps the deck clear of moving vehicles and utilizes the overhead for roadways. Overhead monorail systems have been used successfully at several piers for over 40 years. At all of these installations, one commodity was continually handled over a fixed route, usually relatively short.

The lack of flexibility of this system is evident. It may be, however, that with additional trackage and suitable methods for switching the trolleys from one track to another from a single control point as well as locally, sufficient flexibility will be incorporated into the overhead monorail system to make its operation not only feasible but highly economical.

Automatic, Radio-Controlled Materials Handling System

A system based on the use of electrically operated carriers traveling on monorails, with all movements of the carriers controlled by radio impulses, has actually been designed.

The designers of this system have provided means of controlling the carriers or trolleys by a local loading or unloading crew as well as by a central control dispatcher. A loading crew at any position on the dock would have temporary local control of an empty carrier. After loading of the carrier, the central dispatcher would be signaled by mechanical means and he would take control.

The major obstacle in adapting this idea to a marine terminal already built is the low dock shed doors at such terminals. Other problems would be the fluctuation of the level of the ship's deck at various stages of loading and of the tides. These problems, however, relate only to running the monorail out over the ship so that the trolley could pick up its load directly from or discharge it into the ship's hold. Once the ship's gear has placed the load on the apron, the system could be used to transport materials about the dock shed. With sufficient clearance over the ship's deck a rig such as that shown in Fig. 8-14 would connect the hold directly with the dock shed. Figure 8-15 shows a factory installation of this same type of system, and one can easily see its possibilities for use on the marine terminal.

Such a system would certainly provide a continuous flow of cargo and eliminate delay time on the pier apron. A more efficient distribution of the standard longshore gang could be effected, with the probable result of greater tonnage being handled without an increase in personnel.

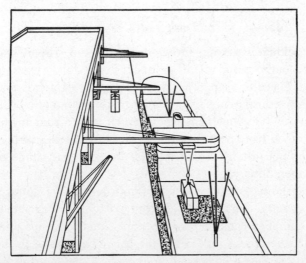

FIG. 8-14. Radio-controlled monorail system: arrangement that would provide a direct connection between the ship's hold and the dock shed without having to use the ship's gear. Courtesy of Trenton Marine Service Corp.

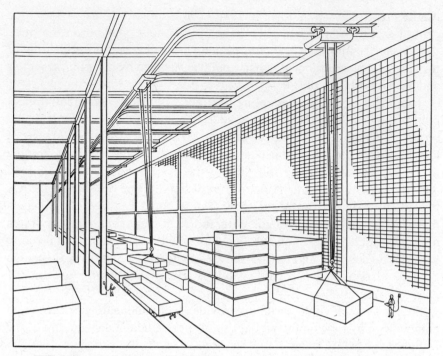

FIG. 8-15.   Typical factory installation of a radio-controlled monorail system.

Factors in the Design of Stevedoring Trucks

Stuffing, stacking, stocking, lifting, breaking out; round loads, square loads, short loads, tall loads; light loads, heavy loads, easy loads, awkward loads: There is probably a greater variety of goods and materials handled in stevedoring than in any other materials handling situation, and as many ways of handling them. Requirements vary from day to day, from ship to ship, and from port to port. No run-of-the-mill, off-the-assembly-line conventional lift truck can handle the immense range of port and stevedore operations.

Maneuverability is exceptionally important in warehouses that are stuffed to capacity, and ships' holds contrive to defeat even the most agile truck. To complicate matters even more, loads are not always neatly stacked on pallets, and forks on a lift truck become useless unless the truck can quickly and readily adapt to the use of one or more attachments.

Stevedore lift truck rated capacities normally range from 6000 to 70,000 lb; the need for individual capacities within this broad range will vary with the geographic location of the ports and the range of materials indigenous to these ports. Generally speaking, the popular capacities are 6000, 8000, 10,000, 15,000, 20,000, 35,000, 45,000, and 70,000 lb, with the lower capacities used for stuffing containers and for general terminal work. Medium-capacity trucks handle steel, lumber, and heavier loads, whereas the upper end of the range is used for container handling (see Figs. 8-16 and 8-17).

Ports on the southern Atlantic seaboard are finding that the 8000-lb truck is being replaced by 10,000-lb pneumatic-tire units that are versatile and maneuverable and can stand the derating encountered when attachments must be substituted for forks. In the Port of Savannah, Georgia, for example, the Clark C500 Y100 is gradually becoming the all-purpose truck for general terminal use; this truck has the agility to work even in ships' holds. It can easily handle loads up to 10,000 lb with forks, and can carry 8000 lb in most instances where an attachment must be employed. These same trucks must be flexible and adaptable enough to handle the assignments of the busy period, which are commonly dictated by the cargo

FIG. 8-16. Container forklift. Courtesy of M/N W. Hewlett, Class of 1985, USMMA.

FIG. 8-17. 80,000-lb diesel pneumatic-tired truck. Courtesy of Clark Equipment Co.

configuration, size, and weight. Forks, roll clamps, bale clamps, carton clamps, rams, side shifters, quick-change couplers, detachable host reels, and other attachments become important handling instruments, and the lift truck with which they work must have the capability to switch devices in minutes as the applications demanded change, thereby minimizing idle truck time, which stevedores can ill afford. There is more at stake than the operator's time when a stevedore truck is idle. Many times a whole crew, who do *not* work for minimum wage, are involved. In this unique industry it is extremely important that lift trucks stay on the job constantly when there is work to do.

The stevedoring industry is different from other industries in many other ways. In many, but not all, ports, lift truck operators tend to be rougher on lift trucks than in most other materials handling operations. Therefore, time and treatment are two essential considerations for the design and manufacture of a lift truck that will survive under these considerations. There are conventional attributes, like durability, reliability and the ability to take a pounding on the docks, that are inherent in a lift truck, but there are seemingly insignificant details that also count toward economies of operation in stevedoring trucks. A good example is a larger Protectoseal filler cap for refueling. Most, or many, ports must also service and fuel larger equipment, such as cranes, which have large-capacity fuel tanks and are fueled faster through a larger nozzle. If the lift truck fleet has the conventional smaller filler cap, then a smaller nozzle or separate source is required for fueling the lift trucks. With the larger filler cap available on stevedore trucks both large and small equipment can be fueled from the same source. This is only one example of the attention to detail gained through working closely with stevedoring and port personnel. Auxiliary fuel tanks also reduce the number of trips back to the refueling station. A common-key ignition system, in which one key will fit any truck in the fleet, is another important time saver.

Many times the removal of a lift truck is accomplished by means of another lift truck pushing it forwards, backwards, or even sideways. But the most innovative method is simply to pick it up by its underside with another lift truck. During this process, there is sometimes a sudden, loud crack, crunch, or sickening sound as the forks rupture the transmission case. The least that will happen is the abrupt sheering of a drain plug. Clark uses steel "belly bands" to protect the undersides of its stevedoring trucks from the disaster just described, or from lower underclearances.

Other important considerations. Gasoline, liquified petroleum (LP) gas, and diesel fuel are the popular fuels for stevedoring trucks, and all three engine options should be available. Electric power also is available and pneumatic electric lift trucks are used in areas where food and citrus cargoes are being handled because of their fumefree characteristics, which also could make them popular in confined areas or for loading passenger ships.

Because it is cleaner burning and less noxious than gasoline and diesel fuel, LP gas is the most widely used fuel on smaller trucks used for stuffing containers and in confined warehouse terminals. Diesel fuel is preferred on larger equipment, especially that with capacities of 15,000 through 70,000 lb.

Stevedoring trucks should be furnished with the standard industry package: sling eyes, single bolt counterweight, and pin-mounted overhead guard for fast, easy removal when it is necessary to operate in extremely low-clearance areas or in ships' holds.

Pneumatic tires are preferred and nearly mandatory for the rough terrain and often muddy areas. However, pneumatic tires are constantly subjected to puncture from debris found shipside and in the yards. A commonly accepted substitute is the punctureproof "solid" or "soft shoe" tire, which delivers the needed traction without punctures.

Low overhead guards for low-clearance areas and triple stage uprights with low down heights complement the stevedoring lift truck and enable it to work in low-clearance areas and retain high stacking capabilities.

One of the least expensive and most indispensible accessories for the stevedore is lights mounted on the mast—an absolute necessity for night work, for stuffing containers, or for working in other dark areas.

Many so-called options have become so necessary on stevedoring trucks that stevedoring companies have learned to incorporate them into the complete stevedoring lift truck. Some items may be eliminated, but all should be seriously considered and evaluated in terms of time and cost saving advantages. These popular industry standards are listed here:

Choice of engine

Low overhead guard and upright down height

"Solid" pneumatic tires

Diesel ignition shut-off

Double auxiliary hydraulic valves

Quick-change coupling

Quick-disconnect hose reels

Hydraulic accumulators

Auxiliary transmission coolers

Protectoseal large-mouth fuel filler

37-amp Marine-type alternator

33-gal Auxiliary fuel tank
Lights
Belly band protection
Slim-line seats
Pin-mounted forks

"Standardization with flexibility" best describes the needs and goals of the port and stevedoring industry. The minimum number of trucks necessary to handle a multitude of requirements is the economic guideline of the stevedoring industry, an industry whose lift trucks are used less, but abused more, than those in other industries.

Conveyors

On the marine terminal, conveyors are used for loading and unloading trucks, railroad cars, and lighters; for loading and discharging the ship through side ports and through the hatches; and for moving cargo to and from the wings in the holds of the ship.

Gravity Conveyors

Gravity conveyors are either simple wooden chutes or made of metal and fitted with steel rollers or wheels. The roller or wheel type conveyors come in sections usually about 5 or 10 ft long and in widths of 18 to 24 in. They may be either steel or aluminum, and are built to withstand rugged usage. The rollers rotate on free-rolling steel ball bearings. The sections are fitted with bars at one end and hook connectors at the other, which makes possible quick setting up and taking down as needed. They are not made for really heavy loads; the weights transported over them should be limited to about $\frac{1}{2}$ ton.

Wooden conveyors, or chutes, as they are called, are somewhat lighter than the steel roller or wheel conveyors. They must be used in place of the steel type with some cargoes, such as explosives. They are built of surfaced hardwood and are sometimes provided with guard rails. In fact, chutes used to handle explosives must have a 4-in. guard rail by law.

Metal chutes built in a spiral have been used successfully to lower case oil and bagged sugar and coffee into a ship's hold by taking advantage of gravity. (See p. 464 for a discussion of the principle of gravity.)

Powered Conveyors

Powered conveyors of many types have been used with great success on marine terminals. These may be classified under four broad headings: (1) endless belt; (2) endless pocket or bucket; (3) screw, and (4) pneumatic elevators or air conveyors. No attempt will be made here to lay down rules or guiding principles on where conveyors should be or should not be used. It is believed evident that the use of conveyors should be considered as a possibility when analyzing any operation. The advisability of using them must be determined from the facts obtained and considered within the framework of the principles of materials handling already discussed. This is, of course, a logical method of approach when considering all available equipment.

Endless belt conveyors. The endless belt conveyor consists of a belt that runs over a drive pulley from which it derives its power. The belt is mounted on a bed in such a manner that a series of wheels take the burden of the load as the load is placed on top of the bed. The underside of the bed is fitted with one or two idler rollers over which the belt passes as it cycles around the bed. Most goods can be sent up inclines as steep as 32° if the belting is made of some good nonslip material. The use of cleats on the belt enables goods to be lifted up a greater incline. These conveyors may be obtained in portable models as well as permanently installed in a fixed position. Figure 8-18 is an illustration of a portable type endless belt

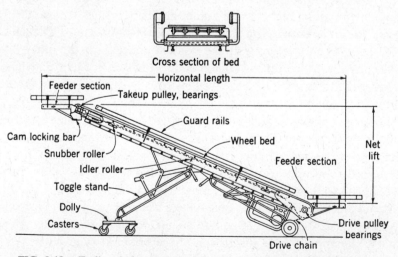

FIG. 8-18. Endless belt conveyor. Courtesy of The Rapids-Standard Co.

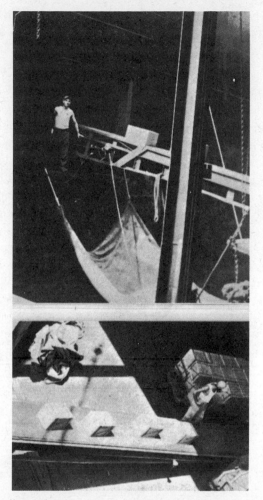

FIG. 8-19.   The endless belt conveyor. View of the ship end (upper) and dock end (lower) of an endless belt conveyor being used to load cartons through the ship's side port. The ship's gear, of course, can continue to work in the lower holds of the same hatch. Courtesy of Cadet-Midshipmen, USMMA.

conveyor showing some of the names of the parts and the mechanism that allows it to be adjusted for working at various elevations all the way down to horizontal. These portable models are extremely versatile materials handling tools. They can be used to mechanize the loading of trucks, for loading ships through side ports, for raising cargo from one deck to another where it might be discharged through a side port, and for many similar tasks (Fig. 8-19). When loading ships through the hatches or discharging from side ports where conveyors might be used, gravity types are the most desirable. (Gravity should always be used, if possible.) Figure 8-20 is another portable model, for heavier work.

Endless pocket conveyors. Probably the most intensive users of the endless pocket type conveyor in the marine field are the banana carrying ships. They load and discharge entire loads using conveyors of this type. These conveyors are large, heavy, specially built, and very costly pieces of equipment. But the speed and economy obtained in the handling of this very special type of cargo, for which speed is important, more than warrants the financial outlay. These ships utilize endless belt conveyors (described above) that bring the cargo from the lower 'tweendeck levels to upper 'tweendeck levels where side port connections with the pier are available. The endless pocket conveyors are used to take the stems from or to the lower holds to or from the pier apron. They consist of heavy canvas pockets that rotate around a bed that is constructed with a joint in the middle of its length so that the conveyor consists of two legs. Making a huge "A," the conveyor is set in place after the ship docks with one leg of the "A" resting in the ship's hold and the other out on the dock. The banana stems are placed in these pockets by longshoremen and taken from the ship to the pier or vice versa, depending on whether the ship is discharging or loading (See Fig. 8-21).

Air conveyors. Machines utilizing air suction methods of picking up and carrying materials through a flexible, noncollapsible tube of large diameter are called *air conveyors*. They are used to discharge a number of

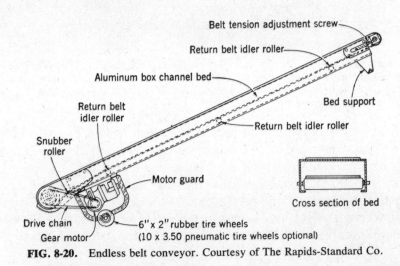

FIG. 8-20. Endless belt conveyor. Courtesy of The Rapids-Standard Co.

FIG. 8-21. The endless pocket conveyor, used to load and discharge banana stems. The dock end is supported on a gantry capable of moving up and down the pier. Courtesy of Port of New York Authority.

commodities and there may be wider use of them in the future. Pneumatic elevators for discharging grain fall into this category. They are a conveyor type in the sense that they utilize an air stream as the conveyor belt. Among the commodities that are discharged by air conveyors are bulk copra, grain, and bulk cement. They may be large, permanently installed units or portable units capable of being moved from pier to pier. Floating grain air conveyors (pneumatic elevators) brought alongside the offshore side of the ship are used to discharge full cargoes of grain into waiting barges or coastal steamers at the rate of 250 tons/hour.

Pallets

The pallet, used in conjunction with forklift trucks and, in some operations, with straddle trucks, has become a very important piece of materials handling equipment. The pallet was developed after the platform skid especially for use with the forklift truck. It is far superior to the skid for marine terminal work, for a number of reasons. It is lighter and vertically shorter, and is capable of being quickly and easily tiered without the use of dunnage between each load.

Pallet Definitions

Before discussing pallets in general, it will be best to define some of the terms that will be used later.

SINGLE FACED: A pallet that has only one deck.

DOUBLE FACED: A pallet that has a top and bottom deck.

REVERSIBLE: A double-faced pallet with both decks capable of supporting a load.

TWO WAY: A pallet that a fork truck may enter from two directions.

SINGLE WING: A pallet on which the outside stringers are set inboard 4 to 6 in. from the ends of the deck boards of the upper deck but flush with the lower deck boards. This gives a nonreversible double-deck pallet adaptable for use with straddle trucks.

DOUBLE WING: A pallet on which the outside stringers are set inboard an equal distance, about 4 to 6 in., from the ends of the deck boards of both decks. This is the construction of most stevedoring pallets and allows for the accommodation of the bar of the bar and spreader type sling.

FLUSH: A pallet on which the outside stringers are flush with the ends of the boards of both decks.

DECKS: The top and bottom surfaces of the pallet.

STRINGERS: The separations and supports for the decks, sometimes called "runners."

CHAMFERS: Beveled edges to permit easy entrance of forks without catching or chipping the end boards.

END BOARDS: Deck boards on the extreme outside edges of the pallet. These are the most vulnerable parts of the pallet. They should be of the best wood, at least 6 in. in width, and renewed as soon as split.

DRIVE SCREWS: Fastening devices for attaching top and bottom deck boards to the stringers. They are either spirally twisted or have annular rings.

Some Pallet Types

There are four basic types of pallets with reference to the type of work for which they are intended. They are the (1) stevedoring, (2) shipping, (3) warehouse, and (4) factory or manufacturing plant pallets. Although this book is mostly concerned with the stevedoring pallet, some comments will be made regarding all four types.

The stevedoring pallet.   The stevedoring pallet is used to handle cargo on marine terminals, and as such, it receives a good deal of rough treatment. Because of their heavy duty, the stevedoring pallets are made of heavier materials than any of the other types. The stringers are made of 3- by 4-in. or 4- by 4-in. lumber and the deck boards are made of 2-in.-thick deals (nominal size). The end boards should be not less than 6 in. in width and made of no. 1 or 2 grade stock board. The other deck boards may be of random widths not less than 4 in. The stringers should be 4- by 4-in. material if the length of the pallet is over 6 ft. The fastenings used on stevedoring pallets may be either stove bolts or drive screws 4 in. long with large heads.

The stevedoring pallet must be capable of accommodating a large number of commodity types and sizes and as a result a number of sizes have been used. For large, bulky packages such as furniture and general household goods, the 4- by 7-ft size has proven convenient. For smaller packages, the 4- by 5-ft size is commonly used.

The shipping pallet.   The shipping pallet must conserve space and weight as much as possible. It is not loaded, unloaded, and handled as much as the stevedoring pallet and can be made with smaller scantlings throughout. Consideration of the accommodation of the shipping pallet in connecting carriers and in stowage in the ship has led to the use of a few standard sizes that are smaller than the stevedoring pallet. The Navy has indicated that either the 48- by 48-in. or 40- by 48-in. size is the most economical size for palletized loads (see Fig. 8-22). The latter are the best for transport in the closed truck body, because the truck requires a 40-in. dimension for greatest use of available cubic. If rail transport or truck transport with an open platform bed is used, the 48- by 48-in. pallet is suitable. The average boxcar width is 110 in., whereas the average truck closed body type has a width of $88\frac{1}{2}$ in. If the shipping pallet's smallest dimension is 48 in., two pallet loads cannot be stowed side by side in the closed body truck; this means a great deal of lost cubic during transport. These figures apply only to closed body trucks and boxcars of the United States. When considering transport in other countries, the dimensions of the connecting carriers at the foreign port must be determined for full coordination of the cargo movement. If open flat-bed trucks are used, the 48- by 48-in. pallet can be accommodated. The stringers on the shipping

•

NAILING

a. Use #6 screw gauge x 2½″ cement coated or chemically etched drive screw nails or #10 wire gauge x 2½″ cement coated or chemically etched annular ring (fetter ring) nails.

b. Boards 3⅝″ to 4½″ require 2 nails at each bearing point.
   4¾″ to 6¾″   "   3 " " " " " .
   7″ to 9″   "   4 " " " " " .
   9⅛″ to 11″   "   5 " " " " " .
   11⅛″ to 12″   "   6 " " " " " .

c. Drill deck boards if necessary to prevent splitting.

d. Stagger nails to prevent splitting.

e. Nailing thru notches not acceptable.

f. When a board exceeding 5″ nominal width covers a notch, one less nail can be used.

LUMBER

a. Use sound square edge lumber free of decay and free of knots with an average diameter greater than ⅓ of the width of the board. No piece shall contain any defect which would materially weaken the strength of the piece.

b. Use the following woods: white ash, beech, birch, rock elm, hackberry, hickory, hard maple, oak, pecan.

c. Use random width boards 3⅝″ or wider, surfaced one side ⅞″, hit or miss. Surfaced faces to be exposed faces. All boards on any one face to be of uniform thickness.

d. Space between top deck boards to be 1″ min. and 1½″ max. Space between bottom deck boards not to exceed 1½″ except 11″ spaces as noted.

e. Boards must be placed so that each notch is completely covered by a single board of not less than 4″ nominal width.

f. Length tolerance plus or minus ¼″.

g. Minimum moisture content shall be not less than 12% and the maximum moisture content not to exceed 25% at time of shipment.

•

```
┌─────────────────────────────────────────────┐
│                                               │
│           DEPARTMENT OF THE NAVY              │
│         Bureau of Supplies and Accounts       │
│                                               │
│   SW RELEASE NO. 52C (Superseding RSX Release No. 52B)  │
│              31 October 1948                  │
│                                               │
└─────────────────────────────────────────────┘
```

FIG. 8-22. Courtesy of U.S. Naval Supply Research and Development Facility, N.S.D., Bayonne, NJ.

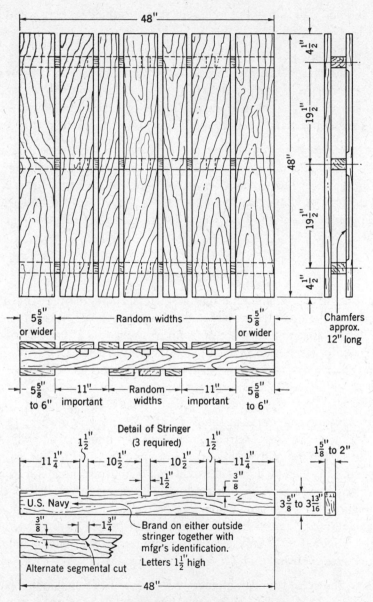

FIG. 8-22. (*Continued*)

507

pallet should be of 2- by 4-in. material, whereas the decking should be of only 1-in. boards.

The warehouse pallet.   The warehouse pallet is used to receive commodities in a warehouse for storage for periods of many months or even years. The materials are palletized as economically as possible, and the pallets are tiered and left in storage. Careful planning of aisle widths and pallet sizes is necessary to obtain efficient use of the available area.

The factory pallet.   The factory pallet is used to handle special materials in large volume as the material is worked during the manufacturing of some product. Factory pallets must be specially designed in size and strength. It is obvious that in every case a study must be made of the entire procedure and the best pallet determined in the light of the data obtained.

Pallet Patterns

Operations on the commercial marine terminal utilizing the stevedoring pallet do not permit extreme care in building pallet loads. It is sufficient to make certain that the tiers are built up in such a way that one tier ties together with another so that sufficient stability is obtained to keep the load together while it is being transported on the pier.

In the building of loads on pallets for shipment on the pallet, or when the operation makes it possible to exercise care in this matter, greater utilization of the pallet area available can be obtained by building the load in accordance with some definite pattern. When building the pallet load, some items may be allowed to extend beyond the deck of the pallet an inch or so. This may be done with steel drums and cases made of wood. It should be avoided with bagged commodities or cargo in cardboard cartons. Figure 8-23 shows some standard Navy pallet patterns. All possible arrangements on the 48- by 48-in. pallet can be reduced to four basic patterns, which, for the sake of common nomenclature, may be identified as *block, row, pinwheel,* and *brick.*

BLOCK: Square or round containers will always be loaded in this pattern on a 48- by 48-in. pallet. This is the least desirable of all the patterns, in that there can be no cross tiering to tie the load together for stability.

ROW: The row pattern is one of the most common and is used with oblong materials for which a multiple of the widths and a multiple of the lengths both approximate 48 in. An excellent cross tie is provided by placing alternate layers at 90° angles to each other.

FIG. 8-23. Basic pallet patterns.

PINWHEEL: The pinwheel pattern is another widely used pattern for material of an oblong nature that does not readily conform to the row pattern. Cross stacking is easily accomplished by reversing alternate layers. The major objection to this pattern is the fact that there is usually a chimney in the center of the load. This pattern is extremely flexible and can be varied.

BRICK: The brick pattern can be used only in a few well-defined instances, namely, with a box whose width is one-half or one-quarter of its length, and whose length is one-half, one-quarter, or one-third of 48 in. Frequently a box will fit both the pinwheel and the brick pattern. However, when the pinwheel pattern is used, four or eight columns are formed, and this gives less stability to the load.

It must be remembered that these patterns cannot be followed under practical operating conditions on most marine terminals. In the first place, the containers are usually placed on the pallet at the pier for transport to the pier apron only. In the second place, union labor agreements prevent the full utilization of a pallet by limiting the load to a set number of containers or a set height with the last tier frequently used to tie the load together. These pallet patterns are for use on shipping pallets, where space utilization is of importance.

Palletized Loads Versus Loose Cargo

Although there is probably no doubt in the reader's mind that palletizing loads reduces the labor involved in transporting cargo, it will help to emphasize this fact if we present a quantitative comparison between palletization and single-unit operations. Figure 8-24 is a chart showing the net savings in man-hours in shipping 100 tons of general cargo on pallets as compared with shipping the same cargo as loose packages.

Not all of the operations indicated in this chart would be performed in a commercial venture; however, we can use the applicable data to obtain an approximation of the difference in man-hours between the two systems in question. Using appropriate information from the chart, we can prepare Table 8-1.

From Table 8-1 we see that loose cargo, in this case, took about $2\frac{3}{4}$ times more man-hours than the palletized cargo. If we knew the hourly

TABLE 8-1

Operation	Man-Hours Required	
	Palletized	Loose Cargo
Load and strapping pallets at contractor's (shipper's) warehouse	50	0
Loading car at contractor's (shipper's) warehouse	8	50
Unloading car at pier	9	24
Loading and stowing on ship	37	47
Unloading ship at discharge port	24	164
Loading truck at dock	8	54
Unloading truck at consignee's	9	57
Total man-hours	145	396

wages paid the longshoremen and other labor in a given case, we could easily arrive at an estimate of the cost difference. Commercial utilization of the palletized load offers some problems that reduce the gross savings indicated by the above figures.

Some Problems of Commercial Palletization

In the above movement of 100 tons, 77 pallets were used. These 77 pallets represent about 3 tons of weight and 554 ft^3 of space. They must be shipped back to the shipper empty unless some form of pallet exchange or pooling is devised. Not all of the savings in man-hours are to the credit of any one of the three parties directly affected by palletizing the cargo. In fact, according to the above figures, the shipper used 8 man-hours more than he would if the cargo had been left loose. Therefore, out of a total saving of 251 man-hours, 8 hours are actually lost at the shipper's warehouse, 211 man-hours are saved at the loading and discharging marine terminals, and 48 man-hours are saved at the consignee's warehouse. Thus, in practice it becomes a problem of coordinating the entire operation and arriving at an equitable distribution of the costs involved in initiating the program. Although problems exist, there is evidence that they are solvable, and as time goes on more and more palletized loads will be shipped.

There are others in the field besides the shipper, shipowner, and consignee who benefit by palletization of cargo and who should be expected to share some of the costs, namely, the underwriter and the contracting stevedore. They might share the cost by reducing their charges to the shipowner and shipper. The underwriter would certainly be a beneficiary of the reduction in pilferage and in individual container damage that accompany palletization.

Palletized Operations Today

For the past 20 years containerization has been changing the face of ports throughout the world. A substantial proportion of containerizable cargo, however, has been untouched by the container revolution. In practice these cargoes are not carried in containers either because no container vessel serves the trade concerned or because the port has inadequate container handling facilities.

An interesting example illustrating modern palletized operations was the experience of Israel with unitization, reported by *Cargo Systems*

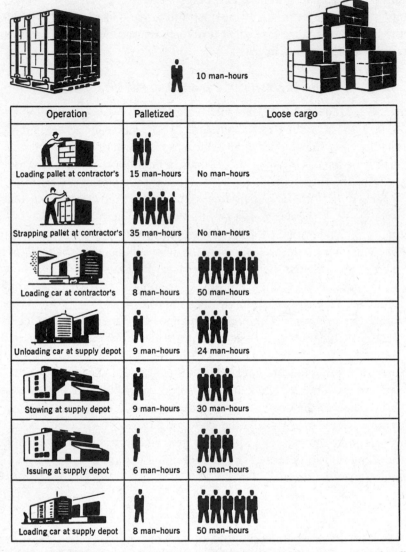

Operation	Palletized	Loose cargo	
Loading pallet at contractor's	15 man-hours	No man-hours	
Strapping pallet at contractor's	35 man-hours	No man-hours	
Loading car at contractor's	8 man-hours	50 man-hours	
Unloading car at supply depot	9 man-hours	24 man-hours	
Stowing at supply depot	9 man-hours	30 man-hours	
Issuing at supply depot	6 man-hours	30 man-hours	
Loading car at supply depot	8 man-hours	50 man-hours	

FIG. 8-24. Palletized unit loads vs. loose cargo. This chart compares two cargo handling methods at each operation in a typical shipment of supplies from Navy contractor to point of use. The number of man-hours required to handle 100 tons of palletized cargo (77 pallet loads) is compared with the number of hours required to handle the same amount of loose cargo (4080 separate packages). The figures used are based on studied conducted by materials handling officers in the United States, Guam, Pearl Harbor, and Trinidad. Ideal working conditions are assumed.

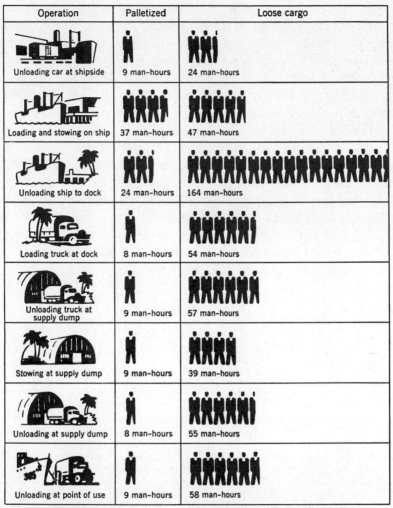

Operation	Palletized	Loose cargo
Unloading car at shipside	9 man-hours	24 man-hours
Loading and stowing on ship	37 man-hours	47 man-hours
Unloading ship to dock	24 man-hours	164 man-hours
Loading truck at dock	8 man-hours	54 man-hours
Unloading truck at supply dump	9 man-hours	57 man-hours
Stowing at supply dump	9 man-hours	39 man-hours
Unloading at supply dump	8 man-hours	55 man-hours
Unloading at point of use	9 man-hours	58 man-hours

Total number of man-hours required for materials handling at all operations:

Palletized cargo....203 man-hours

Loose cargo......682 man-hours

Net saving effected by palletized unit loads...................682 - 203 = 479 man-hours

A plus value........... palletization greatly reduces pilferage potential and damage to individual packages

FIG. 8-24. *(Continued)*

International, the journal of the International Cargo Handling Coordination Association (ICHCA), in its September 1980 and September 1983 issues. Of particular interest was the way in which the Israel Port Authority (IPA) improved its operational efficiency through the widespread implementation of palletization and preslinging (see Table 8-2). By encouraging palletization and preslinging as well as the increased use of containers, the IPA was able to reduce its labor force over a 5-year period by 34%, while cargo tonnage handled increased by 24% (see Table 8-3). The cut in manpower was accompanied by an impressive increase in productivity, thanks to the unitization of general cargoes (see Table 8-4).

Among the cargoes given special attention by the IPA in the unitization project was citrus fruit, which is one of Israel's major exports. Unitization on pallets had been used for some time in other countries for fruit shipments. Special palletized vessels were also used for the IPA project. These vessels had wide open hatches with each hold compartment capable of carrying 300,000 boxes. A multipurpose loader loaded eight pallets in one lift.

TABLE 8-2
TYPES OF CARGO THAT REQUIRED UNITIZATION AND THE METHOD CHOSEN[a]

Cargoes	Unitization System
Timber (hardwood, soft wood)	Bundles or pallets
Bags—import (flour, sugar, rice, chemicals)	Preslinging or 1-ton bags (big bag) or pallets (see Fig. 8-25)
Bags—export (agricultural products, mainly onions, potatoes, carrots, fertilizers, phosphates & chemicals)	Preslinging or pallets
General cargo (leather, single cartons, marmor, barrels, etc.)	Pallets
Frozen meat	Pallets with polyethylene shrinkwrap (see Figs. 8-26 & 8-27)
Logs	Preslinging or special clamps

[a] Reprinted from *Cargo Systems International* (the journal of ICHCA), September 1983, CS Publications, Ltd.

TABLE 8-3
COMPARISON OF TONNAGE HANDLED AGAINST MANPOWER, BEFORE AND AFTER UNITIZATION[a]

Year	Annual Tonnage (10^3 tons)	Number of Workers Employed
1977/78	9,840	4357
1982/83	12,000	2874
Rate of change	+24%	−34%

[a] Reprinted from *Cargo Systems International* (the journal of ICHCA), September 1983, CS Publications, Ltd.

The cargo unitization project implemented by the IPA and very briefly described above required a minimal investment to increase output and at the same time resulted in astonishing increases in productivity.

Unitization of fruit. The unitization of fruit has, relative to that of many other cargoes, been a slow process. According to a United Nations Conference on Trade and Development study published in 1982, only 3.2% of the bananas exported worldwide in 1980 were containerized. This was also true of all palletized fruit. Of the deciduous fruits coming from

TABLE 8-4
HANDLING PRODUCTIVITY—BEFORE AND AFTER UNITIZATION[a]

Type of Cargo	Before Unitization		After Unitization	
	Output of Gang/Shift (Net)	Number of Stevedores Per Gang	Output of Gang/Shift (Net)	Number of Stevedores Per Gang
Bags—import	96	21	300	10
Bags—export	125	18	300	9
Soft wood	63	18	300	9
Hard wood	63	17	120	8
Frozen meat	90	16	250	8
Citrus	250	16	350	8
General cargo	82	20	162	9

[a] Reprinted from *Cargo Systems International* (the journal of ICHCA), September 1983, CS Publications, Ltd.

FIG. 8-25. Courtesy of *Cargo Systems International* (the journal of ICHCA), September 1980, CS Publications, Ltd.

FIG. 8-26. Courtesy of *Cargo Systems International* (the journal of ICHCA), September 1980, CS Publications, Ltd.

FIG. 8-27. Courtesy of *Cargo Systems International* (the journal of ICHCA), September 1980, CS Publications, Ltd.

South Africa to Europe, about one-quarter were palletized during 1982, and it is intended that 100% palletization should be achieved in the near future.

It is also evident that breakbulk handling can respond not only to palletizing, but to the systems approach. One particularly good example of this is an innovative breakbulk cargo handling system, partly originated by the Belgian New Fruit Wharf (BNFW) in Antwerp, that was described in the January 1983 issue of *Cargo Systems International*.

As Figure 8-28 shows, the "fully" automated system is continuous, with loose boxed cargo being placed manually onto the Cargoveyor in the ship's hold and then carried directly, protected from the elements, into the consolidation and dispatch shed by conveyor. On arrival in this area, the bananas are subjected to quality control and then graded. The boxes then progress to one of six palletizing stations, where they are unitized in preparation for distribution by truck or rail wagon. The palletizing machines build pallet loads of either 35 or 40 banana boxes, operating at 55 and 40 pallets/hour, respectively. Three such palletizing machines are employed on the automated line, which means that up to 165 pallets can be dispatched each hour.

Maintaining a constant flow of cargo in this way and unitizing on the land side ensures that close control can be kept on the discharge process,

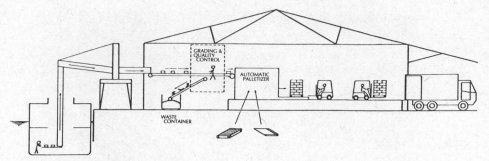

FIG. 8-28. The Belgian New Fruit Wharf, Antwerp, employs an automated, continuous unloading system for discharging fruit produce from ship's hold to road/rail vehicle. The centerpiece of the operation is the Figee Cargoveyor. Courtesy of *Cargo Systems International* (the journal of ICHCA), January 1983, CS Publications, Ltd.

as well as that throughput rates are high and manning low. For example, each palletizer is fitted with photoelectric counters that record throughput volumes. This information, once fed into BNFW's computer, can be used to generate statistical data on discharge volumes and rates, by brand, and consequently to plan delivery by truck or rail.

While banana discharge is efficiently executed by BNFW's Cargoveyor system, other fruits handled at the facility, such as citrus fruits, apples, and pears, are discharged in a more conventional LO/LO manner. This method, employed by many of the major fruit handlers in Europe, involves palletizing fruit cartons onboard the vessel. Once a pallet load is made up, it is lifted onto the quayside, possibly in a multiple pallet load, by conventional crane.

Uncertain future. Although it is clear that the pallet is bound to be employed in fruit handling at one stage or another, its full use along the transport chain is still undecided. Most fruit handling ports are thoroughly in favor of full palletization, since to them it means increased handling efficiency with reduced labor and capital investment costs. Likewise, the advantages to be gained from full containerization from the point of view of the port operator hardly need be spelled out. The variety of vessel types and port facilities in existence, combined with the diverse origins and characteristics of fruit produce, however, prevent any clear resolution of this question in the near future. No doubt unitization methods will be more quickly adopted in the future, especially on those trades to Europe from developed countries, but growth will be less quick in developing countries.

In the United States, Port Newark had been moving boxes of bananas in and out of vessels on conveyor belts. After switching to pallets (see Figs. 8-29 and 8-30), it was found that approximately 3500 boxes of palletized bananas were discharged per hatch per hour. Under the old conveyor system an average of 2500 boxes were discharged each hour. Palletized bananas not only moved faster, but required less labor and rarely suffered damage. The cost of making pallets and strapping the boxes of bananas onto them was insignificant compared with the amount of money saved in labor and damages at discharge.

FIG. 8-29. Pallets are loaded and strapped (with corner protectors) at the banana farms in Ecuador.

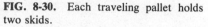

FIG. 8-30. Each traveling pallet holds two skids.

In palletized banana shipping, inexpensive wooden pallets are stacked with 48 cartons of bananas each. At discharge from the ship, the stevedore supplies oversize 4- × 8-ft travel pallets that permit carriage of 96 cartons per load through the warehouse directly to the tailgates of trucks. Forklifts transfer smaller pallets into the trucks. Two further handlings of pallet loads are required before the fruit reaches distributors for marketing.

An added benefit of the pallets is better management of trucks at the terminal, eliminating the congestion characterizing the former conveyor system.

Summary of Advantages of Palletized Loads

a. Less labor required in the handling of palletized materials at all terminals.

b. With palletization, pilferage and damage to individual containers are less possible.

c. Packaging may be eliminated on some products that are palletized properly, resulting in a saving.

d. On some palletized loads, only one or two labels or stencils are required for identification. Without pallets, labeling costs are considerably higher. This is sometimes overlooked as a shipping cost, because labels are applied by a shipping clerk. The cost of preparation of the labels by office help may not be counted into the cost of shipping, of which it is really a part.

   e.   Less personnel problems are experienced, because fewer men are needed when working with pallet loads. The operation is safer. Fewer men have fewer accidents, and thus insurance premiums should be reduced.

Sling Types

An important part of the materials handling equipment on the marine terminal is the *sling*. As used in this section, "sling" means any device attached to the cargo or its container for the purpose of hoisting it on board the ship. There are many sling types: Some might be classified as general slings because they can be used on a large number of container types, and others might be classified as special slings because they are used on only a particular type of container or type of cargo. The choice of sling for any given operation should be made with the objective of maximum safety, minimum delay on the ship and on the dock, and minimum damage to the cargo. Safety is obtained by using slings that give small angles between the parts at the hook and afford an easy and effortless method of attaching and releasing.

Endless Fiber Rope (See Fig. 8-31)

This is one of the most common slings used. Many ships carry their own supply of these slings when operating in trades in which bagged goods such as coffee or raw sugar are carried in large amounts. For a full loading operation, about 25 slings should be available for every set of gear on the ship. Not all of these will be needed at any one time. However, the mate who doesn't carefully control the number of slings given out will, on some runs, find that many more than 25 per gear will be used. About 8 to 10 slings will keep one gear going; more or less may be required depending on the cycle time of the total operation.

   These slings average about 18 ft in length when made; that is, the rope should be cut in lengths of 36 ft when making them. They should be made out of $3\frac{1}{2}$-in. three-stranded manila or sisal. When the short splice is completed, there should be no twists in the sling, or they will be difficult to work with. To prevent turns from being worked into the sling as the splice is made, one turn should be thrown out of the sling just before marrying the two ends preparatory to making the splice. This seemingly trivial point is quite important.

The endless fiber rope sling is well adapted to use with bagged goods, if the bags are rough gunnies; bales; tires; single, large, lightweight, but strongly constructed cases; reels of wire; and similar cargoes. This sling should not be used on bags of refined commodities such as refined sugar or flour, nor on bags of cement. Cement bags are unable to withstand the sling pressures; the bottom bags invariably split and spill the contents, causing considerable damage to the cargo plus making a mess of the ship and contaminating other cargo. These slings should not be used for hoisting several small cases or cartons, or for steel products.

The Fiber Rope Snotter

This is simply a length of rope with an eye splice at each end and generally fitted with a sliding hook. Two of these slings are used on the cargo hook

(a)

FIG. 8-31.   (a) The endless rope sling. Courtesy of Moore McCormack S.S. Co.

FIG. 8-31. (*Continued*)   (*b*) Two endless rope slings. Note danger of damaging some of the bags.

and ride the hook in and out because they are released instantaneously at the end where the load is landed. They are used mainly when handling lumber of the finer grades.

The Wire Rope Snotter

This sling is constructed in the same manner as the fiber rope snotter, but it has more uses. Besides being used on lumber of the rougher grades, it is also used on steel products such as pipe, steel rails, and structural steel shapes. Long wire rope snotters without sliding hooks are used to handle some large bulky cases. When used for this purpose, both eyes are placed on the cargo hook and the case rests in the bites thus formed.

The Web Sling (See Fig. 8-32)

This is simply an endless rope sling with a canvas web sewn between the two parts. It is one of the best slings for handling refined bagged products or bagged cement.

The Platform, Tray, Airplane, or Pallet Sling (See Fig. 8-33)

All of these terms are used to describe the wooden platform type sling. Before pallets were used so widely, this was a simple platform made of

FIG. 8-32. The web sling. Courtesy of Moore McCormack S.S. Co.

FIG. 8-33. (*a*) The airplane or platform sling. This picture shows the use of spreaders and safety nets to prevent cargo from falling off. Courtesy of Peck & Hale, Inc. (*b*) All-metal platform sling with two pallets of bagged coffee. Note absence of protective gratings due to necessity of providing access to pallets for forklifts. Courtesy of M/N T. Waller, Class of 1985, USMMA.

2-in. lumber and fitted with rings at the four corners where four hooks hanging down from a suitable spreader could be engaged. The rings on the platform have now been removed and bars are now used to fit between the upper and lower decks outside of the runner or stringer on the double type pallet. The bars hang from a spreader, which is required to keep sling pressures from bearing against the containers piled on the platform. This is the most widely used sling in ports where forklift trucks are used to handle the pallet. It has always been a popular sling, however. These slings may be used safely for almost all types of cargo, including cartons, crates, cases, bagged goods, cylinders, barrels, drums, carboys, furniture, and short lengths of lumber or pipe. There may be some situations in which the platform sling is used where the web or endless rope sling would actually be better. Such operations might include the handling of some bagged commodities and also of baled goods. The principal reason for the rope slings being superior in these cases is that with them less delay is caused by having to return a number of slings after the landing spot is clogged with empties. An all-metal platform sling is shown in Figure 8-33b.

FIG. 8-34.   Chime or cant hooks. Courtesy of Port of New York Authority.

Chain Sling

Chain slings are used when handling heavy concentrated cargoes such as pigs of iron, lead, tin plate, or copper bars. They are safer than wire rope because they stand up better with such heavy loads.

Cant or Chime Hooks (See Fig. 8-34)

These are used exclusively for hoisting several drums or cask type containers at one time. They are also used for hoisting hogsheads of tobacco but the hook is very broad and offers several inches of bearing surface.

FIG. 8-35.   (*a*) Fiber rope net sling with a pie plate. Courtesy of U.S. Maritime Service.

(b)

FIG. 8-35. (Continued)   (*b*) Unloading bales of paper using a net sling. Courtesy of M/N T. Waller, Class of 1985, USMMA.

Fiber Rope Net (See Fig. 8-35)

The fiber rope net is made of interwoven 2-in. fiber rope in sizes of about 12- by 12-ft square. For loading of some products these nets may be lined with canvas. They are used for bagged goods, mail, single cases, or heavy-duty drums. When lined, they are used for bulk products such as brazil nuts or fertilizer. They should not be used for crates and cases in general, because excessive sling pressures are exerted and damage to the cargo will result. They are also poor slings for lightweight drums for the reason just mentioned, and if used, some leakers are almost certain to develop.

Wire Rope Net (See Fig. 8-36)

The wire rope net has some special uses. One of the most common is for handling rubber bales.

Crate Clamps

The crate clamp is used not only for crates as the name implies, but also on reels of wire cable and rolls of heavy paper. It is an excellent sling for crated automobiles and similar containers because it can be applied and released quickly and because with its use there is no accumulation of slings at either end of the hook's travel.

Side Dogs (See Fig. 8-37)

Side dogs are used exclusively for handling steel plates. When plates are extremely long and limber, more than one pair should be used. Even as many as three pair may be required to prevent the bending of the plates.

FIG. 8-36. Wire rope net sling. Rubber bales are handled in this type of sling.

FIG. 8-37. Side dogs used on steel plate.

FIG. 8-38. Bale hooks. This view shows bale hooks rigged two to an eye, with four sets comprising the entire sling. This is not the same as the type described in the text, which has only one hook to a snotter. Courtesy of Boston Port Authority.

Dog Hooks

Dog hooks are similar to crate clamps. The difference is that the dog hook is attached to the bottom of the crate or case and, because of the dog-leg shape of the hook's shank, they set in place securely. When a strain is taken on the cargo fall, they tend to grip the bottom and will not slip off. They have the advantage over crate clamps in that they are a little less likely to be accidently released, but they are more difficult to set in place than are crate clamps.

Bale Hooks (See Fig. 8-38)

Bale hooks consist of about eight to 10 snotters of 2- or 2½-in. fiber rope spliced into a steel ring. At the lower end of each snotter, which is about 6 to 8 ft long, a hook is attached having a shank of about 8 in. in length. The ring is placed on the cargo hook after the smaller hooks on the several snotters have been engaged on the steel bands of the bales. At least two bands should be engaged on each bale to prevent one band from taking the entire weight and perhaps breaking. Bale hooks cannot be used on bales that are extremely heavy. They are commonly used on wool bales of 200 or 300 lb each.

Bale Tongs (See Fig. 8-39)

Bale tongs are levered tongs similar to ice tongs. They are used to handle cotton bales, and make an excellent sling because they can be applied and removed with great speed.

Newsprint Sling (See Fig. 8-40)

Newsprint is easily damaged by cutting the outer surface or denting the edges of the ends, and the slings for use with this commodity are generally designed to reduce the possibility of this happening, which is high with the endless rope sling. These types of slings are good examples of specialized slings developed because no general sling could be well adapted to handling this commodity. As said before, the endless rope sling is often used for hoisting newsprint, but it is not a good sling for the purpose.

The Pie Plate (See Fig. 8-41)

The pie plate is a circular platform placed in a rope net sling to prevent sling pressures from damaging cargo hoisted in the net. It is an awkward sling and should be avoided if at all possible (also see Fig. 8-35a).

FIG. 8-39. Bale tongs hoisting three cotton bales at one time. Courtesy of New Orleans Board of Harbor Commissioners.

FIG. 8-40. One type of newsprint sling. Courtesy of Port of New York Authority.

FIG. 8-41. The pie plate used with a fiber rope net.

FIG. 8-42. One type of automobile sling. Courtesy of Moore McCormack S.S. Co.

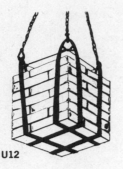

U12

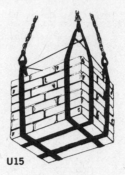

U15

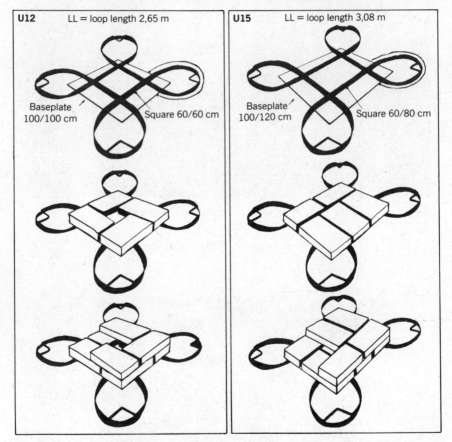

| U12 | LL = loop length 2,65 m |
| U15 | LL = loop length 3,08 m |

U12: Baseplate 100/100 cm — Square 60/60 cm

U15: Baseplate 100/120 cm — Square 60/80 cm

FIG. 8-43. The Unisling. Span Set Unislings are made of synthetic fiber—polyester yarn sealed between two layers of plastic foil. The sling is endless, without a seam, and less dependent on specially shaped lifting hooks. (Note: For woven webbing slings, special hooks are to be used in order to keep the sling flat.) The base plate is welded to the sling and not stitched or clamped. Courtesy of Span Set® Unisling.

FIG. 8-44. The Marino Sling around bags of coffee ready for discharge.

Truck and Auto Slings (See Fig. 8-42)

There are a number of designs of slings for handling automobiles that are not crated. Improvised slings generally cause damage of some type to any car hoisted by them. A specially constructed sling with ample spread between the vertical parts should always be used.

Special Slings

On many pieces of machinery and certain odd-shaped units of considerable bulkiness and/or weight, slings will be found attached when the cargo arrives at the pier. Great care must be taken not to allow such a sling to become lost or damaged during the voyage. If it is removed, note carefully how it was attached.

The Unisling (See Fig. 8-43)

The Unisling unitizes bags and cartons and is designed for one-way use from the shipper to the consignee. It is made of synthetic fiber, namely, a

polyester endless yarn sealed between two layers of polyethylene or plas-
ticized paper. The base plate protects the cargo and ensures quick and
easy stacking. The broad, flat bands guarantee a wide bearing area, reducing the risk of cutting bags. Once loaded aboard a vessel this sling is
unhooked and left intact about the cargo (see Fig. 8-44). It thus facilitates
unloading at the port of discharge.

CHAPTER **9**

PROPER CARE FOR
THE CARGO

(The Ventilation of Cargo Holds)

ELEMENTS OF THE PROBLEM

The body of knowledge covered by this chapter would probably be more accurately described as the air conditioning of cargo holds; however, through common usage on ships through the years, the term *ventilation* is now used to indicate the steps taken to prevent damage to cargoes from condensed moisture within the cargo holds. One good reason for this is because until about 1938 the only *technique* used was ventilation. The technique was *ventilating,* but *air conditioning* was being accomplished. The control offered by the choice of either ventilating or not ventilating is often not sufficient to prevent condensation under conditions encountered at sea. It was left for a shipper of products suffering from improper air conditioning at sea to recognize this need on ships first. Thus, Mr. Oliver D. Colvin, president of Cargocaire Engineering Corporation, was the first to take positive action toward giving ship's officers full control of the phenomena that cause condensation to appear on the ship's hull or cargo. His investigations into the reasons why his canned milk shipments suffered damage by rusting ultimately resulted in the development of a dehumidifier coupled with a forced air ventilation system.

The simple psychrometric principles and the exact methods used to minimize condensation are the topics of this chapter. Before proceeding with a discussion of these principles, the authors wish to comment upon some interesting factors bearing upon the subject matter of this chapter. It has been observed that in the past a number of ship's officers have failed to recognize the need for correct ventilating procedures. Part of the cause of this is their failure to make themselves cognizant of the simple physical principles that would enable them to fully understand the problem with which they are dealing. As a result, the general opinion among many officers with years of experience is that the best thing to do always is to see that the cargo holds are given the maximum amount of air possible. As will be seen later, under some conditions this is the exact opposite of what should be done.

A group of underwriters actually obtained statistical data that substantially proved that such misjudgment actually took place in a segment of the U.S. Merchant Marine. Before World War II, the damage to cargo on the intercoastal run from condensation was determined on a percentage basis. These data were obtained on ships equipped with cowl ventilators only. Following World War II, data were taken on ships with forced ventilation systems. These data showed that the incidence of damage due to condensation had actually increased, despite the improved ventilation system. The improved ventilation system, if used correctly, would have reduced damage, but to the amazement of all concerned the damage had increased.

The only answer to the question of how such a thing could happen is that the users of the improved means, because of their lack of knowledge, were now able to do more damage than previously.

Fundamental Objective of Ventilation

Ventilation has the single objective of preventing moisture damage to cargo that results from condensation within the cargo holds. Since the condensation comes from the air, it seems logical that before discussing the matter further we should learn something about the composition of air and about what other types of airborne damage there are besides moisture damage. Because air is everywhere and because of our intimate association with it, we all accept it as a highly commonplace substance and take for granted that we know all there is to know about it. As implied above, damage caused by condensation is only one of several forms of airborne damage. Components of air other than water vapor also may cause damage to cargo. These will also be dealt with briefly in this chapter; however, it must be emphasized that the only effective and practicable method of combating moisture damage is through air conditioning by whatever means is at our disposal. Other airborne damage types must be eliminated by other techniques, not because air conditioning would not be effective, but because it would not be practicable.

Composition of Air

The principal components of air and their amounts are as follows: nitrogen, 78.03%; oxygen, 20.99%; argon, 0.93%; carbon dioxide, 0.03%; neon, helium, krypton, hydrogen, xenon, ozone, and radon, 0.02%. These amounts are for the total air of the Earth's atmosphere. It must be pointed out that air near the Earth's surface contains some additional substances and that all of these amounts will vary to some degree. The additional substances are dust, water vapor, and under certain conditions accidental components. In a ship's hold, these accidental components are vapors from commodities stowed in the closed space.

Sources of Airborne Damage to Cargo

Nitrogen acts simply as a diluter. It is a very stable gas and does not readily react with other substances so as to produce any problem in the protection of the ship or cargo. Argon, carbon dioxide, and the remaining gases either do not appear in the air at the surface of the Earth or appear only as traces and consequently offer no problem.

Oxygen is a powerful chemical agent and readily combines with other substances. It is well known that oxygen is necessary for combustion; however, within cargo holds it causes more subtle damage through the process of corrosion. It is important to note that moisture must be present also in order for corrosion to take place. A highly practical corrosion preventative measure would be to remove the moisture from the air in areas where it is possible to do so.

Dust exists in all air to some extent. Dust particles are so small that they float through the air for some time before falling to rest. Dust particles average about 0.5 microns (2×10^{-5} in.) in diameter. Excessive dust can cause considerable damage to cargo. This is an obvious fact and one of the first recognized by students of cargo stowage. It should also be recognized that protection from this airborne damage is not possible by ventilation. Protection from dust damage is obtained primarily by careful planning of the stowage so that dusty cargoes are loaded first and discharged last or segregated from cargoes that may be damaged by dust. The classical example of a dusty cargo is Portland cement, but there are many others, such as sulfur, bauxite, coal, copper concentrates, guano, nitrates, and ores of all kinds. Protection from dust for cargoes stowed in the same compartment and at the same level can be obtained by erecting a dusttight bulkhead. This is ordinarily necessary only when a dusty cargo must be handled while delicate goods are stowed in the vessel. The damage caused by dust is known as *contamination*.

Accidental components of air in the hold are the fumes or vapors from liquids, gases, or solids stowed therein. The damage caused by this means is known as *tainting* damage. The vapors of turpentine can cause tea to taste of turpentine. Rubber can give a heavy, pungent, and characteristic odor to silk goods stowed in the same hold. Tainting of this sort will result in damage claims. Protection against tainting damage can be obtained only by segregation, and usually the segregation should provide a space for at least one cargo hold between a highly odorous and a very delicate cargo. Temporary vapor-tight bulkheads are not practicable in the way that temporary dust-tight bulkheads are.

Of all the components of air, water vapor is the most variable in amount, from only a trace in the air over a desert far from any large body of water to as much as 4% by volume under extremely humid conditions. It is contributed to the air by evaporation from water surfaces, soil, and living tissues. Water vapor may also be contributed to air by transfer from hygroscopic substances, and in fact this is the source of the water vapor that causes most of the trouble in a closed hatch. This transfer differs from transfer by evaporation; the controlling factors will be discussed later. The amount of water vapor that a given air sample can hold varies directly with the temperature of the air. We commonly speak of moist and dry air with the implied meaning that the atmosphere takes up vapor much as a sponge takes up water, but the fact is that the vapor occupies space without regard to the other gases. At any given temperature the same amount of vapor can be diffused through a vacuum as through an equal space occupied by air. In other words, the amount of vapor that can occupy a given space depends *entirely* on the temperature. This amount is approximately doubled with each increase of 20°F within the ordinary ranges experienced in the free air. Thus, if the temperature is raised from 0 to 80°F, the capacity of a given space for moisture is increased almost 16 times.[1]

TERMS OF PSYCHROMETRY

Relative Humidity

When air contains all the water it can hold at a given temperature, it is said to be *saturated*. When the dry bulb temperature of saturated air is decreased, the air will reject some moisture in the form of what is known as condensation. When the dry bulb temperature of saturated air is increased, the air will cease to be saturated and will immediately have the ability to hold more water vapor. The ratio of the amount of water vapor actually in the air to the amount that would be in the air if the air were saturated is known as the *relative humidity*.

Specific Humidity

Specific humidity is defined as the weight of water vapor per unit of weight of dry air. In engineering work it is generally given as grains of water per pound of dry air or pounds of water vapor per pound of dry air.

Absolute Humidity

The amount of water in air may be expressed in terms of units of weight of water vapor per unit of volume of dry air. *Absolute humidity* is the term used to describe this unit. In engineering work, absolute humidity is expressed as pounds of water vapor per cubic foot of dry air or grains of

[1] U.S. Department of Commerce, Weather Bureau, *Weather Forecasting*, U.S. Government Printing Office, Washington, D.C., 1952, p. 10.

water vapor per cubic foot of dry air. The psychrometric chart (Fig. 9-1) may be used to determine absolute humidity by first obtaining the specific humidity and then dividing by the volume per pound of dry air. This latter value is also found on the psychrometric chart.

Dew Point

Dew point is defined as that temperature below which the air will be unable to retain the moisture it presently contains. The dew point of any given air sample is entirely dependent on the absolute humidity. The reader should note that under constant barometric pressure, the dew point also varies directly with the specific humidity. This last relationship is reflected in the way that you read the dew point and specific humidity from the psychrometric chart, which is constructed assuming a standard barometric pressure of 29.92 in. of mercury.

Wet Bulb Temperature

Heat causes water to acquire the properties of a gas. When applied to a water surface, heating results in part of the water turning into vapor. The heat that is expended in this process, which is called evaporation, is rendered latent or imperceptible so far as temperature effects are concerned until the vapor is again condensed into visible water particles. When this happens, the heat reappears. This explains why evaporation from a moist surface has a cooling effect, and it logically follows that the more rapid the rate of evaporation, the faster the heat will become latent and the greater the cooling of the surface from which the heat is drawn will be. The wet and dry bulb hygrometer measures the relative humidity based on the principle that the rate of evaporation is directly proportional to the relative humidity of the surrounding air.

To most seamen the wet and dry bulb hygrometer is a familiar instrument; however, a few comments concerning it are in order. It consists of two thermometers mounted close together. The bulb of one is covered with a wick that is dampened just before reading it or kept constantly dampened if mounted on a bulkhead in a permanent location. The wet bulb thermometer will register the dry bulb temperature reduced by the cooling effect of the rate of evaporation. The rate of evaporation, of course, depends on the relative humidity. This is the most accurate method in use for arriving at the relative humidity and dew point of the air, providing the wick is clean, there are no artificial heat sources affecting the thermometers, and there is a light flow of air over the wick to wipe away the high-humidity air film that will develop around the bulb in dead air. If any one of these requisites is violated, the relative humidity and dew point read will be in error. In the absence of any air flow, the wet bulb reading will be too high; the reader should be able to see quite readily the direction of the error in the resulting relative humidity in such a case.

Wet Bulb Depression

The wet bulb temperature will always be either equal to or less than the dry bulb reading. The difference between the wet bulb and dry bulb reading is called the *wet bulb depression*. The wet bulb depression obviously will vary directly with the relative humidity. It will be zero when the air is saturated. The relative humidity may be read from Table 9-1 if data from a wet and dry bulb hygrometer are employed.

Vapor Pressure and Saturated Vapor Pressure

When water enters the air by evaporation, it acts as another gas in the mixture of gases that composes the air. This water vapor exerts a pressure in all directions that is independent of the pressure of all the other gases in the mixture. This is known as the *vapor pressure* of the air, and varies directly with the amount of vapor per unit of volume of dry air, that is, the absolute humidity. This pressure is commonly expressed in inches of mercury and its value contributes to the total length of the barometric column. The maximum vapor pressure under conditions of constant temperature and pressure occurs when the air is saturated. Every change in air temperature or pressure changes the *saturated vapor pressure*.

The table in the upper left-hand corner of the psychrometric chart (Fig. 9-1) gives the saturated vapor pressures for temperatures between 20 and 84°F at a barometric pressure of 29.92. Note that the temperature column headed "Wet Bulb Temp." can be entered with the dew point or dry bulb if the air is saturated, because they are all equal under this condition.

The vapor pressure of air depends on the amount of water vapor per unit of volume of air, but so do the absolute humidity and the dew point. It is important to note and remember that the vapor pressure is an indirect

TABLE 9-1

RELATIVE HUMIDITY TABLES[a]

Difference in Dry and Wet Bulb Readings (°F)

Dry Bulb Reading (°F)	0	1	2	3	4	5	6	7	8	9	10	11	12
30	100	89	78	67	57	47	36	26	17	7			
35	100	91	82	73	65	54	45	37	28	19	12	3	
40	100	92	84	76	68	60	53	45	38	30	22	16	8
45	100	92	85	78	71	64	58	51	44	38	32	25	19
50	100	93	87	80	74	67	61	55	50	44	38	33	27
55	100	94	88	82	76	70	65	59	54	49	43	39	34
60	100	94	89	84	78	73	68	63	58	53	48	44	39
65	100	95	90	85	80	75	70	65	61	56	52	48	44
70	100	95	90	86	81	77	72	68	64	60	55	52	48
75	100	95	91	87	82	78	74	70	66	62	58	55	51
80	100	96	92	87	83	79	75	72	68	64	61	57	54
85	100	96	92	88	84	80	77	73	70	66	63	60	56
90	100	96	92	88	85	81	78	75	71	68	65	62	59
95	100	96	93	89	86	82	79	76	72	69	66	63	60
100	100	97	93	90	86	83	80	77	74	71	68	65	62

Difference in Dry and Wet Bulb Readings (°F)

Dry Bulb Reading (°F)	13	14	15	16	17	18	19	20	21	22	23	24
30												
35												
40	1											
45	13	7	1									
50	22	16	11	6	1							
55	29	24	19	16	10	6	1					
60	34	30	26	22	18	14	10	6	2			
65	39	35	31	28	24	20	17	13	10	6	3	
70	44	40	36	33	29	26	23	19	16	13	10	7
75	47	44	40	37	34	31	27	24	21	19	16	13
80	51	47	44	41	38	35	32	29	26	23	20	18
85	53	50	47	44	41	38	36	33	30	28	25	22
90	56	53	50	47	44	41	39	36	34	32	29	26
95	58	55	52	49	47	44	42	39	37	35	32	30
100	59	57	54	51	49	47	44	42	39	37	35	33

[a] Relative humidity readings taken as percentage. *Example:* What is the relative humidity of air with a dry bulb temperature of 70 and a wet bulb temperature of 60? Answer: 55%.

measure of the absolute humidity, specific humidity, and dew point. This fact is of importance when studying the basis of the moisture equilibrium chart (see Figures 9-3 and 9-14).

Equations for Calculation of Relative, Specific, and Absolute Humidity

Relative humidity can be expressed as a ratio of two masses or two pressures:

$$RH = \frac{A}{A_s} = \frac{e}{e_s}$$

where A = existing absolute humidity.
$\quad A_s$ = absolute humidity if saturated.
$\quad e$ = vapor pressure.
$\quad e_s$ = saturated vapor pressure.

Specific humidity is obviously a ratio of the weight of water vapor in the air to the total weight of the air expressed in the appropriate units. It is also the ratio of the vapor pressure to the total barometric pressure multiplied by a constant. The value of the constant depends on the units used for expressing the specific humidity. The constant is 4500 if the specific humidity is expressed in units of grains per pound.

$$W = \frac{\text{Weight of vapor}}{\text{Weight of air}} \qquad W = K\frac{e}{p}$$

where W = specific humidity.
$\quad e$ = vapor pressure.
$\quad p$ = total air pressure.
$\quad K$ = constant (value depends on units of W).

Absolute humidity is equal to the ratio of the weight of water vapor in the air to the volume of the air sample being considered. Inasmuch as we always obtain the specific humidity from a psychrometric chart based on a pound of dry air, we can find the absolute humidity by dividing the specific humidity by the volume of one pound of dry air.

$$A = \frac{\text{Weight of vapor}}{\text{Volume of air}} \qquad A = \frac{Ke}{pv}$$

where v = volume of one pound of dry air and the other symbols are the same as those in the equation for specific humidity.

The density of dry air is greater than the density of air mixed with water vapor. This fact sometimes comes as a mild surprise to the reader who is accustomed to considering moisture as always adding weight to the dry form of the substance. But water vapor, which acts as just another gas in the total mixture of gases in the air, is actually lighter than the mixture in its dry state; hence, any addition of moisture makes the total mixture have less mass per unit of volume.

HOW TO USE THE PSYCHROMETRIC CHART

The psychrometric chart, Fig. 9-1, is a development of the Carrier Corporation. Whereas tables may be used to solve problems involving psychrometric data, the use of a chart makes the process easier and faster. Inasmuch as it is necessary to have a thorough understanding of the amounts of moisture involved in changing the various psychrometric values of air, it is desirable for the reader to become familiar with the use of this psychrometric chart and the notations that appear on it.

Definitions, Abbreviations, Symbols

DRY BULB TEMPERATURE: *DB*. Temperature of air as registered by an ordinary thermometer.

WET BULB TEMPERATURE: *WB*. Temperature registered by a thermometer whose bulb is covered by a wetted wick and exposed to a current of rapidly moving air.

DEW POINT TEMPERATURE: *DP*. Temperature at which condensation of moisture begins when the air is cooled.

RELATIVE HUMIDITY: *%RH*. Ratio of actual water vapor pressure in air to the pressure of saturated water vapor in air at the same temperature, expressed as a percentage.

SPECIFIC HUMIDITY: *W*. Moisture content of air in terms of weight of water vapor in grains or pounds per pound of dry air.

ABSOLUTE HUMIDITY: *A*. Moisture content of air in terms of weight of water vapor in grains or pounds per cubic foot of air.

ENTHALPY: *h*. Total heat. A thermal property indicating the quantity of heat in the air above an arbitrary datum, in Btu per pound of dry air.

VAPOR PRESSURE: *e*. The pressure exerted by the water vapor contained in the air, in inches of mercury.

VOLUME (as used in psychrometrics):   *v*. Cubic feet of the mixture per pound of dry air.

POUNDS OF DRY AIR: The basis for calculations, this value remains constant during all psychrometric processes.

d = enthalpy deviation, Btu per pound of dry air.

h'_s = enthalpy of air saturated at the wet bulb temperature.

q = heat added or removed in process, Btu per pound of dry air.

m = weight of moisture added to or removed from air, grains or pounds per pound of dry air.

Subscripts 1, 2, and so forth indicate the entering and progressive states in the process.

The dry bulb, wet bulb, and dew point temperatures and the relative humidity are related in such a fashion that if two of these properties are known, all other properties shown in Fig. 9-2 may be read from the chart. When air is saturated, dry bulb, wet bulb, and dew point temperatures are identical. The enthalpy of air for any given condition is the enthalpy at saturation corrected by the enthalpy deviation due to the air not being in a saturated state. Our only interest in the enthalpy value as found on the

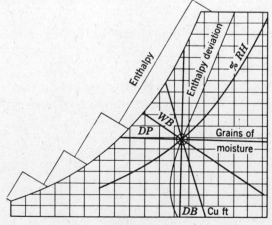

FIG. 9-2.

chart is that it is needed to investigate the approximate heat load involved when renewing the air in a refrigeration system requiring the addition of fresh air from time to time. Not being burdened with the necessity of making precise calculations along these lines, we will not concern ourselves with the enthalpy of added or rejected moisture given on the chart. The sensible heat factor also appearing on the chart will not be used in any of our calculations, and therefore an explanation of this value will be omitted.

NUMERICAL EXAMPLES

EXAMPLE 1

Reading the properties of air.

Given: DB = 70°F. *WB* = 60°F.

Required: %RH, DP, volume (*v*), specific humidity (*W*), absolute humidity (*A*), and enthalpy (*h*).

Solution: Locate point of intersection on the chart of the vertical line representing 70° *DB* and the oblique line representing 60° *WB*. All values are read from this point of intersection.

Interpolate between relative humidity lines on 70°F *DB* line; read *RH* = 56%.

Follow horizontal line left to saturation curve; read *DP* = 53.6°F.

Interpolate between lines representing cubic feet per pound of dry air, read *v* = 13.53 ft^3.

Follow horizontal line to right; read the specific humidity, *W* = 61.4 grains/lb.

Divide 61.4 by 13.53 to obtain absolute humidity of 4.54 grains/ft^3.

Follow wet bulb line to "Enthalpy at saturation" scale and read h_s' = 26.46 Btu. Read enthalpy deviation for point of intersection; *d* = −0.07 Btu. Then the enthalpy of air at the given conditions is $h = h_s' + d$ = 26.46 − 0.07 = 26.39 Btu/lb of dry air.

EXAMPLE 2

Reading the properties of air.

Given: W = 100 grains/lb. *RH* = 60%.

Required: Volume (*v*), absolute humidity (*A*), *DP, DB, WB,* and enthalpy (*h*).

Solution: Locate point of intersection on the chart of horizontal line representing 100 grains/lb and 60% *RH* line. All values are read from this point.

Interpolate between lines representing cubic feet per pound of dry air; read $v = 13.98$ ft^3.

Divide 100 by 13.98 to obtain absolute humidity (A) of 7.15 grains/ft^3.

Follow vertical line downward to dry bulb temperature scale; read $DB = 82.4°$F.

Follow horizontal line to left to saturation curve; read $DP = 67.1°$F.

Follow wet bulb line to the saturation curve; read $WB = 71.8°$F.

Continue out on wet bulb line to "Enthalpy at saturation" scale and read $h'_s = 35.6$ Btu. Read enthalpy deviation for point of intersection; $d = -0.10$ Btu. Enthalpy of air at the given conditions is $h = h'_s + d = 35.60 - 0.1 = 35.5$ Btu.

EXAMPLE 3

(*a*) Calculating the cooling load q based on a pound of dry air entering.

Given: Initial air: $DB = 85°$F; $WB = 77°$F. Final air: $DB = 50°$F, $WB = 50°$F.

Required: Number of Btu removed from the incoming air.

Solution: Read $h_{s1} = 40.58$, $h'_{s2} = 20.30$; $d_1 = -0.09$ Btu, $d_2 = 0$; $h_1 = 40.58 - 0.09 = 40.49$, $h_2 = 20.30$.

$$q = h_2 - h_1 = 20.30 - 40.49 = -20.19 \text{ Btu/lb}$$

(*b*) Calculating the amount of moisture rejected from the incoming air based on a pound of dry air entering.

Read $W_1 = 127.9$ grains/lb, $W_2 = 53.5$ grains/lb.

$$m = W_2 - W_1 = 53.5 - 127.9 = -74.4 \text{ grains/lb}$$

(*c*) Calculating the cooling load based on the total amount of air entering a recirculated air stream.

Given: Initial and final air as given under (*a*). Outside air added at the rate of 1000 cfm.

Solution: Divide 1000 by the volume of 1 lb of entering air, $v = 14.12$ ft^3/lb. $1000/14.12 = 70.9$ lb of dry air. Cooling load,

$$h = 70.9 \times 20.19 = 1431.47 \text{ Btu}$$

EXAMPLE 4

Calculating total amount of moisture involved in changing DP of a given amount of air.[2]

Given: $DB_1 = 90°F$, $DB_2 = 90°F$; $DP_1 = 80°F$, $DP_2 = 35°F$; 100,000 ft^3 of air involved.

Required: The number of pounds of moisture that was removed from the air to cause the change in DP.

Solution: Read $W_1 = 156$ grains/lb of dry air. $W_2 = 30$ grains/lb of dry air. $m = W_2 - W_1 = 30 - 156 = -126$ grains/lb of dry air. $v_1 = 14.3$ ft^3/lb.

Divide 100,000 by 14.3; number of pounds of dry air in state 1 = 7000 lb.

Hence the number of grains rejected amounts to 7000 × 126 or 882,000 grains. But, since 7000 grains equals 1 lb, the amount in pounds is 882,000/7000 = 126 lb.

EXAMPLE 5

Calculating the effect on the DP of air when water vapor is removed from the air.

Note: In solving this particular problem, we will assume a constant DB. When water vapor is absorbed or released by a substance there is also a transfer of heat; hence our assumption of a constant DB will produce some error in our final results. The amount of error will be negligible for our purposes. In this problem we will also assume a constant volume of dry air, which will be the volume for state 1. To eliminate these assumptions would complicate the problem and serve no useful purpose in our study of the subject.

Given: 100,000 ft^3 of air with $DP = 75°F$, $DB = 80°F$.

Required: The new dew point if 100 lb (700,000 grains) of water vapor is absorbed from this body of air.

Solution: Read $W_1 = 132$ grains/lb of dry air. $v_1 = 14.0$ ft^3/lb of dry air.

Hence: 100,000/14.0 = 7092 lb of dry air are being considered. 7092 × 132 = 936,144 grains of moisture in the air before absorption commenced, 936,144 less 700,000 equals 236,144 grains of moisture

[2] Problems of this type must be solved later to understand thoroughly why the condition of a commodity in a nonventilated hatch controls the storage atmosphere's DP.

remaining after absorption ended. 236,144 grains in 7092 lb of dry air is equal to (236,144/7092) grains/lb or 33.3 grains/lb.

Therefore the new *DP* is read on the horizontal line representing a specific humidity of 33.3 grains/lb as 37.5°F.

HYGROSCOPIC MOISTURE TRANSFER

"Hygroscopic" Defined

The word *hygroscopic* is an adjective implying the ability to absorb moisture in the form of a gas. Hence, hygroscopic moisture is moisture absorbed by a substance capable of absorbing water vapor in the gaseous form. Not all substances are capable of this. A substance capable of absorbing hygroscopic moisture is known as a hygroscopic substance. Hygroscopic substances include all substances of an organic nature, such as all grains, wood and wood products, cotton, wool, sisal, jute, paper, sugar, and other products of animal or vegetable origin. Examples of nonhygroscopic substances are all metals and glass products.

Factors Controlling Hygroscopic Moisture Transfer

Hygroscopic moisture leaves the hygroscopic substance and enters the ambient air in a manner that is similar but not identical to the transfer of free moisture to the air. Moisture in the liquid state will leave the parent body and enter the air as a gas so long as the ambient air is not saturated. The speed of evaporation depends on the relative humidity of the ambient air and the temperature of the water. Evaporation will continue even if the water temperature is lowered to the freezing point or below. When water vapor enters the air from ice without passing through the liquid state, the process is known as sublimation rather than evaporation. The important point is that either process, evaporation or sublimation, will continue if the ambient air has a *RH* of less than 100%.

Hygroscopic moisture will leave its parent body if the vapor pressure within the substance exceeds that of the ambient air. This warrants some discussion. The hygroscopic moisture within a hygroscopic substance will possess a vapor pressure somewhat similar to the vapor pressure at the surface of liquid water or on the surface of a block of ice. For any given substance this vapor pressure varies as the *moisture content* of the substance and the *temperature* of the substance. A given combination of moisture content and temperature will produce different vapor pressures in different commodities. It has already been pointed out that the water vapor in the air also has a vapor pressure, which is dependent upon one thing only, namely, the moisture content of the air. Recall that the dew point is also dependent on only this one thing.

It was said above that the hygroscopic moisture will leave the substance if the vapor pressure of the ambient air is less than the vapor pressure of the substance. It follows that if the two vapor pressures are equal there will be no transfer of water vapor. Under these conditions the air and substance are said to be in *moisture equilibrium*. Although it may not seem logical, nevertheless it is true that if the vapor pressure of the air is greater than that of the substance, the water vapor will "flow" back into the substance.

Significance Within the Ship's Hold

If a hygroscopic commodity in a ship's hold has a moisture content and temperature such that the resulting vapor pressure is greater than that of the air in the hold, water vapor will leave the commodity and enter the air until the vapor pressure of the air builds up to the vapor pressure of the commodity, the vapor pressure of the commodity falls to that of the air, or the vapor pressure of thè air will rise and that of the commodity will fall until they meet at some common midpoint. If we constantly ventilate the hold with this low-vapor-pressure air, we rule out the first possibility; the air passing through the hold will continue to remove water vapor from the commodity. The student is apt to jump to the conclusion that the latter procedure will, in fact, lower the vapor pressure of the commodity to that of the air. Although it is true that the commodity's vapor pressure will be lowered, it will not be lowered an appreciable amount unless the process is continued much longer than the longest passage made by any modern ship or air with a much lower vapor pressure than that found at sea is used.

On the other hand, if the hatch is not ventilated for any reason, the possibility of the vapor pressure of the air building up to that of the commodity does exist; in fact, this is what will happen under such conditions. It must be pointed out that under no circumstances will the vapor pressure of the commodity fall to meet that of the air. This can be demonstrated clearly by a numerical example that will be given later.

THE MOISTURE EQUILIBRIUM CHART

Use of Moisture Equilibrium Charts

The moisture equilibrium chart, which is illustrated by Figs. 9-3 and 9-4, can be used to determine the dew point temperature that the air surrounding a hygroscopic commodity will have when in moisture equilibrium with that commodity.

Recall that for any given moisture content of the air the water vapor in the air will have a certain vapor pressure; also recall that for every vapor pressure value there is a corresponding dew point value. Inasmuch as there is a one-to-one correspondence, if we choose any vapor pressure x_1 there will be a corresponding dew point value y_1 and a corresponding specific humidity z_1. Hence, if we know any one of these values, we always can find the other two, providing, of course, that we have suitable tables or a suitable psychrometric chart.

It was noted above that all hygroscopic commodities have vapor pressures depending on the commodity's *moisture content* and *temperature*. (Moisture content of a commodity is expressed as a percentage by weight. If the moisture content of lumber is given as 25%, it means that for every 100 lb of lumber there are 75 lb of dry wood fiber and 25 lb of water in the form of hygroscopic moisture.) The vapor pressure comparison is the criterion used to tell whether hygroscopic moisture is being absorbed by the commodity or by the air or if the air and the commodity are in moisture equilibrium. But vapor pressure can be stated in terms of dew point temperatures. The moisture equilibrium charts give the vapor pressure of the commodity in terms of the dew point temperature that air with an equal vapor pressure would possess. This is done because we are interested in knowing final dew points rather than vapor pressures; that is to say, the resulting dew point has more meaning than the resulting vapor pressure.

The charts can be entered with the air's relative humidity and dry bulb temperature to obtain the dew point of the air. If the *dew point corresponding to the vapor pressure in the commodity* is equal to the air's actual dew point, moisture equilibrium will exist. Note the expression, "dew point corresponding to the vapor pressure in the commodity." Hereafter, in this book, this long and somewhat awkward expression will be replaced by the expression, *commodity's dew point*. It should be clear, however, that the commodity itself cannot have a dew point as defined earlier.

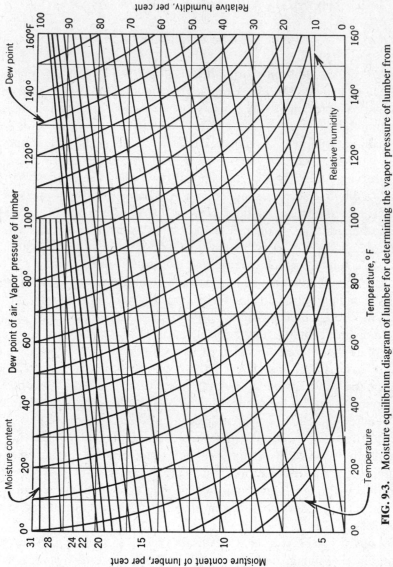

FIG. 9.3. Moisture equilibrium diagram of lumber for determining the vapor pressure of lumber from its moisture content and temperature. Courtesy of Cargocaire Engineering Corp.

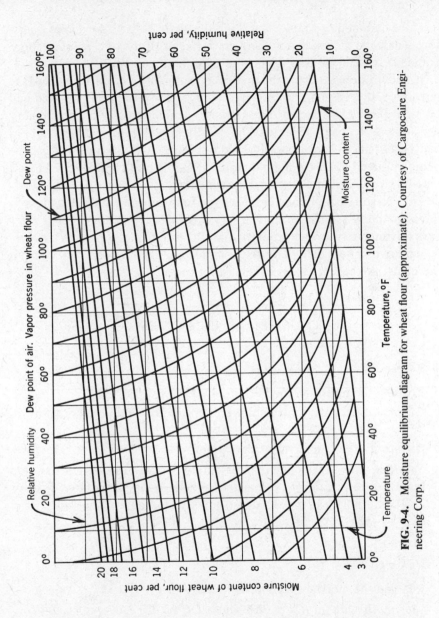

FIG. 9-4. Moisture equilibrium diagram for wheat flour (approximate). Courtesy of Cargocaire Engineering Corp.

If the commodity's dew point is greater than the air's dew point, moisture will leave the commodity and enter the air. If the dew point relationship is reversed, the flow of moisture will be reversed.

It should now be apparent that if a commodity with a high moisture content is stowed in a nonventilated hold, the dew point in that hold will be high. If the exact moisture content and temperature of the commodity are known, the exact level that the dew point will reach within the hold can be obtained from a moisture equilibrium chart. Before this can be demonstrated by means of a numerical example, the method of reading a moisture equilibrium chart must be explained.

The moisture equilibrium chart has the following data recorded upon it: the commodity's moisture content, dry bulb temperature, and vapor pressure (given in terms of dew point); and the air's *RH, DB,* and *DP.* Entering any two known factors for the commodity, the third can be read from the chart; the same is true in the case of the air.

NUMERICAL EXAMPLES

EXAMPLE 1

Reading the moisture equilibrium chart.

Given: Lumber with a moisture content of 20% and a temperature of 70°F.

Required: The commodity's dew point.

Solution: Locate the point of intersection between the vertical dry bulb temperature line for 70°F and the oblique moisture content line for 20%. Interpolate between curved lines to the dew point scale. Read 66°F as the commodity's dew point.

EXAMPLE 2

Reading the moisture equilibrium chart.

Given: Air with an *RH* of 70% and a *DB* of 70°F.

Required: The *DP* of the air.

Solution: Locate the point of intersection between the *DB* temperature line for 70°F and the horizontal line for a *RH* of 70%. Follow a curved line to the dew point scale. Read 60°F as the air's *DP.*

EXAMPLE 3

Estimating the reaction between cargo and air.

Given: The data from Examples 1 and 2.

Required: What will the dew point be in an unventilated hold loaded with the lumber of Example 1 if when secured the air's dew point was as given in Example 2?

Solution: From the discussion about hygroscopic moisture transfer, it should be clear that the hatch will eventually have a dew point of 66°F. In other words, the hatch would have a dew point precisely as dictated by the commodity.

The proof of the truth of the last statement is demonstrated in the following section.

CONTROL OF STORAGE ATMOSPHERE BY THE COMMODITY

Closed Container Hypothesis

In view of the laws controlling the transfer of hygroscopic moisture and the factors controlling vapor pressures in air and in hygroscopic materials, this hypothesis can be made: The characteristics of the air in a closed container are controlled by the temperature and moisture content of a hygroscopic material within the container if the weight of the material is *much greater* than the weight of the air. We show below that this hypothesis is true, using a hypothetical example, and comment upon the significance of such facts for the proper care and custody of cargo on a ship.

The *closed container* we will have in mind is a nonventilated, well-secured ship's hold. That the weight of the material is very much greater than the weight of the air involved will be quite obvious on comparing them in our hypothetical case or under actual conditions on board ship.

Assume: A hold of 80,000 grain cubic. Sixty thousand (60,000) cubic feet is consumed in stowing 600 tons of jute. The jute has a temperature of 80°F and a moisture content of 10%. When the hold is secured, the air has a *DB* of 80°F and *RH* of 90%.

To illustrate the reaction between the air and the jute and arrive at a proof, we must accept certain basic laws of psychrometry. These laws are as follows:

1. Water vapor will always flow from an area of high vapor pressure to an area of low vapor pressure.
2. Water vapor pressure in air depends on moisture content.
3. Vapor pressure in a hygroscopic commodity depends on two factors, namely, (1) the moisture content and (2) the temperature of the commodity.
4. Vapor pressure, dew point temperature, and specific humidity all vary directly with one another and have corresponding values.

To prove the hypothesis we must resort to a quantitative analysis of a given problem. We will use the assumed data given above to illustrate the reaction between the jute and the air. Note that the assumed data are typical of arrangements on a ship.

Proof: On the moisture equilibrium chart, locate the point of intersection between the temperature line for 80°F and the moisture content line for 10%. Read the commodity's dew point as 62°F.

On the psychrometric chart locate the point of intersection between the 90% relative humidity line and the vertical *DB* temperature line for 80°F. Read the following data from this point of intersection: $DP_1 = 76.8°F$; $W_1 = 140$ grains/lb; $v_1 = 14.05$ ft³/lb of dry air.

Laws I and IV above tell us that under the initial conditions, water vapor will flow from the air into the jute. This initial reaction will continue until one of three possible things occurs:

1. The vapor pressure of the air has fallen to the vapor pressure of the commodity.
2. The vapor pressure of the commodity has risen to the vapor pressure of the air.
3. The vapor pressure of the commodity has risen and that of the air has fallen so that they come to equilibrium at some intermediate level between their initial vapor pressures.

In effect, the hypothesis says that possibility (1) will take place. If a quantitative study of the reaction proves this to be true, then we will have proved the hypothesis.

Our quantitative study amounts to a comparison of the quantities of moisture involved. Initially the jute contains $(600 \times 0.1) = 60$ tons of hygroscopic moisture. Sixty tons is equal to 94.08×10^7 grains. The air contains 140 grains/lb of dry air, and there are $(20,000/14.05) = 1423$ lb of dry air in the hold. The total moisture in the air of the hold amounts to $(1423 \times 140) = 199,220$ grains.

We have already noted that initially some of the water vapor will leave the air and enter the commodity. The question we must now answer is how much? The only thing that can stop the flow of moisture is the attainment of equal vapor pressures by the air and the commodity. It is obvious that due to the enormous difference between their moisture contents, even if all the water in the air left it and entered the commodity, the latter's moisture content would be practically unaffected on a percentage basis. This means that its vapor pressure would also be practically unaffected. It is evident then that possibility (1) of the three mentioned above, will take place.

To make this more evident, let us examine the final conditions of the air and the commodity assuming the air is brought to a dew point of 62°F.

At a *DP* of 62°F and *DB* of 80°F we read the following: $W_2 = 83$ grains/ lb of dry air. Assuming the same amount of dry air, we find that moisture equilibrium will be reached when the total amount of moisture in the air is $(83 \times 1423) = 118,109$ grains. Thus it is seen that the amount of moisture that would have to leave the air to make the dew point fall from 76.8 to 62°F is $(199,220 - 118,109) = 81,111$ grains.

This 81,111 grains would, of course, be absorbed by the jute. The factors affecting the jute's vapor pressure would not be appreciably changed by the absorption of 81,111 grains. The jute already contains 940,800,000 grains at 10% moisture content. Adding 81,111 grains would change the percentage only slightly, to 10.0007%.

Under the final conditions in the hold, a commodity's vapor pressure will be the same as at the beginning of the reaction, but the DP of the air will have been lowered from 76.8°F to 62°F.

Significance of the Hypothesis

The significance of the above fact for proper care and custody of cargo is that it offers a complete explanation of the following items to the ship's officers:

1. Why it is possible to restrict all ventilation without danger of moisture damage if hygroscopic commodities are shipped sufficiently dry.
2. Why it is desirable that hygroscopic commodities have low moisture contents.

3. Why stowage of hygroscopic commodities adjacent to a warm bulkhead can cause high-dew-point storage conditions.
4. Why the upper 'tweendecks is the most likely place for heavy condensation.
5. Why *continuous* proper ventilation is necessary for dew point control.

The hypothesis is also significant because it offers an explanation of the causes of continuous heavy condensation and certain other types of moisture damage problems. Finally, it illustrates the extent of the control that the ship's officer has in meeting practical ventilation needs on a ship under operating conditions.

Determining Moisture Content

Much has been said above about the importance of the moisture content of hygroscopic commodities with reference to the problem of moisture damage in a ship's hold. We discuss briefly below the methods used to ascertain moisture content.

Weight Differences

One method for determining moisture content consists of weighing a known volume of the commodity. This sample of the commodity is then placed in an oven and heated to 212°F for about 3 hours. This should remove all the moisture from the sample. The sample is again weighed. The difference between the two weights is equal to the weight of the moisture in the original sample. To obtain the percentage moisture content, divide the weight of the moisture removed from the sample by the original weight of the sample and multiply the quotient by 100. Given as a mathematical equation, this is:

$$\frac{W_1 - W_2}{W_1} \times 100 = \%\text{M.C.}$$

where W_1 = original weight of commodity sample.
       W_2 = dry weight of commodity sample.

If the dry densities of the commodities are available, there is no need to dry the sample in the oven. Instead, after carefully weighing the sample, calculate its density and determine the moisture content from these data in a manner similar to that outlined above.

The Moisture Meter

Another system for measuring moisture content involves the use of a device that measures the resistance to an electric current as it passes through a sample of the material tested. A galvanometer registers the resistance, but instead of indicating voltage or amperage, the dial is calibrated to read percentage of moisture by weight. Since the scales have to be calibrated from empirical data for every commodity type, a different instrument is needed for every commodity. The principal differences between such instruments are the calibration of the scales and the method used to insert into the commodity the electric terminals over which the electric current passes. Figure 9-5 shows the electric moisture meter for the instantaneous determination of moisture content of lumber within the range of 7 to 24%.

Measuring Hygroscopic Transfer

A third method of measuring moisture content is based on the premise that the condition of the air in an enclosed space is controlled by the

FIG. 9-5. Electric moisture meter for instantaneous determination of moisture content of lumber within a limited range. In use, the needles in the handle are driven into the board as shown. The rheostat (selector) in the upper left-hand corner of the instrument is then adjusted until the needle on the milliammeter (upper right-hand corner) is vertical. The moisture content of the board is then read directly from the dial on the rheostat. A modification of this instrument is used for determining the moisture content of grain. Courtesy of Association of Marine Underwriters of British Columbia.

condition of any hygroscopic material contained in the space. We have already proven that this premise is reasonable.

This method requires a container several cubic feet in capacity capable of being hermetically sealed and fitted with a psychrometer. The psychrometer must be so fitted that measurements of the inside air can be taken after the air and commodity have reached moisture equilibrium without allowing any admixture with outside air. The important data obtained from this experiment are (1) the dew point of the inside air and (2) the dry bulb temperature of the commodity (same as that of the air). Since we know that the air's dew point results from the commodity's temperature and moisture content, we can use a moisture equilibrium chart to obtain the moisture content of the commodity.

NUMERICAL EXAMPLES

Obtaining moisture content by allowing moisture equilibrium to be reached.

Given: Some wheat grains are spread thinly on a screen within a hermetically sealed container. After 6 hours, a wet bulb thermometer on the inside of the container reads 65°F and a dry bulb thermometer reads 70°F. It is assumed that the wheat and air are in thermal equilibrium.

Required: Moisture content of the wheat.

Solution: On a psychrometric chart locate the point of intersection between the vertical dry bulb line and slanting wet bulb line. Follow horizontally to the left and read the dew point as 62.3°F.

On the moisture equilibrium chart, locate the intersection point of the curved dew point line and the vertical dry bulb line. Now read the grain moisture content on the oblique line to the left, 13%.

Of the three methods mentioned, the electric moisture meter is the most practical for use by ship or dock personnel interested in making spot checks on moisture contents. The principal advantage of this method is the speed with which a check can be made. The disadvantage is the reliance that must be placed on the correct operation of a delicate instrument. It is necessary to check and calibrate such an instrument frequently to make certain it is giving true readings.

Moisture Contents in Practice

Although it is evident that more moisture content information would in some instances contribute to better out-turns of shipments, there are numerous occasions wherein the shipowner receives no information concerning the moisture contents of hygroscopic cargoes as they are shipped. When the moisture content is known, the information may not be passed on to the officers of the ship or to any dock force personnel. In some cases, even if the information were passed on, its significance would not be understood.

The moisture contents of grain, seeds, nuts, and pulses are of considerable importance, and sometimes shippers of these products are required to certify the moisture content of such items. When the moisture content of grain[3] is sufficiently low, this item is almost completely dormant; it may be kept in sealed containers for years even if heated to abnormally high temperatures. For example, barley may be heated to approximately 200°F without losing its germinating power or undergoing any decomposition. This is a valuable property to have in grain, especially when loaded in a ship's hold, but it is lost when the grain is damp. As the moisture content rises in grain, it ceases being dormant and commences to respire. Respiration is a complicated series of biological and chemical changes influenced by conditions within and without the substance. These changes or processes are frequently referred to as germination, fermentation, turning rancid, sprouting, and so on. The end products of these processes are water vapor, carbon dioxide, and heat.

There is no clear and definite moisture content level below which grain may be considered dormant and above which it respires; however, all grain may be considered practically dormant at moisture contents below 10%. Under normal conditions of storage on land, where ambient temperatures change slowly, the following moisture contents are generally considered safe for several months' storage: soy beans, 16%; maize, 14.5%; wheat, 14%; rice, 12%; flax seed, 11%.[4]

---

[3] As used here and in the rest of this discussion, "grain" will include all types of grain, pulses, and seeds.

[4] Colvin, Hahne, and Colby, "Care of Cargo at Sea," *Transactions of the Society of Naval Architects and Marine Engineers*, 1938, p. 122.

With higher moisture contents than those stated, the rate of respiration increases rapidly. See Fig. 9-6 for the respiration rates of rice and sorghum grains at 100°F. This diagram uses the amount of carbon dioxide produced in 24 hours as an indication of the respiration rate. It also illustrates the difference between the respiration rates of white rice (also known as milled or polished rice) and rough rice (also known as brown rice), which explains the great difference between the stowage techniques used with the two types of rice.

Table 9-2 gives the moisture contents of a number of commodities as they are shipped in practice as determined by S. J. Duly of London. It should be interesting to the ship's officer that in most cases he receives these commodities with moisture contents that are at the limit of the safety level, or, as with Danubian maize, well above it. Hence, it is evident that anything that may cause an increase in moisture content during the voyage is likely to cause some damage to grain cargoes. A discussion of the effects of extreme heat or cold when carrying grain is presented later.

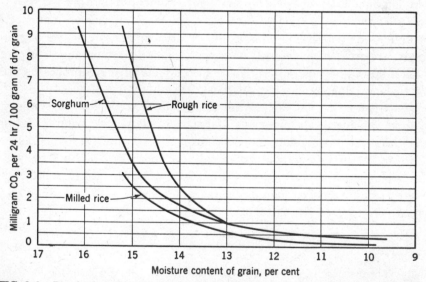

FIG. 9-6. Respiration of grain as a function of its moisture content. Temperature of grain, 100°F; length of test, 96 hours. Courtesy of Coleman *et al., U.S. Dept. of Agriculture Technical Bulletin* 100.

TABLE 9-2
AVERAGE MOISTURE CONTENTS OF COMMON HYGROSCOPIC MATERIALS[a]

Material	Moisture Content, Percent	Material	Moisture Content, Percent
Cacao	8–10	Textile fibers	
Cereals		Cotton	8–11.5
Maize (plate)	11–14.4	Jute	12–17
Maize (Danubian)	17–19.5	Silk	11
Rye (Polish)	17.19	Wool, raw	9.5–20
Wheat	10–14	Hides and skins	
Coffee	8.5–10.6	Dry	14–20
Dried fruits		Dry salted	14–20
Dried apricots	0.28	Wet salted	40–50
Prunes	24–28	Tobacco leaf	
Raisins	17–28	Usually	12–13
Sultanas	18–19	Range	9.5–17
Oil Seeds		Timber	
Copra	3.4–5.6	Kiln dried	11–16
Cottonseed	7.1–9.5	Air dried	16–24
Ground nuts	4.4–5.6	Unseasoned	over 30
Linseed	5.9–12.1	Wood Pulp	
Soya	7.9–12.1	Kraft (sulfate)	10–20
Tea	6–8	Bleached chemical	10–20
		Unbleached chemical	10
		Mechanical	50–55

[a] Information from S. J. Duly, "Condensation on Board Ship," *Journal of the Royal Society of Arts,* February 1938.

The Deterioration of a Maize Cargo[5]

During the 1936–1937 maritime strike on the west coast of the United States, a ship loaded with about 10,000 tons of River Plate maize was tied up in San Pedro, California. Because of the moisture content of the corn that is carried on this particular route, the cargo heats up rather slowly. If

[5] Colvin, Hahne, and Colby, "Care of Cargo at Sea," *Transactions of the Society of Naval Architects and Marine Engineers,* 1938, p. 129.

not unloaded within about 6 weeks, cleaned, and cooled, it will depreciate in value. In this case, the corn could not be unloaded and the temperature in the holds filled with grain continued to go up slowly. The owners of the cargo made tests in all hatches, and when the temperature approached 100°F they decided to take some action to prevent a further rise in temperature.

The method decided upon included the cooling of the bulk corn by use of dry ice. Two tons of dry ice (solid carbon dioxide) were spread over the surface of each of these hot hatches. The carbon dioxide gas that evaporated from the numerous 50-lb blocks was cool and heavy. It sank into the mass of corn, where it displaced some of the trapped air. The respiration of grain and the metabolism of mold germs require oxygen. Heat is produced by respiration and metabolism. Removal of the oxygen undoubtedly slowed down these processes, and the heating slowed down accordingly. However, this desirable result was only of minor benefit in the long run.

The dry ice did lower the temperature of a several-foot layer of grain on the surface. This cooling was noted during the first 2 or 3 days. The amount of cooling that dry ice can perform can be calculated from the following data:

1. Specific heat of dried corn: 0.28 Btu/lb/°F.
2. Latent heat of dry ice: 235 Btu/lb.
3. Final temperature of carbon dioxide gas: 80°F.
4. Temperature of dry ice: −110°F.
5. Specific heat of carbon dioxide: 0.21 Btu/lb.

From the above data it can be seen that 2 tons of dry ice is capable of absorbing about 1,200,000 Btu. With this information and the remaining data (item 1 above) it can be seen that approximately 80 tons of corn could be cooled 26°F, and a smaller amount could be cooled even more. Just how much corn was cooled or the exact degree of cooling cannot be stated, but it can be determined easily that in a hold about 60 ft by 60 ft in area, 80 tons of corn would make a layer about 1 ft deep. This gives us some idea of the effectiveness of the method chosen to cool the cargo. It should be remembered that the corn is continually generating additional heat by its respiration.

It was later observed that the cooling action of the dry ice was sufficient to chill only the surface of the grain cargo. The interior remained between 90 and 116°F. This means that the vapor pressure of the corn below the top layer was very much higher than that of the corn in the very cold top layer. From what has been said about hygroscopic moisture transfer, it can be seen that a good deal of moisture would diffuse toward the cold surface. However, the surface was so cold that no moisture could evaporate from the hot cargo.

When the grain was finally unloaded, the surface layer of the grain was dripping wet; it was estimated that as much as 100 tons of water was collected there. This layer was entirely destroyed because it had been spoiled through germation, molding, and discoloration. The cooling effect of the dry ice vanished rapidly below this surface layer, but the entire mass of grain was hot and moist and had materially depreciated in value.

From all evidence, it seems that if carbon dioxide had to be used, it should have been applied in the form of a warm gas, not as dry ice. The gas should have been injected near the bottom of the cargo, so that it could have displaced the air contained in the grain. The surface layer of the grain should have been ventilated with dry, or relatively dry, air which would have removed any moisture that might have come from the interior.

An experiment modeling the method just suggested was made in another Pacific Coast port at the same time. A powerful portable fan was placed in the open cargo hold and the surface of the grain was ventilated vigorously during the entire strike period of over 3 months. This particular corn cargo did not suffer the heavy damage incurred by the one mentioned above.

During the early part of the first case, the question of whether or not the damp and warm corn cargo should be dried came up. This would probably have been the logical thing to do. A simple system for drying the corn was worked out, but the Argentine cargo owner declined to use it, because drying the grain would mean a loss of weight in the grain. Of course, he did not anticipate the heavy damage that occurred later. Similar situations undoubtedly occur in other cases; grain shippers should realize that a small loss in weight will often mean a much better outturn and may be more profitable. It is quite probable that many cases of heavily depreciated grain cargoes could be traced to a high moisture content when loaded.

The Effects of Hot and Cold Bulkheads

The effect of a heated bulkhead, deck, or overhead on a hygroscopic commodity is to raise the vapor pressure of the commodity and thereby do one of two things: (1) drive moisture out of the commodity into the ambient air, or (2) diffuse moisture from the heated zone of the cargo to the cooler zone.

If the moisture is driven into the ambient air, it must be removed by powerful air currents or the dew point will climb until heavy condensation appears on the overhead. When this condensation is heavy enough to drip back down on top of the cargo, it may cause heavy damage. This is a description of the daily cycle in the upper 'tweendecks of a ship in tropical or summer climes when the ship's ventilation system cannot remove the moisture-laden air before the evening drop in temperature. Sudden drops in temperature brought about by squalls can also cause condensation.

It is quite obvious by now that if the initial moisture content is low enough, heat sources will have no effect. Unfortunately, cargoes are not usually shipped with extremely low moisture contents; we must rely on reliable ventilation equipment properly operated.

If bulk grain is loaded in a space where one of the bulkheads is considerably warmer than the others, the hygroscopic moisture will diffuse from the heated area toward the cool area. This can cause no trouble if the moisture content is low; however, if it is bordering on the critical level of about 14%, the cool grain will receive moisture that will drive its percentage above 14%. At the higher level, the grain may heat and deteriorate. This could be an indirect cause of damage to grain (see Fig. 9-7).

If grain is loaded in a hold in such a way that it comes against an extremely cold steel plate, such as the ship's side in the winter or a poorly insulated reefer bulkhead, the vapor pressure is lowered. Moisture will then diffuse from the warmer zones to the cold zones; thus, the percentage moisture content in the cold zone goes up (see Fig. 9-8). This will cause a localized damage area, probably only a few inches thick, against the steel plate. The higher moisture content may cause the grain to heat, sprout, and spoil. During discharging, the spoiled grain will tend to adhere to the metal.

Limiting Damage Due to Hot or Cold Bulkhead

The extent of damage caused by hot or cold bulkheads will vary as (1) the temperature difference between the bulkhead and the cargo, and (2) the

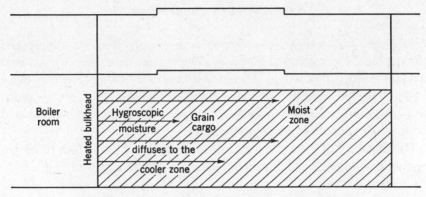

FIG. 9-7.   Effect of a heated bulkhead on adjacent cargo.

nature and condition of the grain cargo. If the cargo is prone to spontaneous heating, as, for example, are soy beans, the sudden increase in moisture content will cause an increase in the respiration of the cargo. This latter change will augment the effect of the hot bulkhead. It is obvious that this entire chain of events and the resulting damage can be made less likely by conditioning (drying out) the cargo before shipment. Some protection may be obtained by insulating boiler room bulkheads and similar places. A slow increase in temperature will not produce effects as damaging as does sudden heating.

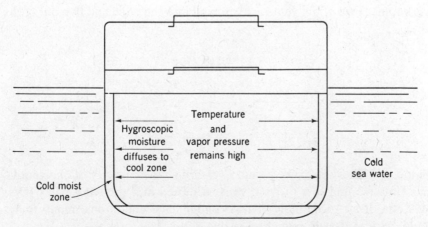

FIG. 9-8.   Effect of a cold bulkhead on adjacent cargo.

Ventilation by Natural Means

Most ships built today are equipped with supply fans, or supply and exhaust fans, for ventilating purposes. The change from the old cowl vent to some method that would guarantee a continuous supply of air, when needed, was long past due when it arrived. To get some idea of how much air a cowl ventilating system can deliver to a hold under ideal conditions, let us make some reasonable assumptions.

The capacity of a duct in terms of the amount of cubic feet of air per minute it will handle can be calculated by a simple equation:

$$Q = A \times V$$

where Q = capacity in cubic feet.

    A = cross-sectional area of the distributing duct in square feet.

    V = velocity, in feet per minute, of the air flowing through the duct.

If we know the wind velocity and the size of a cowl vent, we can calculate the volume of air delivered to the hold. It must be pointed out, however, that turbulence and eddying of the air as the cowl vent changes the air's direction 90° and confines it within its interior will reduce the effective velocity. We will assume that the reduction is 25%.

Making the above assumptions, we can calculate the cubic feet per minute (cfm) of air delivered to a hold fitted with four cowl vents, each having a diameter of 18 in. at the deck level. The wind velocity is 15 mph (1320 fpm). Two of the vents are trimmed into the wind and two out of the wind.

$$A = \pi r^2 = 3.14 \times (18/2)^2 = 3.14 \times 81$$

$$Q = \frac{3.14 \times 81 \times 1320}{144}$$

$$= 2331 \text{ cfm}$$

Reducing 2331 by 25%, we obtain 1749 cfm as the capacity of one ventilator. Thus we find that the two vents trimmed into the wind will deliver 3498 cfm of air to the hold. This is under ideal conditions; many things could cause a drastic reduction in this amount.

A reduction of the actual wind velocity, a change in the wind direction, a change in the ship's course, heavy spray, fog, or rain may stop all ventilation of the cargo holds. If this happens when conditions are such that some means of air conditioning is needed to prevent condensation, damage will certainly result. It is evident that natural ventilation means cannot be relied on unless the ship is carrying cargo for which ventilation is of no importance. However, if ventilation is not required, it would be more economical and sensible to remove all cowl vents and secure the vent coamings by use of wooden plugs and three tightly secured canvas covers.

In contrast to the uncertain capacity of the system just described above, a forced system will consistently deliver 4000 to 10,000 cfm of air, depending on the capacities of the fans and ducts in the system. Used intelligently, such a system can eliminate as much as 80% of all condensation. The remaining 20% can be eliminated if the ship is equipped with a dehumidifier and connecting ducts that are used correctly.

Damage to Cargoes Caused by Moisture

Damage to grain, seeds, nuts, and pulses from moisture has been discussed earlier. The moisture causes a deterioration of the commodity through accelerated respiration, which eventually results in partial or complete loss of such cargoes. The moisture that causes such damage may be free moisture dripping onto the cargo from the deckhead, the hygroscopic moisture already present when the cargo was shipped, or hygroscopic moisture acquired during the voyage.

Other hygroscopic commodities and nonhygroscopic commodities in packages may be damaged by heavy stains from dripping condensation. This type of damage may not materially affect the product itself, but damages may be claimed on the basis of some reduction in salability of the product.

Cargoes of metal products may be heavily damaged by moisture condensing directly on the commodity and causing stains or corrosion.

Some special damage types exist, as in the case of refined sugar. If refined sugar is allowed to absorb hygroscopic moisture and it subsequently releases this moisture, it will become caked. If excessive caking occurs, damages may be claimed; under correct conditions, caking need not occur.

Eliminating Moisture Damage

Ways and means of reducing the detrimental effects of condensation are outlined below. Some of the systems or techniques in use are not very effective, and others are very expensive; the ship operator should investigate all methods with great care. The ship's officer should be aware of all methods and attempt to make them as effective as possible.

Air Conditioning Holds

Proper air conditioning of the ship's holds can eliminate all of the moisture damage due to condensation on the ship or cargo and some of the damage caused by hygroscopic moisture. It should be noted that in the case of refined sugar *proper air conditioning* means the closing off of all circulation in the hold.

Stevedores and other practical cargo handlers have for many years practiced certain basic precautions against moisture damage that do not include ventilation. They have found from experience that they cannot rely on ventilation methods. Whether this situation continues depends entirely on the equipment placed on ships of the future and the knowledge of the officers who use the equipment.

Dunnaging

The most common technique used for minimizing moisture damage is the thorough dunnaging of cargo so that it cannot touch any metal, drainage is provided beneath the cargo, and ventilation air channels through it are provided. This protective measure is necessary whether or not suitable air conditioning systems are present.

Covering Top Tiers

The top tiers of cargo are sometimes covered with a moisture-repellent material, such as a tarpaulin or plastic, to prevent dripping condensation from striking the cargo. The idea here is to catch such moisture and direct it into the frame spaces, where it will fall and eventually find its way into the bilges.

One bad feature of this technique is that the air may become so moist below the covering that heavy condensation will occur there, especially if ventilation with low-temperature air is taking place over the top of the covering. If this happens, the corrective measures could actually be the cause of more damage than would have occurred if nothing had been done.

Hermetically Sealing Containers

In some cases, shippers of products that are of a nonhygroscopic nature but may be easily damaged by condensation pack their products in hermetically sealed containers with small bags of a desiccant sealed within the package. The reaction between the trapped air and the desiccant is such that the dew point of the air is reduced to a level at which no condensation can possibly occur on the metal parts; furthermore, the moisture content of the trapped air is so low that the contents can remain in the package for years without rusting or other forms of corrosion. The material used to wrap the contents must be capable of shedding free moisture from the outside also.

Coating Exteriors

Some products with a highly polished finish are coated with a thin protective covering applied like paint. This covering prevents corrosion and may be removed by wiping the surface with a rag soaked in a paint thinner or specially designed product.

DEW POINT CONTROL

Importance of Dew Point Control

All of the means used to protect cargo from moisture damage resulting from condensation on cargo or hull are unnecessary if the dew point of the air in the ship's hold is kept below the temperature of the cargo and the hull or deckhead. Control of the dew point should be the basis of all our thinking on and efforts toward minimizing moisture damage in the hold.

It has been pointed out that successful dew point control with natural ventilation systems is quite impossible. Forced ventilation systems give some measure of protection. They will fail when the outside air has a moisture content that is higher than that of the air in the ship's hold. When this occurs, the outside air should not be brought in for ventilating purposes. If ventilation is stopped, the dew point inside the hold may climb to the level of the outside air. In other words, control is lost. In periods of heavy fog, rain, or mist, ventilation with outside air should be stopped. This action also brings about a loss of dew point control.

Control of the dew point is only complete when the ship is equipped with a device capable of manufacturing an adequate supply of low-dew-point air, which is mixed with *recirculated air* in the closed holds; this should be done only when it becomes technically impossible to keep the holds opened to the outside air.

Such a device is called a *dehumidifier*. How a dehumidifier is integrated with a forced ventilation system to afford complete mechanical control of the dew point is the topic of a later section.

Basic Rules for Ventilating

If there is a single guiding rule that may be used for ventilation, it is as follows: *When the dew point of the outside air is lower than or equal to the dew point of the air in the hold,* **ventilate.** *When the dew point of the outside air is greater than the dew point of the air in the hold,* **do not ventilate.**

Hard and fast rules can sometimes lead us into difficulties because of particular changes that may take place in outside atmospheric conditions; also, the rule stated above assumes that the operator has sufficient and trustworthy instruments for obtaining the data needed for making a decision. Because of these considerations, four general rules for four specific situations are given below. It is important to understand that these rules should be used only as guides.

Hygroscopic Cargoes Going from a Cold to a Warm Climate

This is not a critical situation. There is little danger of ship's sweat, although there may be the possibility of cargo sweat. Without instruments to check the condition of the air in the holds, a ship's officer could never be certain that he was doing the correct thing. After battening down, it is advisable to keep a close check on the condition of the air in the holds and the outside air. In all probability, it will not be necessary to ventilate initially. At first, the inside air will probably have a dew point that is lower than that of the outside air. The dew point will probably climb upward gradually, but until the inside dew point is higher than the outside dew point there is no purpose in ventilating.

Hygroscopic Cargoes Going from a Warm to a Cold Climate

As a ship proceeds from a warm climate to a cold climate, it will experience a gradual drop in atmospheric dew point as well as a drop in dry bulb temperature. Under these conditions, a danger of heavy condensation on the ship's hull and top decks exists. The air in the below deck spaces will be receiving large quantities of moisture from the hygroscopic commodities, and unless constant and vigorous ventilation is maintained, heavy condensation is certain.

Nonhygroscopic Cargoes Going from a Cold to a Warm Climate

Under these conditions, the temperature of the cargo being loaded will usually be much lower than the dew point of the air through which the ship will pass as it proceeds into warmer climes. It will be necessary to keep the outside air out of the hold; hence, ventilation should be stopped. If ventilation is maintained during the voyage, condensation will occur on the cargo.

Not ventilating will protect the cargo during the voyage, but unless the cargo temperature rises above the dew point of the air at the port of discharge, there will be condensation on the products when they are discharged. However, some damage will have been prevented. If the nonhygroscopic cargoes are steel products, they might be badly rusted if soaked with condensation all during the voyage.

The possibility of large masses of steel rising in temperature during a voyage of only a few weeks is small. Few data concerning heat transfer through cargoes in a ship's hold have been gathered; however, the heat transfer is known to be slow. There will be a rise of about 1°F per day per 25° temperature differential between the cargo and the outside or sea temperature. A cargo of steel products 40° colder than the sea temperature will warm up to the sea temperature in about 20 days, according to Colvin, Hahne, and Colby.[6]

Nonhygroscopic Cargoes Going from a Warm to a Cold Climate

The possibility of trouble in this situation is remote. There is no possibility of cargo sweat, because the cargo is warm and will remain warm during

---

[6] Colvin, Hahne, and Colby, "Care of Cargo at Sea," *Transactions of the Society of Naval Architects and Marine Engineers*, 1938, p. 121.

the voyage. Hull sweat can occur, but the slightest ventilation will prevent this.

Mechanical Control of the Dew Point

All mechanical dew point control systems (as differentiated from simple forced ventilation systems) require three distinct *divisions* of equipment. For maximum control under all conditions, all three division are necessary. These divisions are:

a. *The Hold Fan and Duct System:* This acts primarily as a forced ventilation system. It may be designed with a forced supply and natural exhaust or vice versa; the most efficient systems have a forced supply and a forced exhaust.

b. *The Instrumentation:* This is the means of gathering data from which the operator is able to make intelligent decisions concerning what to do with the system. This division will be explained later in greater detail, but it should be mentioned here that these instruments measure the moisture contents of the air in the holds and the outside air. With reliable information, the operator can judge accurately how to set up the system. With incorrect information or no information, the system may be used to cause more condensation than would result without the use of the system at all.

c. *The Dehumidifier:* This is a machine for removing moisture from the air by adsorption or absorption. Air thus dried is injected into the recirculated air stream within the hold.

The axial flow fans used at each hatch should have a rated capacity of 6000 cfm. The outlets in the holds should be fitted with volume controls, which are adjusted when the system is installed. These volume controls must not be tampered with; otherwise the correct volumes will not be sent to the individual spaces. The duct system should be designed to split off portions of the total 6000 cfm being delivered in proportion to the number and sizes of the holds to which they are destined and route them to these spaces. In other words, a hold of three divisions vertically (Fig. 9-9) would have the 6000 cfm delivered in units of 2000 cfm to each of the spaces. In each space there would be three outlets mounted in a fore and

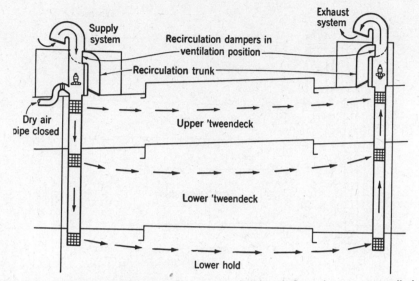

FIG. 9-9. Cargocaire hold fan and duct system, showing air flow when set on ventilation. Courtesy of Cargocaire Engineering Corp.

aft direction. Each of these outlets would deliver an equal amount of air, in this case 666⅔ cfm.

If the dew point of the outside air is lower than or equal to the dew point of the inside air, the system should be set up to ventilate. If the outside dew point is greater than the inside dew point, the system should be set up to recirculate and add dry air (see Figs. 9-9 and 9-10). To obtain maximum penetration of the cargo by the air stream, the cargo must be stowed with great care and with particular provision for vertical and horizontal air channels through the stowed block.

Surprisingly enough, the hold fan and duct system as described above was not used on the Mariner type Maritime-Administration-designed ships. These ships were equipped with a forced supply system but not a forced exhaust system. The exhaust system, in this case, is called a *natural exhaust.* Actually, the exhaust relies on a slight pressure being built up in the hold to eject the air through a covered opening into the atmosphere. The hold then becomes what is termed a *plenum chamber,* or a space under a very small pressure. The pressure would be only a fraction of a pound per square inch. This system is not as efficient as one having a forced exhaust as well as a forced supply.

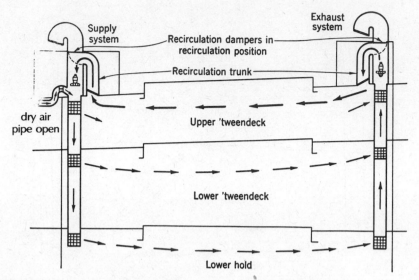

FIG. 9-10. Cargocaire hold fan and duct system, showing the air flow when set on recirculation and adding dry air. Courtesy of Cargocaire Engineering Corp.

Need for Adding Dry Air While Ventilating

When the outside dry bulb temperature is falling rapidly and the system is already on ventilation, the dehumidifier should be put into operation and the dry air used to increase the differential between the hold air dew point and the temperature of the ship's hull. This is a special case, but should be understood so that condensation may be prevented under such conditions.

On runs from South America to the United States during the winter on the east coast, there occurs an excellent example of the conditions described above. When ships leave the Gulf Stream and pass by Cape Hatteras northbound, the dry bulb temperature may fall 40° within 12 to 24 hours. If precipitation does not occur, the system may remain on ventilation. The differential between hull temperature and inside dew point temperature will decrease until there may be less than a degree difference. The addition of dry air to the ventilation air stream may increase the differential so that in the event of further temperature drops or complications, condensation will not occur. See Fig. 9-11 for a graphic illustration of the above discussion.

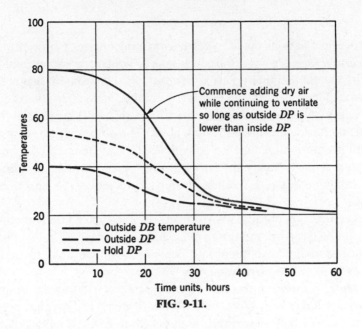

FIG. 9-11.

Need for Recirculation Without Addition of Dry Air

The only time that recirculation of the air in the holds should be maintained without the addition of dry air is when it is desired to raise the temperature of cold nonhygroscopic cargoes stowed below the upper 'tweendeck level. Here is the situation envisioned: The lower hold contains steel products that have been loaded at a very low temperature. Unless the temperature of these products is raised to a level above the dew point of the air at the port where they are to be discharged, they will become wet with condensation when the hatches are opened.

During the voyage, the upper 'tweendecks may reach a temperature above 100°F, while below the water line the steel will remain cold. The steel's temperature will rise, but very slowly, and it may not rise enough to prevent condensation from the cause mentioned above. To obtain some heat transfer to the cargo of steel, the air within the hold may be recirculated without the addition of dry air. Adding dry air will do no harm, but it will accomplish nothing. There is no object in operating the dehumidifier unnecessarily. To make this arrangement more effective, the only upper 'tweendeck outlets and intakes on the ducts should be covered. Do *not* cover the recirculation trunk openings.

Instrumentation

The older installations made by Cargocaire Engineering Corporation use a *hythergraph* for measuring the conditions of hold air, weather, and dry air made by the dehumidifier. This is an aerological instrument that measures two atmospheric factors: (1) relative humidity, and (2) dry bulb temperature. These data are used in conjunction with a psychrometric chart or suitable tables to determine the dew point. The dew point is the important element in determining what to do with the total system. One hythergraph is located at each hold exhaust side, one on the bridge, and one near the dehumidifier. On a five-hatch ship there would be seven hythergraphs.

Newer instrumentation with sensors in each cargo hold records the dew point on a recorder graph in the chart room or wheel house. With the outside dew point also recorded, the mate will be able to determine whether to ventilate or recirculate.

The *hand-aspirated* or *sling psychrometer* is used to determine any instrumentation errors. When taking a reading with the psychrometer, the wick of the wet bulb is wet thoroughly with distilled water using an eye dropper. Do not allow excessive water to remain over the wick. The instrument should be held at arm's length when taking readings and the operator's hand should not be near the metal cover, which should be closed over the bulbs. This prevents the readings from being affected by the operator's breath and body heat. The wet and dry bulb readings are used with a psychrometric chart or suitable tables to determine the dew point.

Probably the greatest weakness in the total system for mechanically controlling the dew point in the hold of a ship lies in the instrumentation. On some ships the system has been maintained in good order and the results are good. However, the system requires a considerable amount of attention to be kept in good operating order. It takes time also to obtain the psychrometric information necessary for the operator to make correct decisions as to the proper setting. For these reasons, the instrumentation has often been neglected. As a result, many ships equipped with excellent and expensive mechanical dew point control systems have failed to use them correctly. If these systems fail to check condensation from any cause, it is probably due to improper operation by the ship's personnel. These machines are constructed according to the laws of physics, and if operated intelligently, they will prevent condensation.

The Dehumidifier

As mentioned in the beginning of this chapter, by 1939 dehumidification technology was taking definitive shape as the means of protecting ship cargoes from moisture damage during transport and storage. Tropical military actions during World War II underscored the need for dehumidification to protect delicate and sensitive equipment. At the end of the war, dehumidification was essential in moth-balling hundreds of naval and cargo vessels. Dehumidification permitted vessels to be "sealed" and stored at near-operating conditions, thus reducing appreciably the time required to return a vessel to full operation.

Cargocaire's first unit was the *Model H*. The desiccant used in this unit was silica gel. Silica gel is a form of silicon dioxide that looks like quartz crystals or rock salt. This mineral has the physical property of being able to *adsorb* water vapor from the air in large amounts. Note that silica gel *ad*sorbs; it does not *ab*sorb. The water vapor leaves the air and clings to the surface of the silica gel. Silica gel is extremely porous and the diameter of the pores is so small that a single cubic inch of silica gel presents a surface area of 50,000 ft^2. Adsorption is a physical transfer of water vapor; it is not a chemical process. Since the moisture simply clings to the surface of the silica gel until it is driven off by heat and no chemical change takes place as in *ab*sorption, the silica gel never has to be replaced, providing that it is kept clean. It was highly important that the silica gel be kept clean; hence, all air going into the dehumidifier was filtered. The filters were required to be replaced regularly. After the water vapor was removed from the silica gel by passing heated air over the crystals, the silica gel was used over again for drying purposes.

The *Model S* Cargocaire dehumidifier was a later design. It was a streamlined model, smaller and lighter than the Model H. The large and cumbersome inlet air dampers were replaced by a relatively light and small four-way valve. Cylindrical silica gel beds replaced the flat beds. Axial flow fans replaced air pumps and heavy conventional fan units. The precooler on the adsorption side was removed. The damper control motor was replaced by a piston that was actuated by compressed air.

The *Kathabar* dehumidifier used a liquid desiccant—a solution of lithium chloride. A later Cargocaire dehumidifier, which replaced the Model S, used triethylene glycol as a liquid desiccant.

Drihold, manufactured by Thermotank Limited of Glasgow, Scotland, used refrigeration to accomplish the dehumidification process. The air passed over coils of a refrigeration system kept at a temperature of 35°F; thus, the air's dew point was reduced to 35°F in passing through the unit. Drihold was the first system to employ the concept of an individual hold dehumidifier.

These early dehumidifiers are described along with more recent types in Table 9-3.

Central System Versus Individual Hold System

The *central system* is a method of dew point control used on most older ships. The dehumidifier is usually located in the machinery space. The air is taken from the outside atmosphere, dried, and delivered through a large dry air pipe to each hold in the ship. However, the dew point of the dried air increases as the outside (weather) air becomes more humid, with the result that the plant produces its wettest air just when cool cargoes require the driest air. This is a deficiency inherent in any ship dehumidification system using a centrally located plant that takes its air to be dried from the outside atmosphere.

The *individual hold system* (Cargocaire's Auto-Mate) locates the dehumidifier in the individual hold; the air is taken from the hold itself and an accumulative drying effect is obtained regardless of weather conditions outside under a true recirculation system.

Cargocaire System

The Cargociare Independent Cargo Hold System (see Fig. 9-12) includes:

a.   The *Dewpointaire®* system which consists of temperature sensors located (1) within the stow of cargo of each cargo hold, (2) in the outside atmosphere, and (3) in the sea chest; and dew point sensors, which are located (1) in the exhaust air duct from each cargo hold and (2) in the outside atmosphere. The sensors are connected to the Dewpointer (see Fig. 9-13), which indicates and records these temperatures and dew points.

b.   The dehumidification system, which consists of *HoneyCombe®* dehumidifiers (see Fig. 9-14), four-way ventilation and recirculation air control, air control valves, and cargo hold air circulating fans. This system enables the mate to take whatever action is called for by the Dewpointer quickly and easily.

Dewpointaire system.   The heart of this system is the Dewpointer control unit (see Fig. 9-13), which is located in the wheel house or chart room.

TABLE 9-3

Model	Manufacturer & Year First Manufactured	Type	Desiccant	Principle of Operation
Cargocaire Model H	Cargocaire Eng. Corp. Amesbury, Mass. (1938)	Central	Silica gel (adsorbs)	Physical process—adsorb phase and reactivating phase
Cargocaire Model S	Cargocaire Eng. Corp. Amesbury, Mass. (1939)	Central (smaller & lighter)	Silica gel (adsorbs)	Physical process—adsorb phase and reactivating phase
Cargocaire TEG Unit	Cargocaire Eng. Corp. Amesbury, Mass. (1947)	Central	TriEthylene Glycol (TEG) (Caire Col) liquid desiccant (absorbs)	Chemical process—rate of absorption = rate of regeneration
Kathabar	Humidity Conditioner Corp. Narbeth, Penn. (1950)	Central	Lithium chloride (Kathene) (absorbs)	Chemical process—rate of absorption = rate of regeneration
Drihold	Thermotank LTD. Glasgow, Scotland (1955)	Ind. hold	None	Refrigeration to reduce dew point
Bry-Air	Marine Dehumidification Company Peabody, Mass. (1962)	Ind. hold	Silica gel (adsorbs)	Rotating desiccant bed
Auto-Mate	Cargocaire Eng. Corp. Amesbury, Mass. (1958)	Ind. hold	Dry desiccant salts in a honeycomb wheel (adsorbs)	Honeycomb principle of wheel rotating—$\frac{3}{4}$ of wheel reactivating with heated air and $\frac{3}{4}$ of wheel adsorbing water vapor from hold via 1500 cfm fan

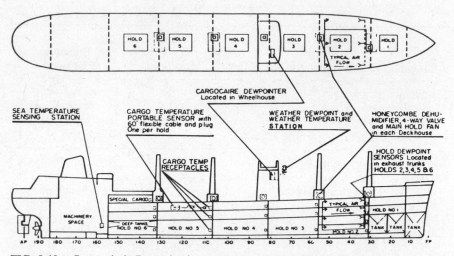

FIG. 9-12. Cargocaire's Dewpointaire system uses independent hold dehumidification to prevent ship and cargo sweat damage. Courtesy of Cargocaire Engineering Corp.

This unit includes two Honeywell multipoint recorders—the upper for ship sweat conditions and the lower for cargo sweat conditions. (Ship sweat is sweat that occurs on the ship structure and damages cargo by dripping. Cargo sweat is sweat on the cargo itself, as occurs with canned goods and metal products.)

Temperature and dew point sensors, located throughout the ship as illustrated in Fig. 9-12, are connected electrically to the recorders, which plot the following temperatures and dew points (some ship systems do not include all of these stations):

Symbol on Chart	Color of Chart Record	Measurement Recorded
D	Brown	Dry air dew point
W	Purple	Weather air dew point
T	Red	Weather temperature
S	Red	Sea water temperature
Numerals 1 thru 7	Green	Hold air dew point
Numerals 1 thru 7	Black	Hold air temperature
Numerals 1 thru 7	Red	Cargo temperature

1. *Temperature Scale, 0-100°F*
2. *Indicator, Print Wheel, Ink Pads*
3. *Stripchart*
4. *Illumination Control*
5. *Legend Plate with Print Sequence*
6. *Power "On-Off" Circuit Breaker*
7. *Auto-Mate Panel with Independent Cargo Hold Controls*
8. *Spare Parts Drawer*
9. *Access Panel To Sensor Cable Connections*

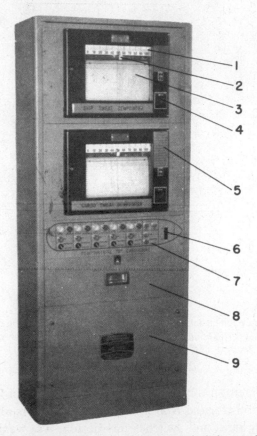

FIG. 9-13. Dewpointer unit in wheel house or chart room includes two Honeywell strip chart recorders that automatically plot temperature and dew point information from a number of strategic locations throughout the ship. Control panel below recording instruments contains individual controls for each cargo hold. Courtesy of Cargocaire Engineering Corp.

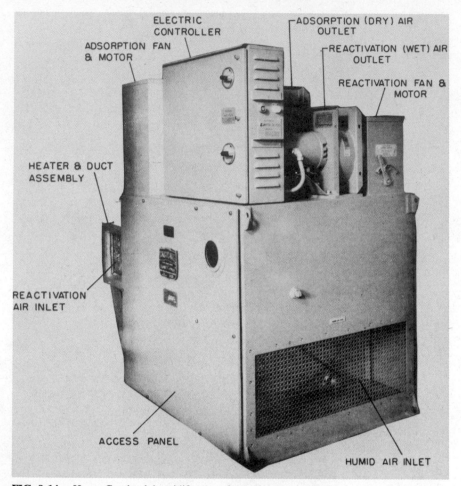

FIG. 9-14. HoneyCombe dehumidifier (one for each cargo hold) is located in deckhouse or fan room. This device takes humid air from the cargo hold exhaust air duct and returns dried air to the hold. Courtesy of Cargocaire Engineering Corp.

Also mounted on the Dewpointer (below the recorders) are the start and stop buttons for the *HoneyCombe* dehumidifiers and the cargo hold air circulating fans, and switches for the four-way ventilation/recirculation valves, one control for each hold.

A *HoneyCombe* dehumidifier (see Fig. 9-14) is located, along with an associated ventilation/recirculation valve and circulating air fan, in a deckhouse or fan room near each hold. Each can be operated either

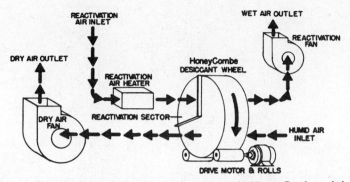

FIG. 9-15. Air flow diagram. Desiccant or drying agent in HoneyCombe unit is continuously reactivated or dried by stream of hot air moving through the smaller reactivation
section of the slowly rotating desiccant wheel. Courtesy of Cargocaire Engineering Corp.

locally or remotely with the push-buttons and valve switches on the Dewpointer. As illustrated in Fig. 9-15, humid air from the cargo hold passes
through the permanent, solid *HoneyCombe* desiccant wheel, which removes the moisture. The desiccant is continuously reactivated or dried as
the whole wheel slowly rotates through a stream of hot air supplied by an
air heater.

System operation.   The only action needed to prevent sweat damage is
to keep the green (hold air dew point) records on the Dewpointer charts
approximately 20°F to the left of the red (weather, sea water, and cargo
temperature) records. This is done by introducing into the cargo hold
either outside air, if sufficiently dry, or air that has been dried by the
HoneyCombe dehumidifier. To enable the ship's officer to interpret the
Dewpointer's records at a glance, the rule "RED RIGHT ALL RIGHT"
is imprinted on every recorder chart.

The overall Cargocaire Independent Cargo Hold System is shown
schematically in Fig. 9-16. Temperatures and dew points from cargo holds
(for simplicity, only hold no. 2 is shown in detail), temperature and dew
point from a weather station, and temperature of the sea water are all
transmitted continuously to the Dewpointer, which records them on
either or both of the two strip chart recorders. Should the ship's officer
see the green chart record of any hold approaching closer than 20°F to the
red chart record for that hold, he can immediately take appropriate action
by means of the push-buttons located on the *Dewpointer.*

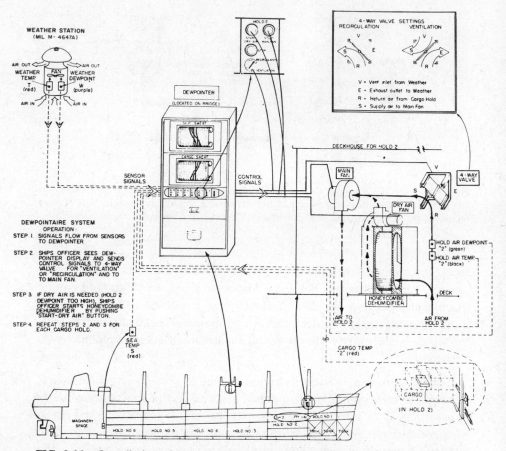

FIG. 9-16. Overall view of Dewpointaire system. Ship's officer quickly checks ship and cargo sweat conditions as recorded on two Honeywell strip chart recorders in *Dewpointer* unit. If dehumidification is needed, he handles it right from the control panel below the instruments. Courtesy of Cargocaire Engineering Corp.

The following is a typical operating procedure.

a. *Initial Startup*. Turn Dewpointer on by throwing circuit breaker on control panel to "On" position. Allow one hour for all system components to stabilize and to provide readable colored records on both strip charts.

b. As soon as hatches are secured for each hold, put the four-way valve switch for that hold in the "Recirculation" position; then press the "Start Main Fan" button for that hold.

c. Compare green and red records with corresponding numerals on both recorders on Dewpointer.

d. If the record for every hold shows a separation of at least 20°F (about 2 in. on the chart) between correspondingly numbered green and red records (with green to the LEFT of red), there is no immediate probability of sweat and no action need be taken.

e. If correspondingly numbered green and red records for a given hold are NOT separated by at least 20°F, observe the weather dew point record (purple and marked "W").

f. If "W" is higher than the green record, weather air is not dry enough for ventilation. Since the four-way air valve switch is already in the "Recirculation" position, and the main fan is already operating, you need only push the "Start Dry Air" button, which starts the *Honey-Combe* dehumidifier.

g. If the weather dew point track ("W") falls 20°F to the left of the red tracks, weather air is adequately dry. Turn the four-way air valve switch to "Ventilation," STOP the dehumidifier, and restart the main fan.

The Cargocaire Independent Cargo Hold System automatically performs many routine jobs that would otherwise burden the busy mate. For example, it automatically records cargo and hold conditions over 100 times a day and compares them with the weather. It does not limit the mate's cargo responsibility, but rather helps him achieve a good out-turn.

Bry-Air System

Bry-Air dehumidifiers dry a mix of moist fresh and returned air with a rotating bed filled with activated alumina desiccant. As the moist air moves through the dehumidification sector of the bed, the desiccant adsorbs the moisture and the dried air is moved to the cargo space, as shown in Fig. 9-17. In the reactivation sector, air from outside the vessel is heated to 250°F and forced through the bed via a reactivation air fan. This air is heated aboard ship by steam or electricity. The hot air dries the moisture out of the rotating bed due to the difference in vapor pressures and thus reactivates the desiccant.

Selection of a Dehumidification System

What dehumidification system is selected depends on the particular needs of the carrier. Before a selection is made, however, there should be a

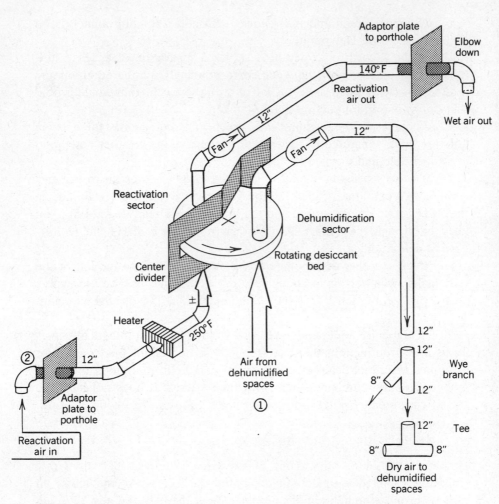

FIG. 9-17. Schematic of Bry-Air D-H system. Air from the dehumidified spaces (1) is drawn through the rotating desiccant bed, which adsorbs moisture from it. The dry air fan then blows the dehumidified air back into the dehumidified areas via ducting. Air drawn in from outside the vessel by the reactivation air fan (2) is heated to about 250°F. This hot air dries the moisture out of the rotating bed and thus reactivates the desiccant. Courtesy of Marine Dehumidification Company.

detailed analysis of all the systems available. The carrier should consult with ship's officers before selecting the system to be installed. Cost is a key concern, but should not be the predominant factor. The predominant factor should be which system will insure the best out-turn of a shipment of cargo.

Summary and Conclusion

Moisture damage or sweat damage probably ruins more cargo each year than any other form of oceanic shipping hazard. This is partly because sweat, or condensed moisture, is so prevalent a condition in the ordinary vessel's cargo holds, and partly because so many commodities are susceptible to damage by moisture. Textile goods are discolored and mildewed, metals and machinery are rusted, food stuffs of many sorts are rendered unpalatable, fibers and grains are heated and fermented, woods of some kinds are stained, and many minerals and chemicals are dissolved or changed in chemical composition. Scores of other commodities too numerous to mention are likewise susceptible to sweat damage.

The *Carriage of Goods by Sea Act of 1936* states that the carrier is responsible for making the holds in which goods are carried fit and safe for their reception, carriage, and preservation. The carrier, however, is not responsible for damage to cargo arising out of such causes as "acts of God," "perils of the sea," and "inherent vice of the cargo." The formation of sweat on the ship's structure or the cargo is the result of natural laws and can be prevented; therefore the carrier is often found liable for cargo damage resulting from sweat.

Ship sweat, the most common source of moisture damage to cargo, is from dripping overhead condensation. The extent of damage depends, of course, on the vulnerability of the commodity and the length of time that the condensation has been dripping before the hatches are opened. Ship sweat or condensation on the ship's structure is a result of the steel becoming cooled to below the dew point of the air in the holds through contact with cold air, rain, or sea water. This causes the air to release its moisture, which condenses on the overhead and on the ship's sides.

Cargo sweat is next in importance as a source of damage, and is the result of condensation of moisture on the cargo itself, usually metal products. Cargo sweat occurs when cargo is loaded cold and stored in a hold where its temperature will rise very slowly. After reaching warmer weather, moisture evaporates from other cargo, lumber in packing cases, dunnage, and cartons, raising the dew point of the surrounding air above the temperature of metal cargo, causing condensation to form on them. This process may take place even though the hold is kept tightly sealed throughout the voyage. If the hold is ventilated with moist, tropical air, the condition will be greatly aggravated.

Thus, it is evident that if the dew point of the air in the holds can be kept below the temperatures of the hull and cargo at all times, no condensation will form. This dew point control is accomplished by means of dehumidification systems.

CHAPTER 10

RESEARCH AND
DEVELOPMENT
IN THE INDUSTRY

Generally speaking, it is true that industrial progress has passed the waterfront by during the first half of this century. The causes seem to be complex and many. But the problem was recognized, and men were encouraged to think about how to change things for the better. It is the purpose of this chapter to list some of the institutions or groups that are responsible for making this progress a reality and to indicate broadly their sphere of influence. A secondary purpose of this chapter is to look into the future of sea transport as a basic requirement for society.

593

BASIC PROBLEM AND GROUPS RESPONSIBLE
FOR SOLUTION

Port Speed and Sea Speed

The most economical marine operation is that which makes the sum of all economic factors affecting port speed and sea speed equal to a minimum value. In the past, the industry has concentrated on the problems of attaining an economical sea speed, and much progress has been made. Ships built during World War I burned approximately 0.8 barrels of oil/mile at a speed of 10 knots. Ships of World War II were capable of burning 0.7 barrels of oil/mile at a speed of 16 knots. These economies have been possible because of better power plant and hull designs.

The only time that a ship is earning money for her owners, however, is when she is under way between ports with a paying cargo in her bottoms. Therefore, it is evident that efforts must now be expended by labor, management, and the government to increase port speed. Port speed is entirely dependent on cargo operations.

The correct steps and the correct direction of these steps is the concern of many groups. In the past several decades, there has been a steady increase of interest in this matter. At present, there is promise of much progress in the next decade. Research and development are the responsibility of labor, management, government, and certain private groups. Every group or individual connected with the transportation of goods by water should be concerned with the success of the combined thinking on this matter. Certain groups have greater responsibility than others, for some are charged with initiating corrective action as well as with investigation of the problem.

It is the authors' opinion that immediate relief from the slowdown in ports can be attained to some degree by two positive steps: First, there must be an increase in cooperation between labor and management. Labor must be willing to give all ideas for mechanization a fair trial. Management must be willing to give labor a fair share of increased compensation in the form of increased salaries and better working conditions. The greatest problem here is to work out a practicable measuring device for ascertaining the increase in compensation and to define what is meant by a *fair share*.

Second, management, government, and labor must work out a system of encouraging *thought* on ways and means of increasing productivity and, what is of equal importance, a system of giving *all ideas* a full and complete test to determine their worth. This is, perhaps, the first step that must be taken to mitigate the problems of the waterfront. Without *ideas* there can be no progress; therefore, it is imperative that a climate that encourages ideas be created by the responsible powers.

It may be that the only way to make really noteworthy progress is to change the design of the ships and the piers so that they may work more closely together and make greater efficiency possible. This, of course, is no more than an hypothesis. The truth must be determined by research and development by the interested groups mentioned later in the chapter. Any change must be gradual, but once the need for change is determined, such change should be initiated.

Maritime Administration

The governmental agency having the greatest control over the commercial maritime affairs of the United States is the Maritime Administration (MARAD), which manages the details of running the maritime affairs of the government.

One of the responsibilities of MARAD is that of developing plans for research and development leading to more efficient ships. Prototype ships are built and tested and plans are made and kept current for the quick construction in large numbers of these advanced design types.

The authors gratefully acknowledge permission from Alexander C. Landsburg, who was Coordinator of MARAD's "Ship Designs for Mobilization" project, to present below a major part of a paper of his that appeared in the *Naval Engineers' Journal* of April 1977. MARAD's conceptual study of three general cargo ship designs appeared in Mr. Landsburg's paper, entitled "Merchant Ships for Wartime Mobilization: Prototype Design and Construction for Readiness." It reads in part as follows:

THE *SS SCHUYLER OTIS BLAND* AND THE MARINER PROGRAM

Development of the mobilization prototypes [sic], *SS Schuyler Otis Bland,* was begun in 1948 under the premises that:

"First the design was prepared to obtain a modern, efficient merchant ship; and second, to make available a ship which could be reproduced on a mass production basis, if future events so warranted."

Major impetus for the project stemmed from the fact that the existing U.S. Flag Merchant Fleet had been built over a relatively short span of years, making block obsolescence inevitable. Changing trade conditions implied that ship construction programs of our competitors were leading toward higher sea speeds, specialized cargo gear, and other more efficient equipment.

The design featured a high 18.5 knot speed using an advanced plant having 865psig–900°F steam conditions rather than the predominant 440psig–740°F. The Ebel burtoning arrangement and mechanically operated hinged hatch covers also were introduced and proved their efficiency on the *Bland*.

Construction plans were developed by the constructing shipyard (Ingalls Shipbuilding Corporation) especially for use in other shipyard [sic] for quick production during a wartime mobilization.

Upon completion in July 1951, American President Lines chartered the *Bland,* which had also seen extensive successful service with the Military Sealift Command (MSC).

The MARINER design incorporated many of the features of the *Bland* and was developed with anticipation of the conflict in Korea. Vice Admiral E. L. Cochrane of the Maritime Administration promoted the MARINER's 20 knot capability as necessary for antisubmarine warfare (see Table 10-1).

With Presidential approval, the Maritime Administration contracted in 1951 for the construction of 35 ships. Only 10 MARINERS were delivered before the end of the Korean conflict in July 1953. Over 500 ships had been reactivated, however, from the Maritime Administration Reserve Fleet by 1952, illustrating the need for fast action. After a brief stay in the Reserve Fleet, most of the MARINER type ships were sold to commercial operators. They have been very successful in increasing the competitive posture of the [U.S.] Merchant Fleet.

Recent Mobilization Designs

The Maritime Administration has produced several preliminary designs for mobilization over the years. The PD-133 "PACER Class" (Productive

TABLE 10-1
MOBILIZATION CARGO SHIP DESIGN CHARACTERISTICS

	Hog Islander	Liberty EC2-S-C1	Victory VC2-S-AP2	Victory VC2-S-AP3	Bland C3-S-OX1	Mariner C4-S-1a	Pacer[a] PD-133	PD-148[a]
Length overall (ft)	401.0	441.5	455.3	455.3	478.2	563.7	565.0	588.0
Length between perpendiculars (ft)	390.0	416.0	436.5	436.5	450.0	528.5	535.0	540.0
Beam (ft)	54.0	57.0	62.0	62.0	66.0	76.0	78.0	85.0
Depth (ft)	32.0	37.3	38.0	38.0	41.5	44.5	45.5	48.5
Design draft (ft)	24.4	27.7	28.6	28.6	28.5	29.8	28.5	29.3
Deadweight (lt)	7,825	10,865	10,900	10,861	10,516	13,409	13,050	10,000
Displacement (lt)	—	14,245	15,200	15,200	15,910	21,093	20,800	21,940
Speed (kts)	11.5	11	15.5	16.5	18.5	20.0	20.0	22.0
Scantling draft (ft)	—	27.7	28.6	28.6	—	31.6	31.5	—
Deadweight (lt)	—	10,865	10,900	10,861	—	14,885	—	—
Bale cubic (ft³)	372,300	499,573	453,210	453,210	561,290	736,723	874,200	880,000
Shaft horsepower (normal)	2,500	2,500	6,000	8,500	12,500	17,500	17,500[b]	24,000[b]
Number crew	41	54	43	43	52	58	40	41
Number passengers	—	0	28	28	12	12	—	5
Gross tonnage	5,735	7,191	7,612	7,612	8,918	9,216	10,000	—
Net tonnage	3,445	4,309	4,553	4,553	5,267	5,366	6,000	—
Stowage factor (ft³/lt) (Bale cubic/deadweight)	47.6	46.0	41.6	41.7	53.4	54.9	67.0	88.0

[a] Preliminary designs.
[b] Maximum continuous.

All-purpose Cargo Emergency Replacement) consisted of a series of designs. The purpose of the project was to have "in being" plans and specifications for a General Purpose Cargo Liner which could be built in quantities of 25 or more to alleviate shipping shortages in non-major emergencies, such as the Korean War, or to replenish the Merchant Marine Reserve Fleet, which consisted of ships nearing the end of their usefulness. The design was developed as a minimum cost ship having characteristics enabling it to operate competitively over a 25-year life. The general arrangement is shown in Fig. 10-1.

The concept of the design followed the principles of versatility and flexibility. A modular deck house was incorporated, and engine room plant alternatives consisted of steam, medium speed diesel, slow speed diesel, and gas turbine plants. An optional parallel mid-body section to extend the length and capacity of the ship increased its flexibility. Consideration was given to building a ship with a straight frame approach similar to that of the Blohm and Voss PIONEER Class, but model tests indicated that this was not desirable economically due to the large increase in power required.

Another MARAD design, PD-148, was developed as a multi-purpose cargo ship intended for operation by the Military Sealift Command (MSC). Table 10-1 presents the design's general characteristics. Basic design requirements demanded a very versatile ship which could handle many different types of cargoes while featuring selective loading. A stern ramp as well as side ports allow a variety of access to the cargo. A fast cargo handling rate remained a major requirement, with no emphasis placed on austerity or quick production.

CURRENT MOBILIZATION SHIP
DESIGN REQUIREMENTS

The Maritime Administration established a "Ship Designs for Mobilization" Project in 1974 to develop modern ship designs for wartime production. The first phase of the project looked at needed shipping capabilities, shipyard production problems, alternative engineering systems, and design constraints. From these studies, *three* concept designs were developed to further identify design requirements.

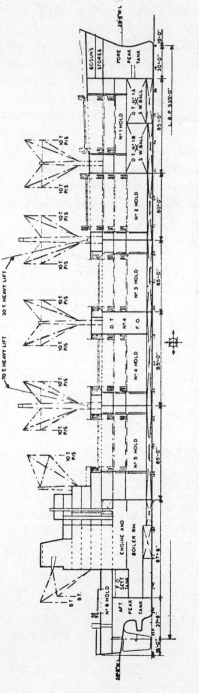

FIG. 10-1. PD-133 PACER design (characteristics listed in Table 10-1).

General Cargo Mobilization Designs

In an effort to define General Cargo Ship requirements, the MARAD Office of Ship Construction developed *three* versatile general cargo designs as candidates for the mobilization fleet. The designs were developed only through enough of the preliminary stages to insure feasibility and allow decisions on what characteristics would be appropriate.

Figure 10-2 and Table 10-2 compare the profiles and characteristics for the three designs. A general arrangement plan for the 550 COMBINATION is found in Fig. 10-3. Many decisions are involved in producing a suitable design, such as ship size, ship speed, type of cargo gear, and cargo to be carried. Various important considerations are described in the following paragraphs.

Ship Size

The first basic and difficult decision is that of desired ship size. Machinery lead time is a critical problem because of the extremely long times required for production of main propulsion equipment and supporting machinery components. In order to produce a vessel rapidly, it would seem better to build it of large capacity so that fewer machinery components would be required. Lead times to obtain the components also would be better matched with the longer times needed to construct large vessels.

TABLE 10-2

VERSATILE GENERAL CARGO MOBILIZATION DESIGNS:
PRINCIPAL CHARACTERISTICS

	500 General	550 Combination	670 RO/RO
LOA (ft)	517.5	574	697
LWL (ft)	497	550	665
LBP (ft)	489	539	655
Beam (ft)	73.5	97.0	105
Depth, main deck at side (ft)	45	61.0	66.5
Draft, design (ft)	28	31.33	32

TABLE 10-2 (*Continued*)

	500 General	550 Combination	670 RO/RO
Deadweight (LT)	9,900	19,650	27,470
Cargo Dead-weight (LT)	7,985	16,800	23,883
Displacement (LT)	16,920	29,860	42,170
Draft, scantling (ft)	30.33	36	32
Deadweight (LT)	11,680	28,850	27,470
Displacement (LT)	18,610	39,060	42,170
Lightship (LT)	6,930	10,210	14,700
Bale capacity (ft^3)	676,790	1,640,000	2,709,750
Liquid cargo (ft^3)	18,340	26,500	—
Deck area (ft^2)	53,140	150,530	251,400
ABS SHP	14,200	21,200	27,000
Powerplant	Diesel elec.	Med. sp. geared diesel	Gas turbine electric
Propellers	1	1	1
Speed at DWL (knots)	20	20	20
Fuel consump-tion (bbl/day)	495	675	1,013
Fuel oil	Diesel	Residual	Residual
Range (nautical miles)	12,000	12,000	12,000
Gross tonnage (tons)	10,530	22,520	33,960
Net tonnage (tons)	6,370	13,630	20,550
Stowage factor (ft^3/LT) (bale cubic/cargo DWT)	84.8	97.6	113.5

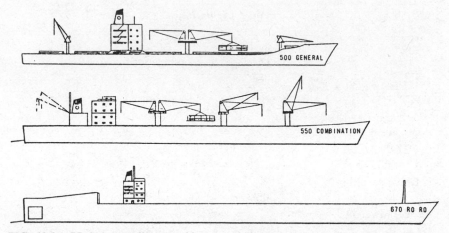

FIG. 10-2. PD-204 versatile general cargo mobilization designs (characteristics listed in Table 10-2).

The second consideration is manning requirements. Crew component on modern merchant ships remains substantially constant with ship size. Large vessels can thus transport the required cargo using fewer crew members than numerous small vessels. In fact, the required number of crew members would be almost identical for the small or large vessel. Since a shortage of qualified officers and seamen will probably occur during the emergency, expected manpower availability should be studied and contingency planning should be undertaken to provide anticipated training requirements.

The third consideration is versatility and deployment. The small vessel has the advantage since there would be a greater number of ships available for deployment in a variety of situations. A large number of small vessels also means less absolute loss per ship sinking.

Cargo types and lot sizes to be carried may also affect the determination of ship size. Large or outsize cargoes can be carried only in large ships or in ships with very large cargo holds. Cargo lot size requirements may influence ship size, although specific needs are not known. Generally the large ships offer significant gains in cargo stowage efficiency.

Shipyard capacity would appear to be the strongest indicator as to what size ship should be built. Vessel dimensions could be constrained to maximize the number of shipways available for their construction. A tabulation of the number of shipways in private U.S. shipyards, by maxi-

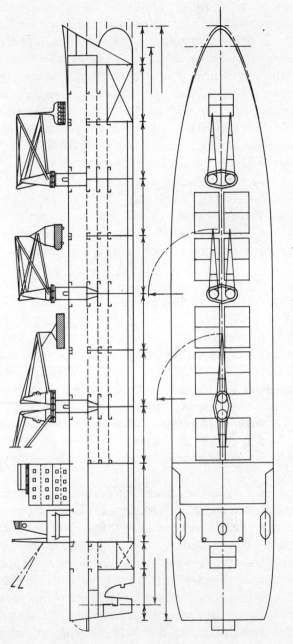

FIG. 10-3. PD-204 550 combination–general arrangement.

mum length and beam capability, indicates, for instance, the number of shipways capable of building a vessel having a particular length and beam, but only shipways accommodating vessels of 475-feet length and 68-feet beam are tabulated. In addition, Navy work would probably occupy many of the larger shipways during wartime. Although mobilization designs would be constructed in existing shipyard facilities to the maximum extent possible, present planning indicates that several new shipyards would have to be built. Few restrictions on ship length or ship beam other than canal constraints would exist in these new shipyards.

Of the many considerations affecting the choice of vessel size, few lead to firm conclusions. Therefore, *three* different size vessels were designed to allow further analysis of needed size. The largest was constrained to pass through the Panama Canal and the smallest through the St. Lawrence Seaway (see Table 10-2). Each design featured its own arrangement and cargo handling approach in order to show a range of systems available.

Ship Speed

Navy guidelines for determining National Defense Features to be installed in subsidized vessels at Government expense recommend a minimum sustained speed of 20 knots for cargo type ships and 16 knots for tankers intended for use as naval auxiliaries. Higher speeds are desirable, as they would permit independent sailings without reliance on convoys. A high percentage of the planned mobilization fleet would be involved in convoy operations, however, and the question of what speed convoys would actually maintain is an important consideration.

It is clear from a powering standpoint that the hull form should be designed for the highest speed desired. Should the machinery availability during construction dictate installation of lower power, the ship may be delivered to run at slower speeds with very little loss in efficiency. A slow speed hull form, however, cannot realize faster speeds without a great penalty in the amount of power that must be installed in the hull. . . . The MARINER hull was designed for 20-knot operation, while the VICTORY was intended for 16 knots. A plant of 28,000SHP would thus be required to drive the VICTORY hull at 20 knots, whereas only 14,000SHP is needed in operating the MARINER. At the VICTORY slower 16 knot speed, however, the much larger MARINER requires only 1,200 horsepower more than the VICTORY, a far smaller penalty.

Optimal Characteristics

With speed and size constraints determined, MARAD's "Least Cost Ship" computer program was used to generate basic dimensions for the three vessels. This program optimizes ship characteristics using minimum required freight rate as the measure of merit. Construction costs and fuel operating costs, however, were the only factors utilized. Consideration of other operating expenses was thought not appropriate for vessels being built and operated during wartime.

A stowage factor of 90 cubic feet per Long Ton was used based on a preliminary study of cargoes to be carried.

Cargo Handling

The primary function of the mobilization design would be to maintain the flow of critical materials and commodities between Allies and to resupply military equipment and supplies necessary in a massive ground war. A multipurpose Cargo Ship was envisioned which would primarily frequent developed ports. Over the beach lightering and other unloading operations in open water are important, but it was not considered necessary for these vessels to be designed primarily for such roles.

The rate of cargo handling required is an important consideration in the design. A major problem during the wars was the long queues of ships in ports waiting to be unloaded. Whether a high cargo handling rate would alleviate port congestion is questionable, but RO/RO capability and high speed cargo cranes are certainly attractive. High speed pedestal cargo cranes (see Fig. 10-4) are popular abroad but are not presently produced in the United States. Such cranes, however, have versatility and could save construction time during mobilization since they could be manufactured at a central facility and then sent to the shipyard for quick and easy installation. A cylindrical or rectangular base is welded directly to the deck of the ship and the crane is ready to go after being bolted to this base and connected to the electric power supply. Where these cranes are used in the designs, the alternative capability to install conventional cargo gear is included by insuring adequate spacing between hatches.

500 GENERAL

The 500 GENERAL, a Lift-On/Lift-Off type vessel, [Fig. 10-2] is a basic General Cargo Ship having *six* holds. Four of the holds are forward of

FIG. 10-4. Twin high speed pedestal cargo cranes.

the "House" and two aft. The larger holds have 62-ft × 42-ft hatch openings and pontoon covers providing a maximum of straight drop stowage for the rapid and efficient handling of cargoes. The tank top in these holds has a 19.3-ft. clearance for out-sized cargoes. Thirty-Long Ton, twin pedestal cranes service Holds Three and Four, while 15-Long Ton pedestal cranes work the other holds. The cranes feature dual speed ranges permitting maximum efficiency in handling cargoes of various unit weights.

Figure [10-4] demonstrates the cargo handling flexibility of the twin 30-ton crane system installed between Holds Three and Four. Note that each crane can be used independently, a) to service adjacent Holds or a coordinated effort can be accomplished by operating the cranes in a double harness, and b) for control by one operator at either hold. The tandem working arrangement enables the handling of tanks or other cargoes with weights up to 60 tons along with controlled pendulation when cargo discharge must be performed at anchorage ports.

550 COMBINATION

The 550 Combination RO/RO and Lift-On/Lift-Off vessel has *eight* cargo holds (Fig 10-3). The *seven* forward holds are serviced by pedestal cranes and the aft hold by conventional gear. This vessel also has a stern ramp for loading RO/RO cargo directly onto the second deck. The ramp was designed to handle 60-ton Tanks and could be constructed for RO/RO access to the Main Deck. Hatch covers on both the Main and Second Decks are flush to accommodate the RO/RO traffic. The Second Deck height of 18 feet at the center line was developed for the out-sized military cargoes, including trucks and helicopters.

The 30-ton capacity twin cranes provide 60-ton lift capacity when worked in double harness and 120-ton lift capacity when *two* twin cranes at each end of a hatch are married. Maximum lift capacities are shown in Table 10-3. The present layout allows deletion of a second set of twin cranes with outreach from the other remaining cranes still capable of unloading all holds on the ship. This is an advantage in the case of a shortage of cranes when the ship is being delivered. The deleted set of cranes could be installed on a future voyage when they become available and/or higher cargo handling rates are required. Conventional booms and King Posts are used aft to work the small No. 8 Hold.

Cargo hatches were sized to accommodate containers and cell guides if an all-container configuration is desired. The containers could also be stowed outboard of hatch openings if required. The stern ramp is suitable for unloading into Army Beach Lighters. The nearly unobstructed RO/RO or Second Deck has only Engine Room uptakes extending through the deck. Provisions have been made underneath the wings of the "House" for drive through of RO/RO cargo on the Main Deck; RO/RO cargo can be stowed easily on this deck because of the flush pontoon hatch covers.

TABLE 10-3
550 COMBINATION CARGO GEAR/HOLD CAPABILITIES

Hold No.	Max. Lift	Remove Cranes at BHD 29 Max. Lift
1	20 LT	20
2	20 LT	20
3	60 LT	20
4	60 LT	20
5	120 LT	60
6	120 LT	60
7	60 LT	60

670 RO/RO

The 670-foot design is most appropriate for Roll-On/Roll-Off cargo. The optimized dimensions are very similar to MARAD designation C7-S-95a. The 670 RO/RO (Fig. 10-2) is an alternative that is intended as an investigation of other arrangement possibilities. The major difference lies in the use of two longitudinal bulkheads extending the length of the ship in order to provide completely unobstructed RO/RO access on and off the decks. Longitudinal sub-division is not necessary inboard of the bulkheads, thereby saving the installation of watertight bulkheads and doors. At the stern, there are *three* watertight doors: the stern door and ramp suitable for off-loading into Army Barges, and the stern side doors useful when several ships are moored together as a ramp for unloading from one ship to another or to the pier.

Movement of the cargo from deck to deck is by a series of internal ramps located at the stern. There are also emergency ramps amidships in case the stern is damaged or deck to deck access is blocked. Emergency ramps can be used to bring the RO/RO cargo up to the Main Deck, from which it can be removed from the ship through *three* emergency hatch covers. Outboard of the longitudinal bulkheads are provisions for storage of container-sized cargo below decks. This cargo would be carried onboard by way of the ramps and dropped into position with straddle carriers or other similar vehicles.

Two decks have 17-foot clearance for outsized military cargoes or stacking of two containers on the deck. Container handling would be done with forklift trucks. Containers can be carried on the tank top, but with great difficulty. With some expertise, *four* containers can be stowed on each deck level in most of the wing holds. The stern ramps would be built to a capacity to support the largest earthmoving equipment anticipated.

Table 10-4 compares the cargo handling features of all three designs. A short study was also performed on the cargo handling rates of the 550 COMBINATION. It is estimated that this vessel could be loaded in approximately 36 hours at an efficiently run pier with skilled longshoremen. The off-loading time would probably take up to 72 hours due to congestion of the piers at the off-loading facility. Removing the twin cranes between Holds No. 4 and No. 5 would increase the loading time by about 22 hours.

TABLE 10-4

CARGO HANDLING COMPARISON

	500 GENERAL	550 COMBINATION	670 RO/RO
Cargo (DWT)	7,895	16,800	23,883
Max. hatch	62' × 42'	42' × 25'-11"	16' × 40'
Max. lift (LT)	60	120	—
Max holds worked	4	7	—
RO/RO	No	Yes	Yes
RO/RO deck height (ft)	—	17.0	17.0
Ramp width (ft)	—	22.0	25.0
Bale cubic (ft^3)	676,790	1,640,000	2,709,750
Deck area (ft^2)	53,140	150,530	251,400
Tank top deck height (ft)	19.25	17.75	11.0
Container capacity (TEU)			
Under deck[a]	317	423	884
On deck[b]	140	258	432

[a] Cell guides, only in hatch opening.
[b] Two high on deck, except Hold #1, no pedestals.

Further Design Work

The 500 GENERAL, 550 COMBINATION and 670 RO/RO concept designs were reviewed by OPNAV, NAVSEA, MSC, and the Army. Strong preference [at the time] was indicated for the middle sized ship, the 550 COMBINATION. The 500 GENERAL was considered limited in its cargo handling approach and capacity, while the 670 RO/RO was thought too large and specialized. Further design work in 1980 based on the 550 COMBINATION has evolved into a contract level design. Five alternative power plants and three cargo handling options are now available.

THE U.S. NAVY

The work within the naval establishment relative to research and development leading to improved cargo operations is carried on by Naval Sea Systems Command and Military Sealift Command. They are especially interested in solving problems that are common to military operations, but some of the ideas developed by them may become useful in commercial cargo operations. In fact, it is probable that the Landing Ship Tank carrier (LST) idea is the father of the current ideas that deal with larger and faster RO/RO ships in both the commercial and military areas.

Three projects with which the U.S. Navy is vitally concerned are the Sea Shed, ARAPAHO, and the Near-Term Rapid Deployment Force.

Sea Shed

A concept designed to meet a military emergency could soon be helping some commercial ship operators gain a competitive edge in the continuing shipping recession. It is the Sea Shed—a large, open-topped container that, in its military version, permits tanks and other heavy and outsize cargoes to be carried in containerships. It has been developed as a result of a $2.4 million project funded jointly by MARAD and the Naval Sea Systems Command. The logic behind the development is compelling: As the U.S. Merchant Marine has increased its commitment to containerization, the capacity to carry outsize equipment and other breakbulk has gone down. The Sea Shed project was initiated as a way to restore this capacity. The units would be stockpiled for deployment aboard containerships in the event of war or other national emergency.

Four prototype military Sea Sheds have been built for MARAD by Tracor Marine, Fort Lauderdale, Florida. They have already undergone their first operational tests. As can be seen from Figs. 10-5 and 10-6, these particular units are of open-truss design and have large side openings.

The commercial shipping potential of the Sea Shed is considerable. Essentially, the Sea Shed functions as a portable 'tweendeck system. This is made possible by a work-through floor composed of two bipanel folding sections. The external panel of each section is connected by a hinge to the Sea Shed main frame and actuated via self-contained winches.

A standardized module has been developed with dimensions of 40 ft × 25 ft × 12 ft high, a weight of 30 tons empty, and a capacity of 150 tons. Sea Sheds of this size can accommodate a wide variety of outsize loads,

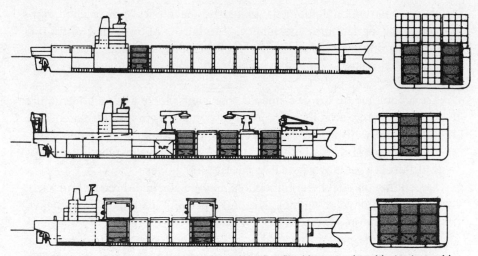

FIG. 10-5. Typical shipboard installations of the Sea Shed in a containership (top), combination carrier (center), and container-convertible product carrier (bottom). This system permits ship operators more flexibility to accept alternative cargoes.

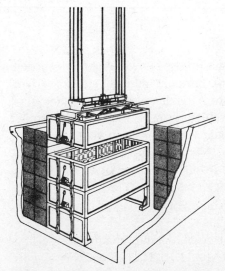

Shed end walls interface with ship's guide structure.

Self-contained winch system speeds opening and closing of floors, requires plug-in connection to the ship service electrical system.

FIG. 10-6. Sea Shed interfaces.

such as construction equipment, industrial machinery, power plant components, or agricultural machinery. Modules are provided with ISO fittings permitting rapid installation, removal, and handling of empty units using 40-ft spreaders. An ample number of D rings and cloverleafs reduce lashing time.

To accommodate the Sea Sheds, ships must have a cellular structure able to take three laterally adjacent 40-ft containers in a single hatch. The modules' end walls are designed to interface with the ship's guide structure. An adapter frame attached to the end wall allows installation in ships with different spacings between adjacent cell guides.

Depending on hold height, modules may be stacked three or four high. An (optional) elevated support structure permits loading of extra heavy items (e.g., stators, compressors) on the tank top.

Modifications needed to enable existing containerships to take the Sea Shed system are relatively minor and can be accomplished during scheduled overhauls.

Other ships for which the Sea Shed offers the possibility of a useful degree of extra flexibility include combination carriers and container-convertible designs of forest product carriers. There is also potential for the system's application in neo-bulk carriers, dry bulkers, and river barges.

Arapaho

The ARAPAHO program envisions the use of Navy helicopters carried aboard commercial containerships to conduct antisubmarine warfare (ASW) operations at sea. In addition to ASW operations, the helicopters could also be used for maritime surveillance, mine clearance, close air support for the U.S. Marine Corps, and evacuation of combat casualties.

For certain missions, such as close air support, ARAPAHO ships could also carry Harrier type aircraft. The ARAPAHO program also envisions the use of bulk carriers, barge ships, RO/RO vessels, and tankers for such carriage.

In October 1982, a U.S. merchant containership conducted the first under-way, shipboard tests of ARAPAHO. The 18,000-ton *Export Leader,* outfitted with a complete aviation facility (flight deck, hangar, fuel system, night lighting, power supply, and damage control), successfully served as a base for six types of military aircraft.

Rapid Deployment Force

The decision of the U.S. government to set up a "rapid deployment force" for speedy movement of troops and their equipment to potential centers of military conflict has led to a number of interesting contracts for merchant vessels.

One of the major components of the rapid deployment fleet will be eight SL-7 high-speed containerships, formerly owned by Sea-Land. These 120,000-shaft-hp vessels are to be converted by National Steel and Shipbuilding Company, San Diego; Avondale Shipyard Inc., New Orleans; and Pennsylvania Shipbuilding Company, Chester. The Navire Co. has been contracted to supply hatch covers, side doors, platforms, a hinged ramp, and a portable ramp for each vessel.

The SL-7s were introduced into the North Atlantic trade in 1972 and represented a major escalation in both vessel size and speed. With a capacity of 1974 TEU and a service speed of 33 knots, they are, with the exception of the liner *United States,* the fastest merchant vessels ever constructed.

The conversion they are to undergo (see Fig. 10-7) will enable them to transport a wide variety of military equipment and supplies, including

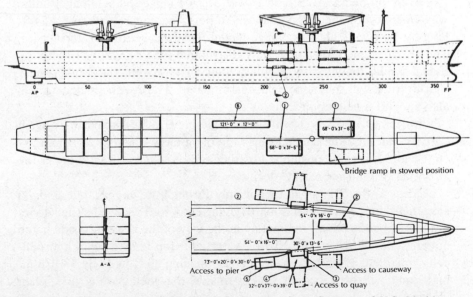

FIG. 10-7. General arrangement of the SL-7 conversions. Courtesy of U.S. Maritime Administration.

tanks and helicopters. It entails constructing an additional deck above the original weather deck level between the forward and midship superstructure blocks. This new deck has been designated the "flight deck," while the main deck below is to be termed the "hangar," or "A deck." The former second deck, originally arranged outboard of the cellular holds only, will be fully plated in and designated the "B deck," while beneath this, two further 'tweendecks are to be constructed, the "C" and "D" decks.

Two hatch openings are to be provided on the flight, A, B and C decks, one arranged to port and forward of a paired-crane installation, the other to starboard and abaft the cranes, which can serve either hatch opening. The D deck will have a single opening.

The flight deck covers will be Navire weathertight, flush folding units, wire-operated by the cranes. The port side cover, consisting of four panels, will give a clear opening 68 ft long by 31 ft wide. The starboard side cover will be identical, but arranged further aft. The panels are designed for a uniformly distributed load of 200 lb/ft^2 or, alternatively, a fully loaded 5-ton truck. The after two panels of the starboard cover are, in addition, intended to be capable of withstanding the dynamic landing load of a helicopter.

Covers on decks A, B, and C will be side-hinged, single-panel units, wire-operated by the ship's cranes. To port, the covers will give a 54-ft-long by 16-ft-wide clear opening, as will those starboard, but further aft. They are designed for a uniformly distributed load of 525 lb/ft^2; alternatively, a fully loaded 5-ton truck, a 65-ton tank, or an 18.5-ton-capacity forklift truck will be able to be accommodated. The covers on deck A will also support a helicopter.

On deck D, the cover will be a single-panel, side-hinged unit, again wire-operated by the ship's cranes, giving a clear opening 25 ft long by 16 ft wide. It is to be positioned on the starboard side, directly under the after half of the starboard C deck cover.

Access to the SL-7s will be not only through the covers, but also via side shell doors, port and starboard, giving entry to B deck. The shell doors are to be positioned approximately in the way of the starboard side hatch covers, and will be of the Navire watertight, top-hinged type, hydraulically operated. Each is to have a clear opening 30 ft wide by 13 ft 6 in. high, and will be recessed slightly inside the shell plating line when closed.

Outboard of each side door will be a bottom-hinged platform that stows by folding up over the closed door. Each platform is to be 32 ft long, with a width at the door of 39 ft 4 in., decreasing to 37 ft 5 in. at the outer edge. When lowered, the outer edge will be supported by a pair of cantilever arms positioned at either side of the opening and above the side door.

Leading off the platform will be a portable Navire ramp; only one such unit will be supplied to each ship. The ramp will work port or starboard and will be able to be used at 90° to the vessel centerline, lying forward parallel to the centerline, or aft skewed 5° outboard. Suitable for a tank grossing 65 tons, the ramp comprises two panels and a finger flap. It is 73 ft long, with a minimum driveway width of 20 ft. The ramp, which will be stowed on the flight deck and positioned on either side by the crane, will be used to discharge cargo into flat-topped barges at sea level. Alternatively, the ramp will be able to land on a convenient section of pier or quayside, by removing one of the two sections if necessary.

The cargo access equipment will be able to function in sea states up to 3, and is designed in accordance with the American Bureau of Shipping (ABS) rules and U.S. Coast Guard standards.

THE U.S. ARMY

The U.S. Army maintains the transportation center at Fort Eustis, Virginia, which carries on research and development into all phases of military cargo operations and related logistical problems.

THE U.S. COAST GUARD

The U.S. Coast Guard does not concern itself with the equipment and methods of cargo operations in general. However, under the provisions of R.S. 4472 (46 U.S.C. 170), the Commandant of the Coast Guard has promulgated regulations for handling, stowage, storage, and transportation of explosives and certain other types of dangerous articles. The Merchant Marine Inspection Division of the Coast Guard also is responsible for examining applicants for licenses as deck officers of merchant ships to determine their proficiency and knowledge of cargo operations.

THE AMERICAN BUREAU OF SHIPPING

The American Bureau of Shipping (ABS) is a classification society, and as such they are interested primarily in the ship itself and its equipment; this includes means for protection of the cargo spaces from the sea. They also have developed rules for certification of cargo gear (see pp. 390–391). These particular rules do not affect the ship's classification; they are an entirely separate set of regulations that were developed at the request of the industry by the Bureau's special subcommittee on cargo gear. One of the reasons for the development of these cargo gear standards was that many of the principal maritime nations have statutory regulations concerning cargo gear. The United States does not have such regulations, but it is necessary for U.S. ships to comply with the regulations, or acceptable equivalents, when they visit ports in countries having such requirements.

Another connection with the cargo operations of the merchant ship held by the ABS is that of inspection of refrigeration facilities of ships. The ABS's rules provide for the following inspections after the installation of refrigerating facilities for cargo:

1. *Loading Surveys.*   A survey made with a view to issuing a certificate stating that the refrigerated holds and the machinery are fit to carry the refrigerated cargo intended.

2. *Semiannual Surveys.*   A general survey of the refrigerating machinery and installation.

3. *Intermediate Surveys.*   Surveys made every 2 years providing for increased inspection and requiring the opening up of certain machinery.

4. *Special Periodical Surveys.* Surveys made every 4 years that are extensions of the intermediate surveys. The inspection of all parts is more detailed and careful.

5. *Continuous Survey.* These surveys may be carried out on vessels that are engaged on voyages of 3 months' duration or less. They provide for making all of the previous surveys periodically on a step by step basis. This prevents the tying up of the ship for any significant length of time.

Because of its functions, it often appears as if the ABS is a government agency. This is not true. The ABS is a self-supporting corporation under the laws of the state of New York; it has no capital stock and pays no dividends. It is financed through fees charged for the services rendered by its surveyors in connection with approval of plans, supervision of vessels under construction, various types of surveys, and the assignment of load lines. The ABS is recognized officially by the U.S. government for such purposes as classifying vessels owned by the United States and for purposes of making load line surveys in accordance with certain laws.

THE NATIONAL CARGO BUREAU, INC.

The National Cargo Bureau (NCB) is governed by a board of directors of 18, including the Commandant of the U.S. Coast Guard and the Maritime Administrator. A group of officers carry out the policies and decisions of the directors and manage the Bureau. The executive, finance, operating, membership, and Pacific Coast advisory committees help to develop policies for consideration by the board of directors. The technical aspects of the NCB are the direct responsibility of the chief surveyor, under the guidance and direction of the officers. The chief surveyor is assisted by deputy chief surveyors at New York, Houston, and San Francisco.

The NCB is a nonprofit membership organization incorporated under the laws of the state of New York, and its purposes are:

1. To provide a private agency to formulate recommendations to the government as to regulations concerning the safe stowage of dangerous goods.
2. To work internationally at industry level to achieve uniformity of safety standards and regulations for the stowage of cargo.
3. To act as an information-gathering agency on the problems of transporting the thousands of commodities offered for water transportation, and to make this information available to the shipping industry and other groups.
4. To offer the shipping industry a low-cost cargo loading inspection service directly at the loading operation.

The cargo loading inspections are made by surveyors who are deemed qualified to advise ship operators and stevedores on the safe stowage of cargo. This service is available, on a cost basis, to shipowners and operators who wish to avail themselves of it. The presence of an impartial, objective NCB surveyor supports the efficiency of the ship operator's personnel. The surveyor supplements the operating and stevedoring personnel. When a surveyor completes the inspection of cargo loading and finds the stowage satisfactory, he issues a certificate of loading.

From the foregoing, it should be clear that the NCB does not concern itself with cargo operation problems except as they are related to the safe stowage of particular commodity types. The NCB is concerned that the stowage does not endanger the cargo itself, adjacent cargo, the ship, or the crew.

THE INTERNATIONAL CARGO HANDLING COORDINATION ASSOCIATION

Early in the year 1951, a British naval architect, A. C. Hardy, and a French economist, F. X. Le Bourgeois, were discussing in a private social conversation the need for an organization dedicated to the continuous investigation of methods and equipment for increasing the efficiency of the movement of oceanborne cargoes throughout the world.

As a result of their conversation, they contacted many of their friends to obtain a reaction to the proposal of setting up some type of nonprofit organization or group along the lines of a technical society for the primary purpose of meeting the above-mentioned need. The response was immediate and enthusiastic. An inaugural meeting met in London on October 30, 1951. This was the beginning of the International Cargo Handling Coordination Association (ICHCA). Since that time, the organization has developed so that at present there are 24 national committees helping to support the objectives of the ICHCA, including the U.S. national committee, which held its first meeting July 28, 1953.

The objectives of the ICHCA are to serve the international public by bringing together all who are interested in improving the efficiency of cargo handling for the common purpose of assisting international seaborne trade. It is hoped, thereby, to coordinate the practical and scientific experience of stevedoring concerns, packers, shippers, shipowners, port authorities, industrialists, and others of kindred professions who acquire knowledge that is of local and possibly limited value when unrelated but, when connected under discussion, examination, and publication by the ICHCA, will tend toward progress and improved efficiency in all aspects of manufacture, packing, transport, handling, and stowage of seaborne goods.

The problems that the ICHCA is dedicated to attacking are those problems that have the greatest effects on the future of the merchant marine of any country. Because of the high labor costs in the United States, some of these problems are more acute here.

The fundamental purpose of the merchant ship is to transport cargo.

Everyone concerned with the industry should always bear in mind that the efficiency of the merchant marine depends on the cargo-operational efficiency of the companies owning the merchant ships and competing on the world's trade routes. However, one should never lose sight of the fact that the efficiency of any given company is a direct function of the efficiency of each ship. And, finally, no ship is operated well unless the men on her are efficient and informed. It is the objective of the ICHCA to foster this total efficiency by general dissemination of information through discussion and publication of the proceedings of symposiums covering the myriad facets of ship operation.

THE MARITIME CARGO TRANSPORTATION CONFERENCE

The Maritime Cargo Transportation Conference (MCTC) was organized under the National Academy of Sciences–National Research Council at the joint request of the Department of Commerce and the Department of Defense. Its purpose was to provide guidance on means and techniques leading to improvement in systems and elements of systems for the sea transportation of dry cargo, as well as to determine critical factors and remedial measures to reduce ship turnaround times (including total time in ports).

The National Academy of Sciences is a private, nonprofit corporation dedicated to the furtherance of science for the general welfare. It was established by Congressional legislation in 1863 for the purpose of providing advice to the government in the fields of the natural sciences and their applied technologies. The National Research Council was established in 1916 to facilitate the scientific and technical advisory services of the Academy, and provided a vigorous program for mustering civilian science in the interest of the war effort. The National Research Council has continued with increasing activity through both World Wars into the present. It deals with scientific problems through committees of outstanding persons who are experts in the various fields of the problems at hand. It organizes and focuses the knowledge and efforts of science in the interests of government and industrial or educational research programs that promise to contribute to national security or general public welfare.

The MCTC consists of a board of advisors constituted of leaders in pertinent phases of maritime cargo handling and transportation who, from their knowledge and experience, provide guidance and review for a small full-time professional staff. Its activities emphasize analysis of current and proposed systems of cargo transportation leading toward improvement in overall performance.

The first study made by the MCTC was an analysis of the time, labor, productivity, and costs involved in transporting a shipload of cargo from the hinterland of the United States to its destination in Europe. This was a preliminary study with the main purpose of defining those areas warranting closer and immediate examination and pointing toward logical methods of organizing future studies. This pilot effect was known as the *Warrior Study,* and its results were used to guide the staff of the MCTC in future research.

THE SOCIETY OF NAVAL ARCHITECTS AND MARINE ENGINEERS

Through the medium of various technical and research committees, the Society of Naval Architects and Marine Engineers (SNAME) carries on a considerable amount of research. Most of this research is into the engineering aspects of building and powering vessels, but one committee, known as the Ship Technical Operations Committee, carries on research into areas dealing with the efficient operation of the vessel. Another recently developed committee, known as the Cargo Handling Committee, is researching methods to make the cargo handling operation more productive and safer.

THE FUTURE OF SEA TRANSPORT AS A BASIC REQUIREMENT OF SOCIETY AND SHIPS OF THE FUTURE[1]

The 1960s was the decade in which unitization began in earnest with the profit-induced desire of speeding up and simplifying cargo handling processess. Ship design was bound to be affected, and thus the LASH, SEABEE, RO/RO, and SPLASH vessels were born, along with modifications of the container vessels, bulk vessels, and general cargo vessels. The whole transport chain took into account door-to-door delivery of unitized cargo via land and sea bridges.

[1] The authors gratefully acknowledge the permission of Cornell Maritime Press and MacGregor Publications, Ltd., to utilize facts and figures from their book *Ships and Shipping of Tomorrow* by R. Schonknecht, J. Lusch, M. Schelzel, and H. Obenaus in this final segment. This book is highly recommended to those readers who desire an in-depth and informative look into the future.

Despite setbacks such as the crisis of the tanker trade in 1974, seaborne world trade can be expected to continue to increase along with the world population up to the turn of the century. It is estimated that a 4 to 5% growth rate will occur each year up to 1990, with a marked increase in the rate of carriage of breakbulk cargo, especially in such unified forms as pallets and containers. Bulk cargoes such as oil, coal, grain, phosphates, and bauxite will continue to occupy the dominant position.[2] The best chances of survival are to be found among those large transport organizations that are active in several fields and that have strong financial backgrounds or are supported by a planned level of employment underwritten by the state.[3]

The following three facts are certain about the future:

1. Technical progress will proceed much more rapidly than it did in the past.
2. The time between technical development and practical application will shorten.
3. Science will play an increasingly important part in the technical and managerial aspects of marine transportation.[4]

Ships of the Future

The categories into which ships of the future will fall already exist. They are listed in Table 10-5. The merchant vessels listed in Table 10-5 do not include passenger vessels, fishing vessels, and ships designed for special missions. In our discussion of future ships we have chosen only those vessels that have been discussed at length in previous chapters of this text. Those vessels are the general cargo ship, containership, barge carrier, RO/RO ship, and the refrigerated cargo ship.

General Cargo Ship

Present. General cargo ships, or multipurpose ships (see Fig. 10-8), form the largest percentage of the world fleet. They are used for all types

---

[2] R. Schonknecht, J. Lusch, M. Schelzel, and H. Obenaus, *Ships and Shipping of Tomorrow,* Cornell Maritime Press, Centreville, Maryland, and MacGregor Publications, Ltd., Hounslow, England, 1983.
[3] Ibid., p. 10.
[4] Ibid., p. 11.

TABLE 10-5

CATEGORIES OF MERCHANT VESSELS[a]

Category	Cargo Variants	Combinations
1. General cargo ship	Universal, multipurpose ship; heavy-lift ship; pallet, timber, and cattle carriers	Semicontainer ship
2. LASH	Lighters	Containers
3. Barge carrier	SEABEE, BACAT	Containers
4. Container	Containers	Refrigerated containers
5. RO/RO	Freight ferry Car carrier Heavy-lift RO/RO	Container, RO/RO Bulk RO/RO Car & passenger RO/RO
6. Refrigerated ship	Fruit, meat, and fish	
7. Tanker	Mineral products, chemicals, bitumen, sulfur, wine & glue	
8. Liquified gas	LPG, LNG, ethylene, and ammonia[b]	
9. Bulk carrier	Universal, ore, cement, and bauxite	Ore–Bulk–Oil (OBO) Ore–Bulk–Container (OBC) Ore–Slurry–Oil (OSO) Bulk/Car Carrier

[a] Information from R. Schonknecht, J. Lusch, M. Schelzel, and H. Obenaus, *Ships and Shipping of Tomorrow*, Cornell Maritime Press, Centreville, Maryland, and MacGregor Publications, Ltd., Hounslow, England, 1983, p. 67.

[b] LPG, liquified petroleum gas; LNG, liquified natural gas.

of breakbulk cargoes, including bags, bales, cartons, crates, cases, barrels, and drums, with the emphasis being on preslung pallets. They are flexible, also being capable of carrying heavy lifts, vehicles, bulk cargoes, refrigerated commodities, and containers. Present-day general cargo ships have several decks, deadweight tonnages of 9000 to 15,000 tons, and economical speeds of between 17 and 22 knots. Technical advances that have brought their in-port times down to a low of 40% include large centerline hatches or two or three rows of hatches side by side (see Fig. 10-9). The time-consuming task of bulling cargo into the wings is thus almost entirely eliminated. Cargo handling equipment includes booms and

Motor cargo ship
(general cargo ship)

FIG. 10-8. Motor cargo ship (general cargo ship). Courtesy of *Ships and Shipping of Tomorrow*, Cornell Maritime Press, Inc., and MacGregor Publications, Ltd.

twin slewing deck cranes that are rated up to a 16-ton capacity. In addition, heavy-lift booms of the Stuelcken type that may be swung between king posts to an adjacent hatch have capacities of up to 120 tons.

The conventional general cargo ship possesses a high degree of flexibility, but does not meet requirements for rapid loading and discharge.

Future. The general cargo ship's developmental possibilities have been nearly fully exploited. In addition to modern booms, slewing cranes, or gantry cranes that can transverse the deck longitudinally, a general cargo will have side ports for use with fork trucks and stern ramps for RO/RO operations (see Fig. 10-10). The general cargo ship of the future could have longitudinal conveyor belts and athwartships conveyors to move units of cargo to their assigned locations on each deck from a computer-controlled elevator. With a floating, automatically operated rack type warehouse, high handling rates could be obtained with small units of cargo (see Fig. 10-11). Other possible ideas include the possibility of converting a general cargo ship to a bulk carrier with 'tweendecks that could swing up or down on hinges.

Containership

Present. The specialized ship known as the containership was unknown prior to 1960. Since then, there have been several stages in the

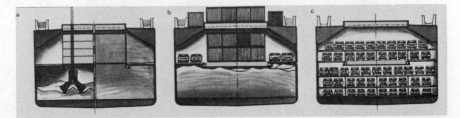

FIG. 10-9. Cross sections through the cargo holds of multipurpose cargo vessels (*a*) carrying bulk cargo; (*b*) carrying bulk cargo, containers, and motor vehicles; and (*c*) carrying motor vehicles. Courtesy of *Ships and Shipping of Tomorrow*, Cornell Maritime Press, Inc., and MacGregor Publications, Ltd.

FIG. 10-10. General cargo ship with universal cargo handling arrangements. The cargo can be handled through hatches, side ports, or a stern ramp, and can also be lifted out of lighters lying alongside the ship. Courtesy of *Ships and Shipping of Tomorrow*, Cornell Maritime Press, Inc., and MacGregor Publications, Ltd.

624

FIG. 10-11. Possible arrangement of a general cargo ship with conveyors for loading and unloading the cargo. Courtesy of *Ships and Shipping of Tomorrow*, Cornell Maritime Press, Inc., and MacGregor Publications, Ltd.

development of this type of vessel. Up to 1968 these vessels were conversions of other ship designs and had capacities of 500 to 700 TEU. The second-generation containerships, designed from the keel up to carry containers, had capacities of 1200 to 1500 TEU, with about 40% of those carried on deck. By 1972 the third generation of containerships included vessels carrying 1800 to 2200 TEU at 27 knots, 2000 TEU at 32 knots, and 2900 to 3000 TEU at 26 knots (see Fig. 10-12). At present, based upon the price of fuel, containerships are built to carry about 3000 TEU at speeds of about 22 knots. About two-thirds of the TEU capacity is carried below decks, with the remaining one-third above deck in three tiers. These vessels do not carry their own cranes, and are therefore not self sustain-

FIG. 10-12. Large third-generation containership for up to 3000 twenty-ft containers. Courtesy of *Ships and Shipping of Tomorrow,* Cornell Maritime Press, Inc., and MacGregor Publications, Ltd.

ing. As a result, all containerships run on liner routes between ports that have appropriate container handling cranes.

Future.   The containership, which today can replace three to five conventional general cargo ships, will in the future be able to replace more than ten. By the turn of the century, 70 to 80% of industrial products will be transported in containers between industrialized countries.[5] Service by containerships will probably be confined to between highly industrialized countries, and those ports that provide the best feeder service will obtain

[5] Ibid., p. 154.

the greatest percentage of this service. The super containership of the future may carry up to 6000 TEU at a speed of 40 knots, once lightweight and inexpensive nuclear plants become available. According to *Ships and Shipping of Tomorrow* (p. 154) the super containership will be moored in deep water with its stern toward shore. After the stern ramp is opened, the overhead crane tracks secured under the deck will be connected to the shoreside track system (see Fig. 10-13). Other overhead crane tracks will lead from the container storage area to a railhead, where national and international trains may be located. The crane tracks will be double so that loading and unloading can take place simultaneously. There will be no human control, as the entire process will be monitored by a central electronic data processing system. Container trains will also be controlled automatically. Only on the containership will there be a human crew (8 to 10 people), who will be replaced regularly using the gold and blue crew concept. Neither crew will concern themselves with the cargo handling process. Their only concern will be the safe and reliable navigation of their 6000-TEU, nuclear-propelled, 40-knot super containership.

Barge Carrier

Present. The barges or lighters carried aboard these specially designed vessels can be considered as floating containers. The LASH

FIG. 10-13. The super containership of the future loading and discharging via overhead crane tracks. Courtesy of *Ships and Shipping of Tomorrow,* Cornell Maritime Press, Inc., and MacGregor Publications, Ltd.

(Lighter Aboard SHip) barge carrier averages 43,000 deadweight tons (DWT) and can carry up to 83 lighters, each with a gross weight of 500 tons (see Fig. 10-14). A lighter crane on this vessel moves forward or aft to lift and position lighters. The lighters are lifted from the water between the forks of a cantilevered stern and moved forward to their hold locations. Barge carriers of the SEABEE type (see Fig. 10-15) utilize a lift platform at the stern to lift two barges (each 1070 tons) from the water to one of three cargo decks. Rail-mounted trolleys carry the barges forward to their stowage positions. These 38,000-DWT vessels can carry 26 such barges. Other types of barge carriers include the 2700-DWT "BACAT" (BArge CATamaran), which can carry thirteen 180-ton lighters, and the BACO-liner (BArge-COntainer-liner), the stowage of which involves floating twelve 800-ton barges into the flooded lower hold through a bow door and carrying 500 containers on deck. As can be seen from Table 10-6, the barge carrier surpasses all other types of vessels in average

FIG. 10-14. The LASH barge carrier. Courtesy of *Ships and Shipping of Tomorrow*, Cornell Maritime Press, Inc., and MacGregor Publications, Ltd.

TABLE 10-6

CARGO HANDLING RATES OF VARIOUS VESSEL TYPES [a]

Ship Type	Average Cargo Handling Rate, Tons/Hour
General cargo ship	70–80
Containership	300–600
RO/RO ship	200–400
Barge (LASH) carrier	1200–1500
Barge (SEABEE) carrier	2500–3000

[a] Information from R. Schonknecht, J. Lusch, M. Schelzel, and H. Obenaus, *Ships and Shipping of Tomorrow*, Cornell Maritime Press, Centreville, Maryland, and MacGregor Publications, Ltd., Hounslow, England, 1983, p. 23.

hourly cargo handling rate. Barge systems are, however, very expensive. Their cargo handling equipment is easily damaged, and weather will often interrupt their cargo operations.

Future. The future of the barge carrier may involve the deletion of the lighter crane or lift platforms for barges. The barges or lighters could be made more seaworthy and then be incorporated into seagoing push tows. These seagoing lighters could be towed as shown in Fig. 10-16 or pushed as shown in Fig. 10-17. The technical solution to the problem of enabling the couplings of the push tow to absorb all the forces in a seaway is not yet available. Since there is no need to enter port with these vessels, they may be nuclear propelled. In-port movement of the lighters or barges may be by steam- or diesel-propelled tugs, enabling compliance with port regulations barring nuclear-propelled vessels.

RO/RO Ship

Present. RO/RO ships have internal ramps for distributing their cargo to various decks. There are internal ramps to the weather deck, where vehicles or containers may be stowed. Loading or discharging of vehicles is via stern ramp, bow ramp, or side ramp. Stern ramps are more prevalent; they are either of the straight type arranged on the centerline, also known as the axial type, or of the type arranged at an angle of 35 to 45° to the ship's centerline, known as the quarter type. Ships with the quarter

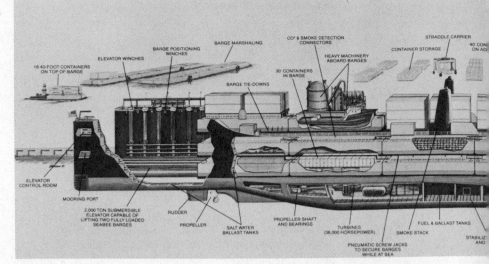

FIG. 10-15. Barge carrier of the SEABEE

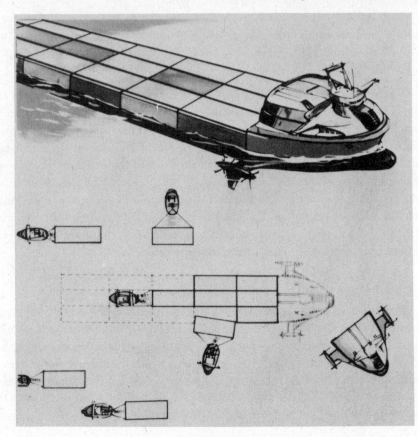

630

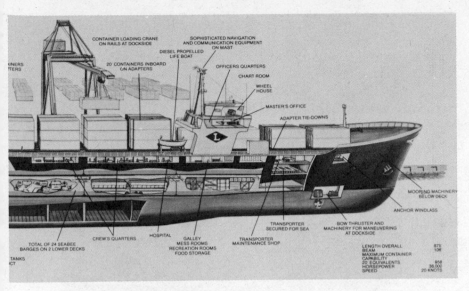

type. Courtesy of Lykes Lines.

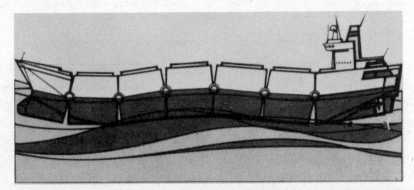

FIG. 10-17. Design for a large ship formed by lighters that are coupled together with movable joints. Courtesy of *Ships and Shipping of Tomorrow*, Cornell Maritime Press, Inc., and MacGregor Publications, Ltd.

FIG. 10-16. (*Opposite*) Seagoing push tow. The closely coupled barges are moved over the ocean leg of the journey by a special push tug, and are then distributed individually or in groups by small tugs. Courtesy of *Ships and Shipping of Tomorrow*, Cornell Maritime Press, Inc., and MacGregor Publications, Ltd.

FIG. 10-18. Combination container–RO/RO ship, originally constructed as steam turbine Principal particulars: length, 212.20 m; breadth, 28.0 m; depth, 19.3 m; draft, 9.3 m; dead-*and Shipping of Tomorrow,* Cornell Maritime Press, Inc., and MacGregor Publications, Ltd.

type ramp are obviously more versatile, and can be used for both liner service and tramping. RO/RO ships vary in deadweight tonnage from 6000 to 42,000 tons, depending on the trade route. Combination container–RO/RO ships can carry containers on deck or in forward container cells as well as vehicles (see Fig. 10-18). The car carrier RO/RO ship is unique in that all it carries is cars. Each of the cars is driven aboard via side ramps or stern ramps and then moved up or down internal ramps to its stowage location. The largest car carriers in service today are 20-knot vessels of 30,000 DWT that can carry up to 6200 medium-sized passenger cars on 13 decks.

Future. RO/RO ships are very expensive. While it is true that the loading or discharging time is only a few hours, thereby ensuring a quick turnaround time, it is also true that there is a substantial cost due to the loss in the use of all available cubic. Future development must involve methods of stowing the cargoes more closely. Automation of the cargo handling process seems virtually impossible, as vehicles must be driven

ship. To economize on fuel, ships of this type have been converted to diesel propulsion. weight, 18,500 tons; propulsion machinery, 25,740 kW; speed, 23 knots. Courtesy of *Ships*

on and off, and must be lashed down and secured for sea. To reduce the cost of RO/RO ships and improve the utilization of the available cubic, two types of vessels are envisioned. Which one is used will depend on the trade route. The RO/RO vessel for the coastal trade will be of 6000 DWT and have a single cargo deck (see Fig. 10-19). It will carry only containers, and will use the LUF (Lift Unit Frame) system to load and discharge the containers. The LUF system, developed in Sweden, consists of a low framework that can accommodate four 40-ft containers or up to six 20-ft containers. A wheeled chassis can be introduced under the framework for transporting containers on or off the vessel's single deck.

The second type of RO/RO vessel will be found on the deep-sea trade routes. The ship will have five or six cargo decks (see Fig. 10-20) or be a combination container–RO/RO ship (see Figs. 10-21 and 10-22). By utilizing a combination of stern, side, and bow ramps leading directly into the 'tweendeck area, the ramps between decks, which cause so much broken stowage, can be eliminated. The remaining cargo areas could be utilized for the stowage of containers. Multihull vessels could also provide a

FIG. 10-19. Container storage area and containership for the LUF system. The containers are stacked on the lift frames in blocks of four or six. Courtesy of *Ships and Shipping of Tomorrow,* Cornell Maritime Press, Inc., and MacGregor Publications, Ltd.

FIG. 10-20. Large RO/RO ship for transoceanic service. Courtesy of *Ships and Shipping of Tomorrow,* Cornell Maritime Press, Inc., and MacGregor Publications, Ltd.

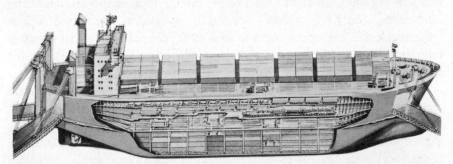

FIG. 10-21. Proposal for a large RO/RO ship with container cells in the lower holds, two stern ramps, and one bow ramp. Courtesy of *Ships and Shipping of Tomorrow,* Cornell Maritime Press, Inc., and MacGregor Publications, Ltd.

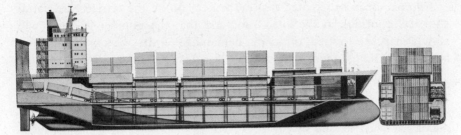

FIG. 10-22.   Proposed design of a combination container–RO/RO ship. The containers are loaded and unloaded into the squares of the hatches and onto the deck by means of cranes, whereas in the lower hold, trailers and vehicles are driven to the individual decks. Courtesy of *Ships and Shipping of Tomorrow,* Cornell Maritime Press, Inc., and MacGregor Publications, Ltd.

better capability for the stowage and rapid handling of RO/RO cargo. For versatility, the future may lie in such combinations of designs as the LO/RO ship (Lift-Off and Roll-On/Roll-Off). These vessels could be equipped with stern ramps, conventional breakbulk cargo holds, deck cranes, and container stowage facilities on deck and/or in cell guides of the forward hold. Bulk–RO/RO ships and BORO (Bulk, Ore, ROll units) are also being considered. Whatever the future holds, it will have to deal with the high capital cost of this type of vessel and with achieving maximum utilization of its available cubic.

Refrigerated Cargo Ship

Present.   In 1980 there are more than 900 refrigerated ships, with a total refrigerated hold capacity of 10 million metric tons. These vessels, with their insulated cargo holds and efficient refrigeration plants, are capable of carrying such temperature-sensitive cargo as fruits, meats, and fish. General cargo ships and refrigerated cargo ships have, despite the rush to larger vessels, remained at about 12,000 DWT (see Fig. 10-23).

FIG. 10-23.   The refrigerated cargo ship. Courtesy of *Ships and Shipping of Tomorrow,* Cornell Maritime Press, Inc., and MacGregor Publications, Ltd.

Future.   It is anticipated that in the near future the refrigerated cargo
ship will continue in the liner trade and that the number of ships will
slowly increase to meet the rising demand for fruits, vegetables, meats,
and fish. Although there will probably be no change in the size or carrying
capacity of these vessels, there should be gradual improvement in the
loading and discharging processes. The loading of individual cartons or
crates on conveyor systems through vessel side ports should give way to
palletized units loaded by forklifts through side ports or in the conventional LO/LO manner via cranes. The systems approach using automatic

FIG. 10-24.   Sail-assisted multipurpose cargo vessel. Courtesy of *Ships and Shipping of
Tomorrow,* Cornell Maritime Press, Inc., and MacGregor Publications, Ltd.

palletizers and computerized accountability, illustrated in Fig. 8-29, seems most feasible for the near future. Within the vessel itself it seems that the conveyor belt system as shown in Fig. 10-11 might be worthy of a feasibility study. The competition between the refrigerated container and the refrigerated cargo vessel will, no doubt, continue. However, it appears when analyzing all the materials handling principles that the refrigerated cargo ship will best serve the demands of this liner trade.

TABLE 10-7

PRINCIPAL CHARACTERISTICS OF NEXT-GENERATION CARGO LINER[a]

	Minimum Ship	Base Ship	Maximum Ship
Length overall	670' 0"	715' 0"	805' 0"
Length between perpendiculars	620' 0"	655' 0"	755' 0"
Beam, molded	105' 6"	105' 6"	105' 6"
Depth to main deck at side	69' 6"	69' 6"	69' 6"
Draft, design	31' 0"	31' 0"	29' 6"
Parallel middle body	None	45' 0"	135' 0"
Block coefficient	0.569	0.597	0.637
Displacement at design draft, LT	32,965	37,083	42,765
Deadweight at design draft, LT	18,700	22,011	26,030
Container capacity,			
Above deck	193	213	253
Below deck	694	768	916
Total	887	981	1169
RO/RO capacity (TEU w/trailers)	230	230	230
Breakbulk capacity, grain cubic feet	379,900	379,900	379,900
Shaft horsepower	22,700	27,100	44,100
Type of machinery	Diesel	Diesel	Diesel
Number of propellers	1	1	1
Service speed, knots	20	21	23
Fuel oil capacity, tons	5,080	5,080	5,080
Number of crew	30	30	30

[a] Information from: Maritime Administration, U.S. Dept. of Transportation, MARATECH, *R & D Technology Transfer Journal*, Vol. 4, No. 5, July 1983, p. 3.

CONCLUSION

Today vessel operators are changing over from steam to diesel propulsion, due to the high cost of fuel. As further alternatives, coal-fired and nuclear ship propulsion are being researched by the maritime industry. Results of a recent study indicate that wind propulsion by sail assist (see Fig. 10-24) is both economically and technically feasible and can reduce the cost of ocean shipping for small- to medium-sized vessels. Although the capital costs of sail-assisted vessels are comparable to those of conventional vessels, sail-assisted ships using a wing sail rig configuration burn some 15 to 25% less fuel at sea; their profitability is also enhanced by weather routing. Another MARAD Administration study recently developed a general cargo liner based strictly on commercial needs of a number of companies. This liner is now envisioned as a combination RO/RO–container carrier with options for design variations to suit individual shipowners and purposes (see Table 10-7). The study also identified the immediate research and development needs of the U.S. merchant fleet as they relate to solving the major technical and economic problems involved in designing and constructing this liner.

It appears that the basic ship types will still be around at the turn of the century. Specialized vessels such as heavy-lift vessels capable of lifting loads between 500 and 1000 tons, hydrofoil vessels, air cushion vessels, submarine tanker and cargo vessels, and wing-in-ground effect machines may be numerous by the turn of the century. Whatever type of vessel provides the best cost-benefit ratio for a given type of service will be adopted. It is also certain that there will be greater communication, cooperation, and coordination between the ship designers and the ports these vessels will serve. This is vital if the specific objective of achieving a rapid and cost-saving turnaround of vessels with efficient materials handling is to be realized.

APPENDIX A

STOWAGE FACTORS OF COMMONLY CARRIED BREAKBULK COMMODITIES

Commodity	Most Common Package(s)	Average Stowage Factor
A		
Abalone	Bags/cases	80/90
Acetate of lime	Drums/casks	80/90
Acetone	Drums/casks	80/90
Achiote (annatto)	Cases	60/64
Acid oil	Deep tank	42
Agar-agar	Bales	70/75
Ajwan seed (ajowan)	Bags	77/80
Alabaster (gypsum)	Bags	43/46
Albumen, dry	Tins in cases	57
Albumen, moist	Tins in cases	46
Alburnum	Bags	57
Alcohol	Barrels/drums	75
Alkanet	Cases	49
Alkyl benzene	Drums/bulk	50/41
Allspice (Jamaica pepper or pimento)	Bags	120/130
Almonds	Cases/bags/hogsheads	73/74/120
Aloes	Cases	40
Alum	Cases/casks	53/58
Alumina	Bulk	18/24
Aluminum	Ingots	29
Aluminum ore	Bags/bulk	40/28
Aluminum scrapfoil	Bags/bales	50
Alunite	Bulk	26
Amber	Cases	37
Amblygonite	Bulk	30
Animal meal	Bags	54
Aniseed	Bags	120
Aniseed oil	Tins in cases	50
Annatto	Barrels	91

Commodity	Most Common Package(s)	Average Stowage Factor
Antimony oil	Bags/cases/bulk	20/25/18
Apples	Barrels/cases/cartons	95/85/95
Apricot kernels	Bags/cases	85/90
Arachides (monkey nuts)	Cases/bags/unshelled	68/70/100
Arachis oil (ground nut oil)	Barrels/bulk	60/62
Aragoes	Cases	39
Archill	Casks/bales	41/88
Areca nuts (betel nut)	Bags	59
Arrack	Barrels/cases	72/55
Arrowroot	Cases/bags	70/53
Arsenic	Cases/kegs	24/33
Asafetida	Cases	43
Asbestos	Cases/bags	55/120
Asbestos cement	Bags	65
Asbestos ore	Bags	70
Asbestos powder	Bags	83
Asparagus (canned)	Cases	45
Asphalt (devil's drug)	Drums/barrels/bulk	45/50/33
Automobiles	Crates/unboxed	150–300

B

Bacon	Cases	59
Balata	Bags	57
Balsam copaiba	Kegs/bottles/cases	60/65/55
Bambara groundnut	Bags	67
Bamboo blinds	Bales	125
Bamboo poles	Bundles	70
Bamboo reeds	Bundles	67
Bamboo splits	Bales/bundles	100/200+

Commodity	Most Common Package(s)	Average Stowage Factor
Bananas	Cartons	135
Barbed wire	Reels	58
Barilla	Casks	34
Barjari	Bags	54
Barks	Bales	150
Barley	Bags/bulk	60/54
Basil	Bales	120
Bauxite	Bulk	29
Bdellium	Cases	78
Beans		
Alexandria	Bulk	48
Black	Bags	52
Castor	Bags	60
Cocoa	Bags	78
Dried	Bags	50
Horse	Bags	60
Jarry	Bags	100
Lima	Bags	65
Locust	Bulk	84
Soya	Bulk	50
Bean oil	Cases/barrels/bulk	53/63/39
Bêche-de-mer (trepang)	Bale	143
Beer	Casks/hogsheads/ cases/cartons	55/57/70/75
Beeswax	Barrels/cases	73/68
Belladonna	Bags	93
Benjamin	Bundles	53
Bennin seed	Bags	54
Beryllium ore	Bags	20
Bidi leaves	Bags	250+
Billets	Bags	12

Commodity	Most Common Package(s)	Average Stowage Factor
Birie leaves	Bags	250+
Bismuth metal	Boxes/ingots	11/7
Bismuth ore	Bags	31
Bitumen	Barrels/drums	45/50
Black lead	Kegs	49
Blackwood	Logs	53
Bleaching powder	Drums	78
Blende	Bulk	22
Blood, dried	Bags	43
Bones	Bulk/bags	94/65
Bone grist	Bags	67
Bone manure	Bags	65
Bone meal	Bags	43
Borax	Casks/bags	54/43
Boric acid	Barrels/cases	65/57
Boussir	Bags	65
Bran	Bags	110
Brazil nuts	Bags/bulk	92/83
Bricks	Bulk	25
Bristles	Cases	70
Brunak (poonac)	Bags/bales	79/76
Burlap	Bales	64
Butter	Cases/cartons/ kegs/tins	52/54/63/73
Butterfat, dried	Cases	77
Butternuts	Bags	103

C

Calaba beans	Bags	80
Calaena ore	Bags	17
Camphine	Drums	63
Camphor	Cases (tin lined)	63

Commodity	Most Common Package(s)	Average Stowage Factor
Camphor oil	Drums	70
Canada balsam	Casks	53
Canary seed	Bags	53
Canella alba	Bales	130
Canes	Bundles/bales	100/120
Canned goods	Cases/cartons	55/60
Caoutchouc capsicums	Varies	153
Caraway seed	Bags	60
Carbon black (lampblack)	Kegs/bags/ paper cartons	117/113/110
Cardamoms	Cases/bags	95/103
Carnarina	Bags	48
Carob	Bulk	87
Carpets	Bales/rolls	170/150 (approx.)
Casein	Bags/cases	68/65
Cashew kernels	Cases	70
Cashew nuts	Bags	75–90
Cashew nut oil	Bulk/drums	38/88
Casings	Casks/kegs	73/63
Cassia	Cases/bundles/bales	163/115/160
Cassia buds	Cases	133
Castor oil	Barrels	61
Castor seeds	Bags	83
Catechu (Cutch)	Cases/bags	54/63
Cattle meal	Bags	73
Caustic soda	Drums	34
Caviar	Cases	53
Celery seed	Bags	75
Cement	Casks	42
Cement, boiler	Double bags	64
Cement, color	Drums	63
Cement, Portland	Casks/bulk/drums	38/22/37

Commodity	Most Common Package(s)	Average Stowage Factor
Cerasein	Barrels	67
Chalk	Barrels/bulk	43/38
Charcoal (wood)	Bags/bulk	100/180
Chasam (chussums)	Bales	77
Cheese, processed	Cases	57
Chestnuts	Cases/bags	123/190
Chickpeas (grams)	Bags	57
Chicle	Bags	57
Chicory	Bags/tins in cases	58–66/75
Chilies, dry	Bags/bundles	103/225
China bark (cinchona)	Bales	110
China clay (kaolin)	Bulk/casks/bags	39/44/48
China grass (rhea fiber)	Pressed/bales	48/70
China root	Cases/bags	94/103
China wood oil (tung oil)	Barrels/tins in cases/ bulk	64/57/38
Chiretta	Bundles	350–400
Chrome ore	Bulk/cases	13/16
Chussums	Bales	77
Chutney	Casks	39
Cigars and cigarettes	Cases	175
Cinchona (Peruvian bark)	Bales	140
Cinnabar	Bags	19
Cinnamon	Bundles/cases	135/100
Citron	Bags/cases	70
Citronella	Drums	77
Civet leaves	Bags	240
Clay	Bulk	20
Clover seed	Bags/barrels	53/50
Cloves	Bales	115

Commodity	Most Common Package(s)	Average Stowage Factor
Clove oil	Drums	63
Cobalt	Bulk/bags	19/12
Coca	Bags	85
Cochineal	Cases (tin lined)	63
Cocoa (cacao)	Bags/cases	75/88
Coconuts	Bags/bulk	95/98
Coconut cake	Bags/bales	79/76
Coconut, desiccated	Cases	75
Coconut fiber (coir)	Bales	100
Coconut oil	Casks/drums/bulk	67/60/39
Coconut shell, broken	Bags	63
Coffee	Bags	70
Coke	Bulk	70–100
Colocynth	Bales	105
Colombite ore	Bags	21
Colombo root, meal	Bags	95
Colombo root	Bales/bags	100/110
Colza oil (rape oil)	Barrels	61
Colza seed (rapeseed)	Bulk	35
Condensed milk	Tins in cases	45
Copal, gum	Cases	80 (approx.)
Copper	Ingots/barrels	11/14
Copper concentrate	Bulk/bags/slabs	15/25/12
Copper cuttings	Bales	115
Copper matte	Bulk/bags	17/20
Copper pyrites, yellow sulfur	Bulk	21
Copperas	Casks	55
Copra	Bags/bulk	102/73

Commodity	Most Common Package(s)	Average Stowage Factor
Copra cake	Bags	63
Copra expeller pellets	Bulk	63
Copra meal	Bags	83
Coquilla nuts	Bags	63
Coral	Bags	90
Corestock	Crates	140
Coriander seed	Bags	123
Cork	Bales/hard pressed	300–420/220
Cork shavings	Bales/bags	280–300
Corn (maize)	Bulk/bags	48/53
Corn flour	Bags	48
Corozo nuts	Bags	63
Corundum ore	Bags/bulk	43/23
Costusroot	Bags	115
Cotton	Bales	79
Cottonseed	Bulk/bags	64–100
Cottonseed cake	Bales/loose	52/58
Cottonseed oil	Barrels/bulk	61/39
Cotton waste	Bales	150
Cowgrass seed	Sacks	62
Cowrie shells	Bags	47
Cowtail hair	Bales	73
Cream of tartar (Argal)	Barrels	63
Creosote	Barrels/drums/bulk	59/60/67
Cubeb (tailed pepper)	Bags	80
Cube gambler	Package	115
Cubic niter– nitrate of soda	Bags	37
Cubic niter– saltpeter	Bags/bulk	37/34
Cudbear	Bales	87
Cuminseed	Bags	80–105+

Commodity	Most Common Package(s)	Average Stowage Factor
Curios	Cases	150–180
Currants	Cases	48
Cuttlefish	Cases	150
D		
Damar	Cases/bags	67/65
Dari jowaree	Bags	55
Detergents	Bulk liquid	45
Dhall dal	Bags	50–75
Divi-divi	Bags	117
Dodder (Hellweed)	Bags	105
Dogs' droppings	Bags	67
Dog skins	Bales	105
Dogtail seed	Sacks	88
Dom nuts	Bags	55–120
Dragon's blood	Cases	85
Dried blood	Bags	57
Dripping	Casks	73
Dur-dur	Bags	87
Durra	Bags	57
E		
Ebony	Bulk	58
Egg albumen	Cases/crates	95/110
Eggs, desiccated	Cases	74
Egg yolk liquid	Tins in cases	43
Egg yolk powder	Tins in cases	78
Emery	Bags	25
Esparto grass	Bales	130–170
Ethyl hexanol	Bulk	45
F		
Farina	Bags/cases	57
Feathers	Pressed bales	65–90

Commodity	Most Common Package(s)	Average Stowage Factor
Feldspar	Bags	35–40
Felt	Rolls/bales	60–150
Fennel seed	Bags	95
Ferromanganese	Bags	16
Fescue seed	Sacks	130
Fiber root	Bales	185
Figs	Cases/baskets	47/50
Firecrackers	Cases	95
Fish manure (meal)	Bales/bags	62/65
Fish oil	Barrels/bulk	58–62/39
Flavin	Cases	44–48
Flax	Bales	85–155
Flax seed	Bags/bulk	60/50
Flour	Barrels/sacks/bags	60/49/45
Fluorine	Cases	85–105
Flourite (Fluorspar)	Mats	34
Foggrass seed	Sacks	170
Formaldehyde	Barrels	55–60
Fossil wax	Drums	67
Frankincense	Drums	63
Fruit, dried	Cases	73
Fuller's earth	Bags	40
Furs	Cases, boxes, or bales	100–150
Fustic	Bulk	70–90

G

Galangel	Bales	95–150
Galena (lead ore)	Bags/bulk	16/13
Galls (gallnuts)	Cases/bag	78–100/83
Gallnut extract	Cases	45
Galvanized iron	Packages/coils	26/30
Gambier	Bales/cases/baskets	97/110/120

Commodity	Most Common Package(s)	Average Stowage Factor
Gamboge	Cases	68
Garlic	Bags	95
Garnet ore	Bags	19
Gasoline	Cases/drums	51/63
Gelatin	Cases/bags	65/60
Geneva (Hollands Schiedam)	Cases	60
Gentian root	Bales	123
Ghee	Tins in cases	53
Gingelly (Sesame)	Bags	59
Ginger	Cases/casks	73/61
Gingerroot	Bags	88
Ginseng	Baskets	115
Glass	Crates	50
Glaxo	Cartons/bags	80/95
Glucose	Cases/barrels	43/47
Glue	Bales/drums/casks/bags	175/120/115/125
Glue refuse	Casks	48
Glycerin	Cases/drums/bulk	42/50/29
Gold slag	Bags	15
Granite	Blocks	17
Graphite (plumbago)	Kegs/cases/bags	49/47/43
Grass cloth	Cases	100–110
Grass seeds	Bags	50–100
Greases	Barrels/drums	55/65
Guano	Bulk/bag	40/43
Guinea corn	Bags	52
Gum, gum resin	Bags	50–65
Gum olibanum (bdellium)	Cases/bales	68/66
Gunny sacks	Bales	50–65
Gutta-percha (gutta)	Bags/cases	85/80
Gypsum	Bags/bulk	44/38

Commodity	Most Common Package(s)	Average Stowage Factor
H		
Hair, human	Cases/bales	70/62
Hay	Bales	120–160
Hematite	Bulk	12–15
Hemp	Pressed bales	92–105
Hemp seed	Bags/bulk	70/53
Henequen (sisal hemp)	Bales	90–130
Herbs	Cases/bags	190/120
Herring	Barrels	45–62
Hessians	Bales	50–65
Hominy, chopped	Bags	53–60
Honey	Barrels/drums	47/40
Hoof tips	Bags	90–100
Horns	Loose	100
Horn shavings	Bags	150–180
Horse hair	Bales	70–100
I		
Illipe nuts	Bags	63
Ilmenite sand	Bulk	13
Indigo, paste	Casks	77
Indigo, powder	Cases	62
Infusorial earth	Bags	90
Iron, pig	Bulk	11
Isinglass	Casks/bales	91/98
Istle (Ixtle)	Bales	95
Ivory	Bundles/cases	33/57
Ivory nuts	Bags	63
Ivory scrap	Barrels	67
J		
Jaggery	Bags	47
Jarrah wood	Bulk	30

Commodity	Most Common Package(s)	Average Stowage Factor
Jarry beans	Bags	100
Jelatong	Cases	73
Jowaree dari (jowar)	Bags	54
Juniper berries	Bags	81
Jute	Bales	66
Jute butts	Bales	59
Jute caddles	Bales (pressed)	88
Jute cuttings	Bales	56
K		
Kainite	Bag/bulk	39/36
Kapok	India bale/pressed bale	63/145
Kapok seed oil	Bulk	39
Kauri gum	Bags/cases	70/63
Kerosene	Cases/drums/barrels	51/63/64
Kieselguhr	Bags	52
Kola nuts	Bags	75
Kolai	Bags	57
Kraft liner board	Reels	57
Kyanite	Bulk/bags	20/33
L		
Lac dye	Cases	77
Lampblack (carbon black)	Kegs/bags/cartons	117/113/110
Lard	Cases/pails/tierces	57/73/63
Lard oil	Barrels	63
Latex	Bulk/drums	37/53
Lead concentrates	Bulk	13
Lead, pig	Pigs/ingots	9/11
Lead, sheet	Cases/rolls	21/18
Lead, white	Kegs	26
Leather	Bales/cases/rolls	85/70/up to 200

Commodity	Most Common Package(s)	Average Stowage Factor
Lentils	Bags	53
Licorice	Bales	96
Licorice extract	Barrels	63
Licorice extract paste	Cases	47
Lignum vitae	Bulk	27
Lily bulbs	Cases	53
Lime, hydrated (calcium hydrate)	Bags/casks	41/46
Lime, borate of	Bags	52
Lime, cirate of	Bags	83
Lime juices	Cases/barrels	55/75
Linen, flax	Bales	100
Linen, tow	Bales	100
Linoleum	Rolls	70
Linseed	Bags/bulk	59/56
Linseed cake	Bags/bulk	50/55
Linseed oil	Barrels/bulk	67/38
Liqueurs	Bottles in cases	70
Locust beans	Bulk	84
Locust meal	Bags	85
Logs	Bulk	75–130
Logwood	Bundles	90
Lubricating oil	Cases/barrels/drums	47/59/56
Lucerne	Bags	107
Lupin seed	Sacks	65
M		
Macaroni	Cases	85
Mace	Cases/bags	90/73
Madder	Bales	65–80
Magnesite, dead burned	Bulk	25
Magnetite	Bulk	16

Commodity	Most Common Package(s)	Average Stowage Factor
Mahogany (acajou)	Logs/boards	28/43
Mail	Bags	100–150
Maize (hominy)	Bulk/bags	48/53
Malt	Bags/tanks	93/67
Manganese ore	Bulk/bags	17/24
Mangrove bark	Bags/bales	77/75
Manioc, mandioca meal	Cases/bags/bulk	90/95/63
Manjeet	Bales	65–80
Manola	Bags	78
Manure	Bags/bulk	40–55/43
Marble	Blocks/slabs/crates	16/19/21
Marjoram	Bales	200–240
Matches	Cases	100–130
Mathie seed (methey)	Bags	95
Mats, matting	Mats, rolls, or bales	100–220
Meat meal	Bags	83
Metal borings and cuttings	Bulk	25–50
Methylated spirits	Barrels/drums	73/71
Mica talc	Cases/bags	40/50
Middlings, semitin pollards	Bags	75
Milk, malted	Bottles in cases	93
Milk, powdered	Cartons/cases/bags	63/80/68
Millet durra	Bags	51
Mirabolans	Pockets/bags	71/73
Mirabolan extract	Bags/cases	47/36
Mohair	Bags/bales	250/125
Molasses	Casks/bulk	55/27
Molasses, concentrated	Baskets	41
Molybdenum ore	Bags	20
Monezite	Bags	50

Commodity	Most Common Package(s)	Average Stowage Factor
Morocco leather	Rolls/bales/cases	200/100/110
Moss	Bales	300–600
Mother of pearl	Cases	46
Mowa (mowrah)	Bags	65
Mowa or mowrah cake	Bags	59
Musk (musquash)	Bales/cases	100/150
Mustard seed	Bags	56–70
Mustard seed oil	Cases/drums	50/54

N

Nails	Cases/bags/kegs	30/27/33
Neat's-foot oil	Barrels/drums	61/73
Nickel ore	Barrels/bags/ingots	24/20/12
Niger seed	Bags	63
Niger seed cake	Bags	55
Niger seed oil	Barrels	62
Nitrates	Bags	40
Nutmegs	Cases	60–70
Nuts	Bags	Varies
Nux vomica	Bags	63

O

Oak	Staves	53
Oakum	Bales/pressed bales	95/73
Oatmeal	Bags	68
Oats, clipped	Bags/bulk	74/66
Oats, unclipped	Bags/bulk	83/76
Ochre puree	Drums/barrels/bags	50/60/80
Oil cakes	Bags	55–70
Oiticica oil	Drums	33
Oleostearin	Bags	68
Olives	Kegs	69

Commodity	Most Common Package(s)	Average Stowage Factor
Olive oil	Barrels/drums	60/62
Onions	Cases/crates/bags	80/82/88
Oranges	Cases/cartons	88/63
Orange oil	Barrels/cases	50/70
Orchella	Bales	95
Ores	Bulk/bags	Varies
P		
Paint	Drums	19
Palm kernels	Bulk/bags	55/65
Palm nut oil	Barrels/bulk	59/38
Palm oil	Barrels/bulk	59/39
Paper	Rolls	85
Paper pulp (dry)	Bales	47
Paper pulp (wet)	Bales	54
Paraffin wax	Bags/barrels/cases	53/73/71
Patchouli patch	Bales/bulk	160/400
Peas	Bags/bulk	53/48
Peacake	Bales	51
Peanuts	Bags	63
Pea pulp	Bags	67
Pepper (black)	Bags	86
Pepper (white)	Bags	75
Phormium	Bales	95
Phosphate rock	Bulk	34
Piassaba (piassava)	Bales	105
Pimento	Bags	125
Pineapples	Cases/crates	58/73
Pistachio nuts	Cases/drums	73/63
Pitch	Barrels/bulk	48/33
Plantains	Bunches	130
Plywood	Crates/bundles	77/73

Commodity	Most Common Package(s)	Average Stowage Factor
Podophyllum ernodi	Bags	95
Poppy seed	Bags	71
Poppy seed cake	Bags	57
Pork	Tierces	61
Potash	Bags/bulk	37/33
Potash, caustic	Drums	28
Potassium, chlorate of	Kegs	56
Potatoes	Barrels/bags/cartons	73/60/62
Prawns	Bags/cartons	85/100
Preserved meats	Cases	60
Prunes	Casks/cases	55/51
Pumice	Bags	110
Puree, yellow ochre	Barrels	55
Putchok (Patchuk)	Bales	135
Pyrethrum flower seeds	Bales	70
Pyrites	Bulk	25
Q		
Quebrache extract	Bags	40
Quebracho	Bundles	70
Quebracho extract	Bags	42
Quercitron	Bales/bags	85/87
R		
Rabbit skins	Bales	170
Raffia grass	Bales/pressed bales	170/125
Rags	Bales/pressed bales	175/65
Railway iron	Rails/ties	14/45
Raisins	Cases	51
Ramie	Bales/pressed bales	83/53
Rapeseed	Bags	63
Rapeseed cake	Bags	52
Rapeseed oil	Bulk	35

Commodity	Most Common Package(s)	Average Stowage Factor
Rattans	Bundles	140
Rattan core	Bales	145
Rattan, split, peeled	Bales	150
Rhea fiber	Bales/pressed bales	83/53
Rice, broken	Bags	54
Rice, cargo	Bags	49
Rice dust (boussir)	Bags	65
Rice meal	Bags	67
Rice paper	Cases	83
Rope, coir	Coils	95
Rope, manila	Coils	83
Rope, sisal	Coils	83
Rope, wire	Coils	30
Rope seed	Bags	63
Rosin	Barrels	59
Rubber	Cases/bales/bags/ sheets/crepe	69/66/66/ 60/120
Rubber latex	Bulk/drums	37/53
Rum	Hogsheads/casks/cases	66/70/58
Rush hats	Bales	140
Rutile sand	Bags/bulk	28/20
Rye	Bulk/bags	50/55
Rye grass seed	Sacks	130

S

Safflower	Bales	68
Safflower oil	Barrels	63
Safflower seeds	Bags	70
Saffron	Cases	72
Sago	Bags	54
Sago flour	Bags	51

Commodity	Most Common Package(s)	Average Stowage Factor
Salmon	Tins in cases/ barrels	51/47
Salt	Bulk/barrels/bags	37/51/39
Saltpeter	Bags	37
Samp	Bags	52
Sand	Bulk/cases	19/60
Sandalwood	Bales	117
Sandalwood powder	Cases	95
Sapanwood	Bulk	105
Sardines	Cases	46
Sarsparilla root	Bales	170
Sausage skins	Casks/kegs	73/63
Scheelite ore	Bags	26
Scrap iron	Bulk/bales	35/50
Semolina	Bags	61
Senna leaves	Bales/bags	120/200
Sesame	Bags	59
Shea nuts	Bags	70
Sheepskins	Bales	140
Sheep wash (sheep dip)	Drums	53
Shellac	Lined cases	83
Shells	Cases	46
Shooks	Bundles	97
Shrot	Bags	67
Shumac	Bags	71
Silk cocoons	Bales/pressed bales	215/70
Silk, piece	Cases	120
Silk, punjam	Cases	170
Silk, raw	Bales	105
Silk, waste	Bales/pressed bales	160/60
Silk, waste	Bundles	115

Commodity	Most Common Package(s)	Average Stowage Factor
Skins	Bales/pressed bales/loose	150/87/185
Slag, basic	Bags/bulk	30/32
Slates	Loose/cases	19/25
Soap	Cases/kegs	65/43
Soda ash	Barrels/bags	46/43
Sooji (suji)	Bags	93
Sorghum	Bulk	48
Soy	Casks	47
Soya beans	Bulk/bags	50/65
Soya bean cake	Bags/bulk	73/68
Soya bean oil	Cases/barrels/bulk	53/63/39
Spaghetti	Cases	95
Spelter	Ingots	10
Spermaceti	Cases/cartons	65/70
Sponges	Bales/pressed bales	160/120
Starch	Cases/barrels	52/67
Staves	Bundles	98
Stearin	Bags/casks	52/63
Steel	Bars/billets/plates/pigs	14/12/11/11
Straw braid (straw plait)	Bales/cases	215/180
Sugar, dry	Hogsheads/bags/bulk	54/47/43
Sugar, green	Bags/baskets	41/49
Sulfur (brimstone)	Bulk/barrels/bags	31/47/36
Sultanas	Cases	51
Sunflower oil	Barrels	63
Sunflower seeds	Bags	105
Sunflower seed cake	Bulk	59
Superphosphate	Bags/bulk	38/40
Syrup	Barrels/hogsheads	50/44

Commodity	Most Common Package(s)	Average Stowage Factor
T		
Talc	Bags	39
Tallow	Barrels/drums/cases/bulk	64/73/46/39
Tamarina	Barrels/cases/bags	43/39/50
Tamarind seed	Bags	65
Tankage	Bags	63
Tapioca, flake	Bags	69
Tapioca, flour	Bags	55
Tapioca, pearl	Bags	63
Tapioca, seed	Bags	63
Tar, coal	Barrels	50
Tar, pine	Barrels	56
Tea	Chests, boxes, or cases	105
Tea dust	Boxes	90
Tea shooks	Cases	62
Teak	Boards/logs/planks	77/85/90
Tejpatta	Bales	275
Thyme	Bales	165
Thymol oil	Cases	53
Ties	Creosoted timbers	45
Tiles, bulk	Bulk	40
Tiles, fire clay	Crates	50
Tiles, roofing	Crates	80
Timothy seed	Bags	69
Tin	Ingots	9
Tincal	Casks/bags	55/44
Tin clippings	Pressed bales	39
Tinfoil	Bundles/cases	23/31
Tin ore	Bags	20

Commodity	Most Common Package(s)	Average Stowage Factor
Tin plates	Packs/boxes	12/17
Tires	Loose	100
Tobacco	Hogsheads/cases/ cartons	190/120/135
Tokmari seed	Bags	57
Tomatoes	Crates/boxes	70/75
Tonka beans	Barrels/boxes	90/70
Tortoise shell	Cases	142
Tow	Bales	100
Tripolite	Bulk	75
Tung oil	Barrels/tins in cases/bulk	64/57/38
Tungsten ore	Bags	17
Turmeric	Bags	83
Turpentine	Barrels/tins in cases/drums	63/55/60
Twine	Bales	65
U		
Umber	Bulk	41
Uranium	Bags	17
V		
Vaionia	Bags/bulk	103/93
Vanadium ore	Bags/bulk	23/25
Vanilla	Cases/bags	63/53
Varnishes	Cases/barrels	57/63
Vaseline	Barrels/cases	60/53
Vegetable fiber	Bales	175
Vegetable oil	Bulk	39
Vegetable wax	Boxes/bags	46–70/57
Veneers	Crates	135

Commodity	Most Common Package(s)	Average Stowage Factor
Vermillion	Cases	45
Vinegar	Barrels	63
W		
Walnuts	Bags	120
Walnut meat	Cases/bags	85/150
Waste	Bales	185
Wattles	Bags	47
Wax	Cases/barrels/bags	53/73/58
Whalebone	Cases/bundles	67/113
Whale meal	Bags	70
Whale oil	Drums/bulk	75/41
Wheat	Bulk/bags	48/53
Whisky	Barrels/cases	69/63
Whitening (whiting)	Barrels/bags	43/38
Wines	Casks/cases	61/65
Wire, barbed	Reels	58
Witherite	Bulk	19
Wolfram	Bags	17
Wood pulp	Bales	50
Wool	Bales, pressed	58
Wool grease	Barrels/drums	60/64
Wood chips, softwood	Bulk	105
Wood chips, hardwood	Bulk	82
Y		
Yams	Bags or crates	92
Yarrow seed	Sacks	90
Z		
Zinc	Ingots	10
Zinc ash	Bags	40

Commodity	Most Common Package(s)	Average Stowage Factor
Zinc concentrates	Bulk/ingots	21/15
Zinc dross	Cases	21
Zinc dust	Cases or drums	23
Zinc, white	Drums or kegs	23
Zircon sand	Bags/bulk	25/30
Zirconium ore	Bags/bulk	20/18

APPENDIX B

PROBLEMS PERTAINING TO MARINE CARGO OPERATIONS

SECTION I:
VERTICAL WEIGHT DISTRIBUTION
OR STABILITY PROBLEMS

Common formulas used:

1. KG or V.C.G. $= \dfrac{\text{Total vertical moments}}{\text{Total weights}}$

2. $GM = KM - KG$

3. Shift of weight: $GG' = \dfrac{\text{Weight } (W) \times \text{Distance } (D)}{\Delta}$

4. Adding or removing a weight: $GG' = \dfrac{W \times D}{\Delta \pm W}$

5. Free surface correction—use tables provided for type of ship or for individual tank:

$$GG' = \frac{r \cdot l \cdot b^3}{12V}$$

$r = \dfrac{\text{Specific gravity (S.G.) of liquid}}{\text{S.G. of salt water}}$

l = length of tank.
b = breadth of tank.
$V = \Delta \times 35$.

Problem #1

A ship in light condition displaces 2000 tons, the KG being 10 feet. The KM in the loaded condition is 12 feet. 3500 tons of cargo are loaded 9 ft above the keel and 400 tons of fuel 15 ft above the keel. Find the GM.

Answer

	W (tons)		V.C.G. (ft)		Moment (ft-tons)
Lt. ship	2000	×	10	=	20,000
Cargo	3500	×	9	=	31,500
Fuel	400	×	15	=	6,000
Totals Δ =	5900				57,500

$$KG = \frac{57,500}{5900} = 9.74 \text{ ft}$$

$$KM = 12 \text{ ft}$$
$$- \ KG = \ \ 9.74 \text{ ft}$$
$$GM = \ \ 2.26 \text{ ft}$$

Problem #2

A ship displacing 15,000 tons has a KG of 30 ft. Two thousand tons of fuel oil is burned from the double bottoms; this mass had a KG of 2 ft. What is the ship's new KG?

Answer

$$D = KG \text{ (ship)} - KG \text{ (F.O.)} = 30 - 2 = 28 \text{ ft}$$

$$GG' = \frac{W \times D}{\Delta - W} = \frac{2000 \times 28}{15,000 - 2000} = \frac{56,000}{13,000} = 4.3 \text{ ft}$$

New KG = 30 ft + 4.3 ft = 34.3 ft

OR

W (tons)		V.C.G. (ft)		Moment (ft-tons)
15,000	×	30	=	450,000
−2,000	×	2		−4,000
13,000				446,000

$$\text{New } KG = \frac{446,000}{13,000} = 34.3 \text{ ft}$$

Problem #3

A vessel displaces 16,000 tons and has a *KG* of 18.5 ft. What will her *KG* be after 600 tons is loaded into a deep tank, with the center of gravity of the 600 tons at a *KG* of 10.5 feet?

Answer

$$D = KG \text{ (ship)} - KG \text{ (load)} = 18.5 - 10.5 = 8 \text{ ft}$$

$$GG' = \frac{W \times D}{\Delta \pm W} = \frac{600 \times 8}{16,000 + 600} = 0.3 \text{ ft}$$

$$\text{New } KG = 18.5 - 0.3 = 18.2 \text{ ft}$$

OR

W (tons)		V.C.G. (ft)		Moment (ft-tons)
16,000	×	18.5	=	296,000
+ 600	×	10.5	=	+ 6,300
16,600				302,300

$$\text{New } KG = \frac{302,300}{16,600} = 18.2 \text{ ft}$$

SECTION II:
LONGITUDINAL WEIGHT DISTRIBUTION
OR TRIM PROBLEMS

Common formulas used:

1. $\text{L.C.G.} = \dfrac{\text{Total long. moments}}{\text{Total weights}}$

2. $\text{Change in trim} = \dfrac{W \times D}{MT1}$

3. $\text{Mean sinkage} = \dfrac{W}{\text{T.P.I.}}$

4. Trim lever = difference between L.C.G. and L.C.B.

5. Change in trim = $\dfrac{\text{Trim lever } (GG') \times W}{MT1}$

6. If tipping center is amidships:

$$\text{Change in draft} = \frac{\text{Change in trim}}{2}$$

7. If tipping center is not amidships:

$$\text{Change in fwd. draft} = \frac{\text{L.C.F.} - \text{F.P.}}{\text{L.B.P.}} \times \text{Chg. in trim}$$

Then:

Chg. in trim − Chg. in fwd. draft = change in draft aft

Problem #1

A ship with a $MT1$ of 1200 pumps 200 tons of fuel oil from a forward tank to an after tank, a distance of 60 ft. The drafts before the shift were as follows: forward, 23 ft 08 in.; aft, 23 ft 04 in.; and mean, 23 ft 06 in. What are the total change in trim and the final drafts?

Answer

① Change in trim = $\dfrac{W \times D}{MT1} = \dfrac{200 \times 60}{1200}$

= 10 in. total change (by stern)

② Change in draft = $\dfrac{10}{2}$ = 5 in Fwd. and aft (assuming tipping center is amidships)

	Fwd.	Mean	Aft
Initial draft	23 ft 08 in.	23 ft 06 in.	23 ft 04 in.
Change	− 5 in.	0 in.	+ 5 in.
Final draft	23 ft 03 in.	23 ft 06 in.	23 ft 09 in.

Problem #2

A ship has a $MT1$ of 1000. 150 tons are loaded 100 ft aft of the tipping center. The drafts before the loading were: fwd., 19 ft 02 in.; aft, 19 ft 04 in.; and mean, 19 ft 03 in. The T.P.I. is 50. What is the total change in trim and the final drafts?

Answer

① Mean sinkage $= \dfrac{W}{\text{T.P.I.}} = \dfrac{150}{50} = 3$ in.

	Fwd.	Mean	Aft
Initial draft	19 ft 02 in.	19 ft 03 in.	19 ft 04 in.
Sinkage	+ 3 in.	+ 3 in.	+ 3 in.
	19 ft 05 in.	19 ft 06 in.	19 ft 07 in.

② Change in trim $= \dfrac{W \times D}{MT1} = \dfrac{150 \times 100}{1000} = 15$ in.

③ Change in draft $= \dfrac{15}{2} = 7.5$ in. down by stern

	Fwd.	Mean	Aft
	19 ft 05 in.	(Stays	19 ft 07 in.
Change	− 7.5 in.	same)	+ 7.5 in.
Final draft	18 ft 09.5 in.	19 ft 06 in.	20 ft 02.5 in.

Problem #3

A vessel has drafts of 26 ft 10 in. fwd. and 31 ft 02 in. aft. Her $MT1$ is 1850 ft-tons and her T.P.I. is 68. Find the final drafts if 410 tons are loaded at a point 205 ft abaft the tipping center. Assume tipping center is amidships.

Answer

① Mean sinkage $= \dfrac{W}{\text{T.P.I.}} = \dfrac{410}{68} = 6$ in.

	Fwd.	Mean	Aft
Before loading	26 ft 10 in.	29 ft 00 in.	31 ft 02 in.
Mean sinkage	+ 6 in.	+ 6 in.	+ 6 in.
	27 ft 04 in.	29 ft 06 in.	31 ft 08 in.

② Change in trim = $\dfrac{W \times D}{MT1} = \dfrac{410 \times 205}{1850} = 45$ in.

③ Change in draft = $\dfrac{45}{2} = 22.5$ in. = 1 ft 10.5 in.

	Fwd.	Mean	Aft
Draft	26 ft 16 in.	(Stays	31 ft 08 in.
Change	−1 ft 10.5 in.	same)	+1 ft 10.5 in.
Final draft	25 ft 05.5 in.	29 ft 06 in.	33 ft 06.5 in.

SECTION III:
STABILITY AND TRIM PROBLEMS

Problem #1

Given:

	W	V.C.G.	L.C.G.–F.P.
Light ship	10,000 tons	35 ft.	350 ft.
Fwd. cargo spaces	5,000	20	200
Aft cargo spaces	4,000	25	550
Passengers/baggage	500	45	400
Fuel oil & water	2,500	10	300
	22,000 tons		

KM at 22,000 tons = 28.5 ft.
*MT*1 at 22,000 tons = 2200 ft-tons.

Mean draft at 22,000 tons = 29 ft 00 in.
L.C.B. is 344.1 ft from F.P.
Free surface correction = 0.6 ft.
   Required: Final *GM* and drafts.

Answer

Stability

	W	V.C.G.	Vertical Moments
Light ship	10,000 tons	35 ft	350,000 ft-tons
Fwd. cargo	5,000	20	100,000
Aft cargo	4,000	25	100,000
Passengers/baggage	500	45	22,500
Fuel oil & water	2,500	10	25,000
	Δ = 22,000 tons		597,500 ft-tons

$$\text{New } KG = \frac{597,500}{22,000} = 27.2 \text{ ft}$$

$$
\begin{aligned}
KM &= 28.5 \text{ ft} \\
- KG &= 27.2 \\
\hline
GM &= 1.3 \\
\text{Free surface correction} &= -0.6 \text{ ft} \\
\hline
GM &= 0.7 \text{ ft}
\end{aligned}
$$

Trim

	W	L.C.G.	Long. Moments
Light ship	10,000	350	3,500,000
Fwd. cargo	5,000	200	1,000,000
Aft cargo	4,000	550	2,200,000
Passengers/baggage	500	400	200,000
Fuel oil & water	2,500	300	750,000
	Δ = 22,000 tons		7,650,000 ft-tons

① L.C.G. = 7,650,000/22,000 = 347.7 ft

② L.C.G. 347.7 ft (Trim Aft) − L.C.B. 344.1 ft = 3.6 ft lever arm

③ 3.6 ft × 22,000 tons = 79,200 ft-tons

④ 79,200/$MT1$ of 2200 = 36 in. total change in trim

⑤ 36/2 = 18 in. = 1 ft 06 in.

⑥

	Fwd.	Mean	Aft
Draft	29 ft 00 in.	(Stays	29 ft 00 in.
Change	−1 ft 06 in.	same)	+1 ft 06 in.
Final draft	27 ft 06 in.	29 ft 00 in.	30 ft 06 in.

Problem #2

Given:

	W	*KG*	L.C.G.–F.P.
Light ship	6,000 tons	30 ft	300 ft
Dry cargo	8,000	25	200
Liquid cargo	1,500	20	350
Deck cargo	500	50	300
Fuel oil & water	1,000 tons	15	400
	17,000 tons		

KM at 17,000 tons = 30 ft.
$MT1$ at 17,000 tons is 1700 ft-tons.
Mean draft at 17,000 tons = 27 ft 00 in.
L.C.B. at 17,000 tons is 260.2 ft from F.P.
Free surface correction = 1 ft.

Required: GM and drafts.

Answer

Stability

	W	V.C.G.	Vertical Moments
Light ship	6,000 tons	30 ft	180,000 ft-tons
Dry cargo	8,000	25	200,000
Liquid cargo	1,500	20	30,000
Deck cargo	500	50	25,000
Fuel oil & water	1,000	15	15,000
	Δ = 17,000 tons		450,000 ft-tons

$$\text{New } KG = \frac{450,000}{17,000} = 26.5 \text{ ft}$$

$$
\begin{aligned}
KM &= 30 \text{ ft} \\
- KG &= 26.5 \\
\hline
GM &= 3.5 \\
\text{Free surface correction} &= -1.0 \text{ ft} \\
\hline
GM &= 2.5 \text{ ft}
\end{aligned}
$$

Trim

	W	L.C.G.	Long. Moments
Light ship	6,000 tons	300 ft	1,800,000 ft-tons
Dry cargo	8,000	200	1,600,000
Liquid cargo	1,500	350	525,000
Deck cargo	500	300	150,000
Fuel oil & water	1,000	400	400,000
	Δ = 17,000 tons		4,475,000 ft-tons

① L.C.G. = 4,475,000/17,000 = 263.2 ft

② L.C.G. 263.2 ft − L.C.B. 260.2 ft = 3 ft trim lever (trim by stern)

③
④ Total change in trim = 3 ft × 17,000/1700 = 30 in.

⑤   30/2 = 15 in. = 1 ft 03 in.

⑥

	Fwd.	Mean	Aft
Draft	27 ft 00 in.	(Stays	27 ft 00 in.
Change	−1 ft 03 in.	same)	+1 ft 03 in.
Final draft	25 ft 09 in.	27 ft 00 in.	28 ft 03 in.

SECTION IV:
WEIGHT CONCENTRATION PROBLEMS

Common formulas:

1.
$$t = \frac{v}{V} \times T$$

Where t = tons to be loaded in desired compartment.
   T = total tons to be loaded.
   v = compartment volume.
   V = total volume.
All volumes are given as bale cubic.

2. Deck load capactiy = $C = \dfrac{\text{Tons} \times 2240}{W' \times L'}$ or $\dfrac{\text{(lb)}}{W' \times L'}$

Where W' = width of compartment.
   L' = length of compartment.

3. Height limitation = $h = \dfrac{C \times f}{2240}$

Where C = deck load capacity.
   f = cargo stowage factor.

Note: Deck load capacity is always expressed in lb/ft^2.

Problem #1: Amount of Weight To Load

What is the number of tons that can be loaded in the upper tweendecks
(U.T.D.) if 13,000 tons are to be loaded?

Given:

Deck	Bale Cubic
U.T.D.	235,130 ft³
L.T.D.	297,452
L.H.	188,056
	720,638 ft³

Answer

$$\frac{v}{V} \times T = t$$

$$\frac{235,130}{720,638} \times 13,000 = 4242 \text{ tons}$$

Problem #2: Deck Load Capacity

Could you load boiler plate that covers an area of 8 ft × 40 ft and weighs 100 tons in the no. 4 hold?

Given: American Racer Class no. 4 Hold.

Deck Capacity	
U.T.D.	360 lb/ft²
L.T.D.	528 lb/ft²
LH	528 lb/ft²

Answer

$$\frac{\text{Tons} \times 2240}{W' \times L'} = \frac{100 \times 2240}{8 \times 40} = \frac{224,000 \text{ lb}}{320 \text{ ft}^2} = 700 \text{ lb/ft}^2$$

The answer is no, unless dunnage is used to increase the area ($W' \times L'$) and thereby reduce the weight per area.

Problem #3: Height Limitation

How high can we load railroad steel with a stowage factor of 14 in the no. 4 lower hold of the vessel above?

Answer

$$h = \frac{c \times f}{2240} = \frac{528 \times 14}{2240} = \frac{7392}{2240} = 3.3 \text{ ft}$$

Problem #4: Maximum Use of Available Cubic with Two Cargoes Differing in Density

How high would you have to stow steel billets with a stowage factor of 12 in a compartment 14 ft high with a deck capacity of 600 lb/ft² so you could fill the rest of the compartments with bales of cork stowing in at 300 ft³/ ton?

Answer

Let x = cubic feet of cork

y = cubic feet of steel billets

$$\text{Density of cork} = \frac{2240}{300} = 7.47 \text{ lb/ft}^3$$

$$\text{Density of steel billets} = \frac{2240}{12} = 186.67 \text{ lb/ft}^3$$

$$x + y = 14$$

$$
\begin{array}{r}
7.5x + 186.7y = 600 \\
-7.5 \times (x + y = 14) = -\underline{7.5x - 7.5y = -105} \\
179.2y = 495 \\
y = 2.76 \ \text{ft}^3 \text{ of} \\
\text{steel billets}
\end{array}
$$

$$x = 14 - 2.76 = 11.24 \text{ ft}^3 \text{ of cork}$$

The steel billets will be 2.76 ft high and the cork 11.24 ft high.

Problem #5: Maximum Use of Available Cubic with Two Cargoes Differing in Density

Given: Steel billets with a stowage factor of 12 are to be stowed in a compartment 12 ft high with a deck load capacity of 400 lb/ft². General cargo with an average stowage factor estimated at 160 is to be stowed over the steel.

Required: How high should the steel be tiered to allow the free space over the steel to be filled with the general cargo and not exceed the deck load capacity?

Answer

First solve for the density of the steel billets and the general cargo.

$$\text{Density of the steel} = 2240/12 = 186 \text{ lb/ft}^3$$

$$\text{Density of the general cargo} = 2240/160 = 14 \text{ lb/ft}^3$$

Let x = cubic feet of general cargo and y = cubic feet of steel. Then:

$$x + y = 12$$

$$
\begin{aligned}
14x + 186y &= 400 \\
-14 \times (x + y = 12) = -14x - 14y &= -168 \\
\hline
172y &= 232
\end{aligned}
$$

$$y = 1.35 \text{ ft}^3 \text{ steel}$$

$$x = 10.65 \text{ ft}^3 \text{ general cargo}$$

Therefore, the steel should be tiered 1.35 ft high; thus, the general cargo would be 10.65 ft high.

Now, what deck space will be required given 50 tons of steel billets?

$$\text{Tons} \times f = \text{Total cubic}$$

$$50 \times 12 = 600 \text{ ft}^3$$

$$\text{Cubic} = \text{Height} \times \text{Area}$$

$$600 = 1.35 \times \text{Area}$$

$$\therefore \text{Area} = \frac{600}{1.35} = 444.4 \text{ ft}^2$$

SECTION V:
STOWAGE FACTOR PROBLEMS

Common Formulas:

1. $f = \dfrac{2240}{D}$

   where D = density in lb/ft^3.

2. $f = \dfrac{2240 \times V}{W}$

   where V = volume in ft^3.
   W = weight in lb.

3. $T = \dfrac{V \times (1 - L)}{f}$

   where L = estimated broken stowage.
   T = tons.
   V = volume of space.

4. $p = \dfrac{V \times (1 - L)}{v}$

   where p = number of pieces.
   v = volume of piece.

Problem #1

Given: A hold of 60,000 bale cubic. A cargo consisting of cases weighing 400 lb and measuring 2.5 ft by 2 ft by 2 ft is to be stowed. Estimated broken stowage is 10%.

Required: The number of tons that can be stowed in the hold.

Answer

Solving for the stowage factor of this cargo:

$$f = \frac{2240 \times 10^*}{400}$$

$$f = 56$$

* 10 = 2.5 ft × 2 ft × 2 ft = cubic of each case

Therefore:

$$T = \frac{60,000 \times 0.9}{56}$$

$$T = 964 \text{ tons}$$

Problem #2

Given: The same data as for Problem #1.
Required: The number of cases that could be stowed in the hold.

Answer

$$P = \frac{60,000 \times 0.9}{10}$$

$$P = 5400 \text{ cases}$$

SECTION VI:
BOOM PROBLEMS

Common formulas:

1. Tension on topping lift span $= \dfrac{\text{Length of span}}{\text{Length of mast}} \times W$

2. Thrust on boom $= \dfrac{\text{Length of boom} \times W}{\text{Length of mast}} + \text{S.H.P.}$

3. S.H.P. (stress on the hauling part) =

$$\frac{(0.1 \times S \times W) + W}{\text{No. of parts at mov. block}}$$

Where S = no. of sheaves.
　　W = weight to be lifted.
　　0.1 = 10% loss due to friction.

No. of parts at mov. block = the mechanical advantage.

Problem #1

Given: A cargo boom is 32 ft long and plumbs a point 25 ft from the foot of a mast. The topping lift span is made fast on the mast 25 ft above the heel of the boom. A gun tackle purchase is being used to lift a load of 12 tons, with the hauling part led through the heel block to the winch at the foot of the boom. (Make an allowance of 10% of the load for friction.)
Required: Tension on the topping lift and the total thrust on the boom.

Answer

$$\text{Tension on span} = \frac{\text{length of span}}{\text{length of mast}} \times \text{weight} = \frac{25}{25} \times 12 = 12 \text{ tons}$$

$$\text{S.H.P.} = \frac{(0.1 \times S \times W) + W}{\text{No. pts. at mov. block}} = \frac{(0.1 \times 3^* \times 12) + 12}{2}$$

$$= \frac{3.6 + 12}{2} = \frac{15.6}{2} = 7.8 \text{ tons}$$

* Includes heel block. (Gun tackle = 2 sheaves + heel block = 3 sheaves.)

$$\text{Thrust on Boom} = \frac{\text{length of boom} \times W}{\text{length of mast}} + \text{S.H.P.}$$

$$= \frac{32 \times 12}{25} + 7.8 = (1.28 \times 12) + 7.8 = 15.36 + 7.8 = 23.16 \text{ tons}$$

Problem #2

Given: The height of the mast is 40 ft, the length of the topping lift span is 40 ft, and the length of the boom is 57 ft.

Required: If lifting a 1-ton weight with a twofold tackle, find (a) the tension on the topping lift, and (b) the thrust on the boom.

Answer

$$\text{Tension of T.L. span} = \frac{\text{length of span}}{\text{length of mast}} \times W$$

$$= \frac{40}{40} \times 1 = 1 \text{ ton}$$

$$\text{S.H.P.} = \frac{(0.1 \times 5 \times 1) + 1}{4} = 0.375 \text{ tons}$$

$$\text{Thrust on boom} = \frac{\text{length of boom} \times W}{\text{length of mast}} + \text{S.H.P.}$$

$$= \frac{57 \times 1}{40} + 0.38 = 1.81 \text{ tons}$$

INDEX

683